Foodborne Microbial Pathogens

FOOD SCIENCE TEXT SERIES

The Food Science Text Series provides faculty with the leading teaching tools. The Editorial Board has outlined the most appropriate and complete content for each food science course in a typical food science program and has identified textbooks of the highest quality, written by the leading food science educators.

Arun K. Bhunia

Foodborne Microbial Pathogens

Mechanisms and Pathogenesis

 Springer

Arun K. Bhunia
Purdue University
West Lafayette, IN
USA

ISBN: 978-0-387-74536-7   e-ISBN: 978-0-387-74537-4

Library of Congress Control Number: 2007934676

Dedicated to my lovely wife Banashri and our two adorable children;
Arni and Irene

Preface

Ever since my days in veterinary school, I was fascinated with the field of microbiology. I always wondered how such a small microscopic organisms are capable of causing infections in other living organisms; big or small, young or old, and healthy or immunocompromised. The subject captured my imagination. Many of the same microorganisms that cause diseases in animals also infect humans. In recent days, pathogens of animal origin impose even greater concern with increasing threat of avian influenza to cause pandemic, and spread of deadly bovine spongiform encephalopathy (mad cow disease) and many bacterial pathogens such as *Listeria, E. coli* O157:H7, *Salmonella, Yersinia*, and *Campylobacter*. I am especially intrigued by the cunning strategy pathogens employ for their survival in a host and their exploitation of host cellular machinery to promote their own invasion into the host. Pathogenic mechanism is complex and unraveling that process requires great minds. Today, microbiologists, cell biologists, and immunologists employing many sophisticated molecular tools are unraveling that secret at a very fast pace. Thus it requires a great deal of efforts to compile and update information in a textbook and it was rather a monumental task. My goal with this book was to paint a bigger picture of pathogenic mechanism of foodborne pathogens, which are responsible for many of modern day outbreaks, and diseases worldwide, and narrate the subject with easy to comprehend illustrations. When I began teaching an advanced graduate level food microbiology course that dealt with pathogenic mechanism of foodborne pathogens in mid nineties, there was hardly any textbook that covered different foodborne microorganisms and the depth of materials needed for the course, especially the mechanism of infection for foodborne pathogens. That necessitated the collection and review of great deal of literature to provide updated materials to my students. That was the beginning and was also the inspiration and motivation to write a textbook on the subject. In the last two decades there had been a tremendous progress in the area of food microbiology especially the study of molecular mechanism of pathogenesis and a great deal of efforts was placed to compile those information in the first edition of the current textbook. In this book, an introductory chapter highlights the significance of foodborne pathogens, epidemiology, and the reason for increasing cases of foodborne illnesses. In Chap. 2, a brief review on biology of microorganisms and the importance of structural components as those relate to pathogenesis is provided.

In addition, diseases caused by viruses, parasites, mycotoxins, and seafood toxins have been included. In Chap. 3, a comprehensive review on the digestive system, mucosal immunity, and the host immune system have been described. This chapter provides the basic foundation for the understanding of the complexity of disease production by foodborne pathogens. First of all, foodborne pathogens' primary site of action is the digestive tract; therefore one must have adequate knowledge to understand the interaction of pathogens with host gastrointestinal tract and second, host innate and adaptive immune responses dictate the progression of a disease. Moreover, some pathogens exploit host immune system as part of their disease producing mechanism. Therefore, it is essential to have some basic understanding of immune system in order to understand the disease process. As it is often said – "It takes two to tango"; or "One need two hands to clap," thus I believe, the knowledge of biology of a pathogen, and the corresponding host immune response go hand in hand to comprehend the full picture of pathogenesis process. In Chap. 4, general mechanism of foodborne pathogens have been included to provide the overall big picture of mechanism of infection and intoxication. In Chap. 5, a brief review on the animal and cell culture models as necessary tools to study pathogenesis is discussed. In Chaps. 6–15, sources, biology, pathogenic mechanism, prevention and control, and detection or diagnosis strategies for individual foodborne bacterial pathogens are described. In addition to traditional foodborne pathogens, descriptions of some of the key pathogens with bioterrorism implications such as *Bacillus anthracis*, and *Yersinia pestis* have been included to provide unique perspective.

In this book, I am pleased to generate both digital and hand-drawn artworks to illustrate the pathogenic process, and I hope these illustrations will aid in better understanding of the mechanism of pathogenesis with greater enthusiasm. I also hope this textbook would be a valuable resource not only for food microbiology graduate or undergraduate students but also for the medical microbiologists, microbiology professionals, and academicians involved in food microbiology and food safety-related research or teaching.

I would like to convey my gratitude to my current and former postdoctoral research associates and graduate students particularly Kristin Burkholder, Jennifer Wampler, Pratik Banerjee, and Ok Kyung Koo for their assistance in collecting literatures and reading the draft chapters. The comments and inputs provided by the students of FS565 over the years were extremely helpful in developing the course contents for this book. Finally, my sincerest and humble gratitude goes to my professional colleagues for their generous time and efforts in reviewing the chapters and providing expert comments and critiques: Chap. 1 (Prof. M. Cousin, Purdue University); Chap. 2 (Prof. M.G. Johnson, University of Arkansas); Chap. 3 (Prof. R. Vemulapalli, Purdue University); Chap. 4 (K. Burkholder, Purdue University); Chap. 6 (Dr. P. Banada, Purdue University); Chap. 7 (Prof. A. Wong, University of Wisconsin and Prof. J. Mckillip, Ball State University); Chap. 8 (Dr. G.R. Siragusa, USDA-ARS, Athens, GA and Dr. V. Juneja, USDA-ARS, Wyndmoor, PA); Chap. 10 (Prof. B. Reuhs, Purdue University); Chap. 11 (Prof. S. Rickie, University of Arkansas and Dr. M. Rostagno, Purdue University); Chap. 12 (Dr. R. Nannapaneni, Mississippi State University); Chap. 13 (Prof. J.S. Virdi, University of Delhi South Campus, India); Chap. 14 (Dr. G.B. Nair, Center for Health and Population Research, Dhaka, Bangladesh); Chap. 15 (Prof. C. Sasakawa, University of Tokyo, Japan).

West Lafayette, IN   Arun K. Bhunia

Contents

Introduction to Foodborne Pathogens

Introduction

Food microbiology can be divided into three focus areas; beneficial microorganisms, spoilage microorganisms, and disease causing microorganisms (Fig. 1.1). Beneficial microorganisms are those used in food fermentation to produce products such as cheese, fermented meat (pepperoni), fermented vegetables (pickles), fermented dairy products (yogurt), and ethnic fermented products such as sauerkraut, idli and kimchi. In fermented products (produced by natural or control fermentation), microorganisms metabolize complex substrates to produce enzymes, flavor compounds, acids, and antimicrobial agents to improve product shelf-life and to prevent pathogens growth and to provide product attributes. Microorganisms with their enzymes also breakdown indigestible compounds to make the product more palatable and easy to digest. In addition, the beneficial microorganisms also serve as probiotics to impart direct health benefit by modulating the immune system to provide protection against chronic metabolic diseases, bacterial infection, atherosclerosis, and allergic responses. Examples of beneficial microorganisms are *Lactobacillus acidophilus*, *Lactobacillus rhamnosus*, *Lactococcus lactis*, and *Pediococccus acidilactici*.

Food spoilage microorganisms are those which upon growth in a food, produce undesirable flavor (odor), texture and appearance, and make food unsuitable for human consumption. Sometimes uncontrolled growth of many of the beneficial microorganisms can cause spoilage. Food spoilage is a serious issue in developing countries because of inadequate processing and refrigeration facilities. Examples of food spoilage microorganisms are *Brocothrix*, *Lactobacillus*, *Bacillus*, *Pseudomonas* spp., and some molds. The microenvironment created in a spoiled food generally discourages the growth of the pathogenic microorganisms, which are considered poor competitors.

Foodborne pathogenic microorganisms (Table 1.1) when grown in a food may not alter the aesthetic quality of products and, thus may not be easy to asses the microbial safety of a product without performing multiple microbiological tests. Foodborne pathogens are responsible for food intoxication (ingestion of pre-formed toxin), toxicoinfection (toxin is produced inside the host after ingestion of

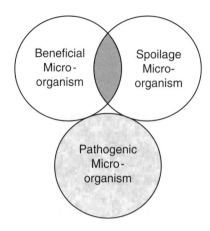

Fig. 1.1 Three branches of food microbiology focus areas: Beneficial and spoilage microorganisms have some overlapping activity (*shaded area*) while pathogens make a distinct group

Table 1.1 List of foodborne pathogens involved in outbreaks from contaminated food and water

Bacterial	Viral	Parasitic
Aeromonas hydrophilia	Astrovirus	*Cryptosporidium parvum*
Bacillus anthracis	Hepatisis A virus	*Cyclospora cayatanensis*
Bacillus cereus/subtilis/ lichniformis	Hepatitis E virus	*Entamoeba histolytica*
Brucella/abortus/melitensis/suis	Nororvirus	*Giardia intestinalis*
Campylobacter jejuni/coli	Rotavirus	*Isopspora belli*
Clostridium botulinum		*Taenia solium/saginata*
Clostridium perfringens		*Toxoplasma gondii*
Escherichia coli		*Trichinella spiralis*
Enterobacter sakazakii		
Listeria monocytogenes		
Mycobacterium paratuberculosis		
Salmonella enterica		
Shigella spp.		
Staphylococcus aureus		
Vibrio cholerae		
V. cholerae non-01		
V. parahemolyticus		
V. vulnificus		
V. fluvialis		
Yersinia enterocolitica		

bacteria) or foodborne infection (ingestion of infective pathogen). Food intoxication is generally caused by *Staphylococcus aureus, Clostridium botulinum*, and *Bacillus cereus*. Toxicoinfection is caused by *Clostridium perfringens*, enterotoxigenic *Escherichia coli* (ETEC), and *Vibrio cholerae*. Foodborne infection is caused by bacterial pathogens such as *Salmonella enterica*, *Campyloacter jejuni*, enterohemorrhagic *Escherichia coli* (*E. coli* O157:H7), *Shigella* spp., *Yersinia enterocolitica*, and *Listeria monocytogenes* and viruses and parasites. Among the foodborne pathogens, Norovirus tops all infections with over a million people infected each year in the US. Protozoan infections are mostly associated with water and fresh produce such as fruits and vegetables. A large volume of these commodities are imported from countries where food production and processing are performed under inadequate hygienic practices, which may contribute to the increased incidences. Molds produce mycotoxins that are mutagenic, carcinogenic, or hepatotoxic and prolonged exposure to mycotoxins may result in serious and sometimes fatal diseases. The onset of symptoms due to mycotoxin intoxication is delayed and is not as dramatic as foodborne bacteria or virus-induced diarrhea, vomiting, or neurological disorders. Thus often significance of mycotoxin ingestion is overlooked because bacterial or viral pathogens affect large populations, cause high morbidity and mortality, and their surveillance statistics are updated routinely. The importance of molds in food rarely peaked consumers or regulatory agencies interest; however, in recent years increased emphasis has been placed on understanding the properties, synthesis, and pathogenesis of foodborne molds and their mycotoxins.

Microorganisms are ubiquitous and can survive and grow in extremes conditions in nature, in food and in human or animal hosts. Some bacteria grow in high temperatures and are called thermophiles (50–70 °C); some even grow in hot geysers such as extremophiles (>70 °C). The most thermophilic and extremophilic microorganisms are nonpathogenic to humans. Some microorganisms grow at refrigeration to ambient room temperatures (1–25 °C) and are called psychrotrophs while some are mesophiles, which grow at 30–37 °C. Both psychrotrophs and mesophiles are capable of causing diseases in humans. Altogether, only a small fraction of all microbes actually cause diseases in humans or animals.

Based on the oxygen requirements, bacteria are grouped as aerobic, obligate anaerobic or facultative anaerobic. Foods are prepared and stored under controlled environments where oxygen concentrations can dictate what type of microorganisms will survive and grow. Anaerobes can grow only in the vacuum packaged or canned foods where oxygen is removed mechanically or by heating. Aerobes grow on the surface of food where oxygen is abundant. A similar scenario applies for pathogens that cause disease in the gastrointestinal tract, where oxygen gradients vary from the upper part of the small intestine to the lower part of large intestine. The upper part is highly oxygenated while the lower part is devoid of oxygen. Again oxygen concentrations vary from the center of the lumen to the proximity of the epithelial lining where the oxygen concentration is higher because of cellular respiration. These environments dictate the types of pathogens that will colonize or be present in different parts of the intestine. Besides microorganisms being introduced into our body via food, water or air; microorganisms also exist since birth as commensals in the digestive tract, skin, nasal passages, and urinary tract and these microbes are generally beneficial to the host. However, under certain conditions, commensals can be opportunistic pathogens. Food is a complex milieu that contains salts, acids, ions, aldehydes, flavoring agents, and antimicrobial preservatives. Pathogenic bacteria transmitted through foods, in most cases, are well adapted to the harsh environments of food and maintain their pathogenic status. During transition to the host, pathogens may express a completely separate set of genes that ensure their survival and disease producing ability in the host.

Food safety is essentially an ongoing problem in today's rapidly growing food industry. Food is globally sourced and also distributed globally; thus, contamination of a food with a pathogen will present a greater economic and social impact than ever before. Food also is being considered as vulnerable to malicious contamination with infective agents; thus, food defense is becoming an essential part of our education. Knowledge and understanding of pathogens in foods and their survival mechanisms in foods as well as in the human host should be important areas of focus to prevent food-related illnesses and increase the well being of the population.

What is a Pathogen?

A pathogen is an organism that is able to cause cellular damage by establishing in tissue, which results in clinical signs with an outcome of either morbidity (defined by general suffering) or mortality (death). More specifically, a pathogen is characterized by its ability to replicate in a host, by its continued persistence of breaching (or destroying) cellular or humoral barriers that ordinarily restrict it, and by expressing specific virulence determinants to allow a microbe to establish within a host for transmission to a new susceptible host. Pathogens could be classified as zoonotic, geonotic, or human origin based on their transmission patterns and movement among different hosts and vectors. Zoonotic diseases are characterized by transmission of infective agents from animals to humans; geonotic diseases are acquired from soil, water, or decaying plant materials; and human origins are exclusively transmitted from person-to-person. Examples of zoonotic pathogens are *Escherichia coli* O157:H7, *Staphylococcus aureus, Salmonella enterica* serovar Typhimurium, *S.* serovar Enteritidis, *Campylobacter jejuni, Yersinia enterocolitica, Mycobacterium tuberculosis*; a geonotic pathogen is *Listeria monocytogenes*; and pathogens of human origin are *Salmonella enterica* serovar Typhi, *Vibrio cholerae, Shigella* spp., and Hepatitis A.

Poverty, competition for food, crowding, war, famine, and natural disaster help pathogens to survive and spread in the environment. Domestication of animals also allowed pathogens to come into contact with the humans and, thus, acquires a new host. In recent years, there is a great concern of possible pandemic outbreak of bird-flu (avian influenza virus) in humans due to the transmission of the virus to the bird-handler or people coming into contact with the infected fowls. Bird-flu virus (strain H5N1) was responsible for several fatal infections in Asia and other parts of the world (see Chap. 2).

What are the Attributes of Pathogenicity?

Some pathogens are designated as primary pathogens, which regularly cause disease. Some are classified as opportunistic pathogens that infect primarily immune-compromised individuals. Nevertheless, both primary and opportunistic pathogens share similar attributes; pathogens must enter and survive inside a host, must find a niche for persistence, must be able to avoid the host's defense (stealth phase), must be able to replicate to significant numbers, must be transmitted to other host with high frequency, and must be able to express specialized traits within the host. For example, *Salmonella enterica* invades host cells when proper levels of O_2, pH, and osmolarity are maintained. This sends appropriate signals to the PhoP/Q regulon for expression of specific invasion-associated genes.

Several factors affect pathogens growth and survival inside a host such as O_2, CO_2, iron, nutrients, pH, bile salts, mucus composition, balance of natural microflora, quorum sensing, and physiological status such as stress hormones like epinephrine or norepinephrines.

Pathogens are clonal and they are generally derived from a single progenitor. They are often selected by the environmental selective pressures such as heat, extremes of pH and antibiotic treatments. However, emergence of a new or a highly virulent form of pathogens suggests possible transfers of virulent genes are occurring between microorganisms. These novel genetic variants have arisen due to a point mutation, genetic rearrangement in the chromosome, and gene transfer between organisms through horizontal (*Shigella* to *E. coli*) or vertical (*E. coli* to *E. coli*) modes. Plasmids, bacteriophages, and pathogenicity islands (i.e., a large piece of DNA carrying virulence gene cluster (see Chap. 4)) serve as the vehicles for gene transfer.

Involvement of a specific gene or genes in pathogenesis can be studied in the laboratory by using various molecular tools such as mutation, genome sequencing and cloning. In mutation, a specific gene is disrupted by chemical mutagenesis, transposon mutagenesis, site-specific mutations (homologous recombination), or in-frame deletion. Genetic complementation is often done to restore the gene function and to confirm the involvement of that gene in pathogenesis. Transposon (Tn) elements are used routinely to induce mutation. The major benefit of using transposon is that they carry antibiotic markers that help in locating a specific gene on the chromosome. However, transposon could act as a transcriptional terminator. If Tn lands on the promoter or the first gene of an operon, it will eliminate the transcription of the genes located downstream. Also, if the transposon is inserted in a house-keeping gene, which is essential for bacterial growth or survival, that microorganism cannot be selected from that experiment. Transposons carry insertion sequences at the end and are generally specific for either Gram-positive (Tn*916*, Tn*917*, Tn*1545*) or Gram-negative (Tn*5*, Tn*7*, Tn*10*) bacteria. In recent years, in-frame deletion has been used widely to create a mutant strain by employing a method called SOE (splicing by overlap extension). Genome sequence of many microorganisms is now available; thus, a series of polymerase chain reaction (PCR) methods is used to selectively remove a target gene by SOE to create an in-frame deletion mutant. Gene sequencing is also a powerful tool for studying pathogenesis. Sequencing of a gene or its product (amino acid sequence) and subsequent matching with the database will reveal the identity and function of the gene. Cloning is also an important strategy for studying pathogenesis. In cloning experiments, a suitable vector is constructed with a gene of interest and, subsequently, transferred to an avirulent strain by electroporation or conjugation. The function of the gene in the new strain is analyzed by in vitro cell culture or in vivo animal bioassays.

Sources of Foodborne Pathogens

Many foodborne pathogens are ubiquitous in nature and generally found in soil, water, animals and plants. Pathogens are introduced into a processing plant through the raw materials. Humans and plant equipment may also bring the organisms to a plant. Recontamination of processed food also frequently occurs and contributes to foodborne outbreaks and illnesses. Pathogens can survive for prolonged period on inanimate objects and serve as a source. A list of foodborne organisms and their survival on inanimate surface have been included in Table 1.2. When considering a product safety, it is important to know the type of microorganisms or toxins likely to be present, their numbers, and concentrations. In addition, their response to the heat (heat-labile or heat-stable), pH, salts, and other processing conditions is important to consider. The numbers and types of microorganisms present in a finished food product are influenced by the original source of the food, its microbiological quality in the raw or unprocessed state, the sanitary conditions under which the product was handled or processed and the conditions for subsequent packaging, handling, storage, and distribution. In a raw agricultural product, generally the maximum microbial load is on the surface, whereas, it is negligible or there is none inside.

Meats, Ground Meat, and Organ Meats

Raw beef carries a large number of *E. coli* since they are natural inhabitants of intestines of mammals, therefore, during slaughter the carcass may be contaminated with fecal bacteria. Fresh meats also can be contaminated with

Table 1.2 Persistence of foodborne pathogens on inanimate surfaces

Organisms	Duration of survival
Bacterial pathogens	
Campylobacter jejuni	Up to 6 days
Escherichia coli	1.5 h–16 months
Listeria species	24 h–several months
Salmonella serovar Typhi	6 h–4 weeks
Salmonella serovar Typhimurium	10 days–4.2 years
Shigella species	2 days–5 months
Staphylococcus aureus	7 days–7 months
Stretococcus pyogenes	3 days–6.5 months
Vibrio cholerae	1–7 days
Viral pathogens	
Astrovirus	7–90 days
Adenovirus	7 days–3 months
Norovirus	8 h–7 days
Influenza virus	1 day–2 days
Rotavirus	6 days–60 days
Hepatitis A virus	2 h–60 days

Adapted from Kramer et al. 2006. BMC Infect. Dis. 6:130

Salmonella and *Staphylococcus aureus* (skin, hide, feathers). Ground beef is made from meat, trimmings, and fat and has a large surface area that favors growth of aerobic bacteria. The knives of meat grinder, if not properly sanitized or washed, may be the source of contamination. Furthermore, one heavily contaminated piece of meat may be sufficient to contaminate an entire lot of ground meat. Pathogens such as *Clostridium perfringens, Bacillus cereus, L. monocytogenes*, and enterohemorrhagic *E. coli* (EHEC) are associated with these types of products. Liver, kidney, heart, and tongue may carry Grampositive cocci, coryneforms, *Moraxella*, and *Pseudomonas*. Lymph nodes are secondary lymphoid tissue where pathogens are brought in for destruction and also serve as a major source for pathogens. Mechanically deboned meat/poultry/fish generally carry lower microbial loads because of less handling and minimal human interventions. Also, electrical stimulation converts glycogen to lactic acid, thus, resulting in lowered pH, which suppresses bacterial loads. During rigor mortis, release of cathepsin (facilitates muscle tenderization) and lysozyme also reduces the bacterial counts.

Vacuum Packaged Meats

In the vacuum packaged product, air is removed to increase the shelf-life. Generally the shelf-life for vacuum packaged meats is about 15 weeks. Initially, the oxygen permeable packaging allows the growth of *Pseudomonas* spp., but lactic acid bacteria (*Lactobacillus, Leuconostoc*, and *Carnobacterium*) take over after the *Pseudomonas* spp. remove oxygen and increase the CO_2 level inside the package, which in turn encourages the growth of facultative anaerobes (lactic acid bacteria) and anaerobic microbes such as *Clostridium* species (*Clostridium botulinum, C. perfringenes*). Later, lactic acid bacteria such as *Lactobacillus* and *Brocothrix* grow as well as other pathogenic bacteria such as *Yersinia enterocolitica* and *S. aureus*.

Poultry

Salmonella enterica serovars Enteritidis, Typhimurium, Infantis, Reading, Blockley; *Clostridium perfringens*; *Campylobacter jejuni*; and *E. coli* are associated with poultry. Workers may also be the source of other *Salmonella* serovars like – Sandiego and Anatum. Fresh poultry may be the source of *Pseudomonas*, coryneforms, and yeasts. Ground turkey also may carry fecal streptococci.

Seafoods

Microbial loads in shrimps, oysters, and clams depend on the quality of the water from which they are harvested. If the sewage is drained, the microbial quality deteriorates. During handling, fecal coliforms, fecal streptococci, and *S. aureus* may be incorporated into the product. *Salmonella* also is found in oysters possibly due to contaminated water. Seafood also is the source for *Pseudomonas* spp., *C. perfringens, L. monocytogenes, Vibrio parahemolyticus, Salmonella enterica* serovar Enteritidis and Typhimurium, *Campylobacter jejuni, Yersinia enterocolitica*, and Enteroviruses (Hepatitis A). Smoked salmon and shrimps also are found to carry pathogenic *L. monocytogenes*.

Fruits and Vegetables

Contamination comes from contaminated irrigation water, animal feces when applied as a fertilizer or from grazing domesticated or wild animals in the orchard, untreated manure, contaminated seeds, insect vectors, food handlers, processing environments, and slicing instruments. Fecal coliform, fecal streptococci, *Pseudomonas* spp., *Erwinia* spp., *C. perfringens, C. botulinum, Salmonella* serovars, *S. aureus*, and lactic acid bacteria are found on vegetables. Protozoan species such as *Giardia intestinalis, Entamoeba histolytica, Isospora belli*, and *Cyclospora cayatanensis* are also associated with produce and herbs. Several fruits and vegetables related outbreaks such as *E. coli* O157:H7 in apple cider, spinach, lettuce and sprouts; *Salmonella enterica* in cantaloupes, cilantro, tomatoes, sprouts and almonds; *Cyclospora cayetanensis* in raspberries; Hepatitis A in strawberries and green onion (scallion); and *L. monocytogenes* in cabbage suggest that further processing or improved quality assurance must be employed with these minimally processed products. Postharvest contamination has been considered the major concern in produce safety; however, recent outbreaks with lettuce and spinach suggest that preharvest crop contamination with pathogens possibly plays a major role (Fig. 1.2). Indeed, a recent spinach outbreak in 2006 with *E. coli* O157:H7 indicated that the source of this pathogen was cattle which grazed in the pasture located near the spinach growing field. Experimental evidence showed that some pathogens like *E. coli* O157:H7 and *Salmonella* are able to survive and grow inside the veins of plant tissues due to the abundant supplies of carbohydrates and moisture.

Dairy Products

Cow's udder, hide, and milking utensils may carry predominantly Gram-positive bacteria such as aerobic sporeformers (*Bacillus* spp.), psychrotrophic *Pseudomonas* spp., and others including *Mycobacterium* and *Clostridium*

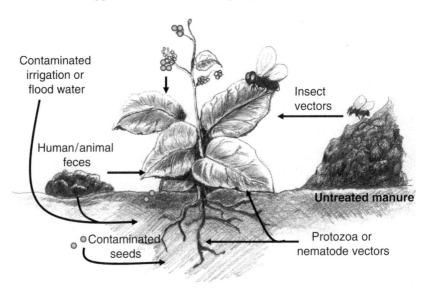

Fig. 1.2 Mode of transmission of foodborne enteric pathogens in fruits and vegetables (redrawn based on Brandl, M.T. 2006. Annu. Rev. Phytopathol. 44:367–392)

species. Sometimes, foodborne outbreaks occur due to consumption of raw milk, home made ice cream containing fresh eggs or cheese made with unpasteurized cow's milk. In 1980–1981, 538 cases of *Salmonella* infection occurred with cheddar cheese and raw milk and certified raw milk. In 1980–1982, 172 cases of *Campylobacter* infection occurred with raw milk and certified raw milk. In 1995, a *Salmonella* outbreak occurred with ice cream when a tanker carrying raw liquid egg also transported pasteurized milk without proper cleaning and sanitization between products. In 1985, 2000, and 2006, *L. monocytogenes* infection occurred due to consumption of Mexican-style soft cheese. This ethnic product is made from unpasteurized milk. *Yersinia enterocolitica* outbreaks also were associated with pasteurized and unpasteurized milk.

Delicatessen Foods

Microbial load and type of pathogens in delicatessen foods such as salads, sandwiches and deli foods, depends on the ingredients, meats, and vegetables used to make them. Food handler's direct contact also can lead to increased incidences of *S. aureus*. Spices in these foods may carry spores of *Clostridium, Bacillus*, and molds. Delicatessen foods also carry psychrotrophic *L. monocytogenes*. Dehydrated foods such as soups of chicken noodle, chicken rice, beef noodle, vegetable, mushroom, and pea may carry as high as 7.3 $\log_{10}$ g^{-1} ml^{-1} of *C. perfringens* as well as coliforms.

Foodborne Pathogen Statistics and Socioeconomic Impact

Worldwide, foodborne pathogens cause numerous sufferings and deaths. In Africa, Asia, and Latin America, there are about 1,000 million cases of gastroenteritis per year in children under the age of 5, which leads to 5 million deaths. In Mexico and Thailand, half of the children aged 0–4 years suffer from the *Campylobacter*-induced enteritis. In Europe, 50,000 cases/million population suffer from acute gastroenteritis. In the Netherlands about 300,000 cases/million population occur yearly. In Northern Ireland and the Republic Ireland, about 3.2 million episodes of gastroenteritis are reported each year. In Australia, 5.4 million cases of foodborne gastroenteritis occur each year. In England 20% of population, i.e., 9.4 million people suffer from acute gastroenteritis each year and the primary contributing microorganisms are identified as Norovirus, *Campylobacter* species, rotavirus, and nontyphoidal *Salmonella* species.

In the US, there are an estimated 6 million cases with 350,000 hospitalizations and 9,000 deaths associated with foodborne infections each year. Foodborne pathogen statistics show declines in incidences from 1996–1998 to 2005 for some pathogens but increased for others: the incidence of *Shigella* decreased by 43%, *Yersinia* species by 49%, *Listeria monocytogenes* by 32%, *Campylobacter* species by 30%, EHEC O157:H7 by 29%, and *Salmonella* Typhimurium by 42%; however, the incidence of *Salmonella* Enteritidis and *S.* Heidelberg increased, each by 25%, and *S. javiana* by 82%. Interestingly, the number of outbreaks and product recalls continued to increase thus placing a huge economic burden on producers and processors (Table 1.3).

Table 1.3 Number of outbreaks, cases, and mortality due to foodborne pathogens in the USA from 1998 to 2002

Pathogens		Outbreaks	Cases	Deaths (%)
Bacterial	*Salmonella enterica*	585	16,821	20 (22.7)
	Escherichia coli	140	4,854	4 (4.5)
	Clostridium perfringens	130	6,724	4 (4.5)
	Staphylococcus aureus	101	2,766	2 (2.3)
	Shigella spp.	67	3,677	1 (1.1)
	Campylobacter spp.	61	1,440	0
	Bacillus cereus	37	571	0
	Vibrio parahemolyticus	25	613	0
	Clostridium botulinum	12	52	1 (1.1)
	Listeria monocytogenes	11	256	38 (43.2)
	Yersinia enterocolitica	8	87	0
	Vibrio cholerae	3	12	0
Viral	Norovirus	657	27,171	1 (1.1)
	Hepatitis A	50	981	4 (4.5)
	Rotavirus	1	108	0
	Astrovirus	1	14	0
Parasitic	*Cyclospora cayetanensis*	9	325	0
	Trichinella spiralis	6	33	0
	Cryptosporidium parvum	4	139	0
	Giardia intestinalis	3	119	0
Seafood toxins	Scombroid toxin	118	463	0
	Ciguatoxin	84	315	1 (1.1)
	Shelfish toxin	5	36	0
Confirmed etiology		2,167	68,981	76 (86.4)
Unknown etiology		4,480	59,389	12 (13.6)
Total (1998–2002)		6,647	128,370	88 (100)

Synthesized from Lynch et al. 2006. Morbid. Mortal. Weekly Rep. 55:1–42

Food animals and poultry are the most important reservoirs for many of the foodborne pathogens. Therefore, meat, milk, or egg products may carry *Salmonella enterica, Campylobacter jejuni, Listeria monocytogenes, Yersinia enterocolitica*, or *E. coli* O157:H7. Control of pathogens in raw unprocessed products is now receiving major emphasis to reduce pathogen loads before arrival at a processing plant. On-farm, pathogen-controlling strategies will help achieve that goal. However, the presence of pathogens in ready-to-eat (RTE) product is a serious concern since those products generally do not receive any further treatment before consumption. In fact, many recent food-borne outbreaks resulted from consumption of undercooked or processed RTE meats (hotdogs, sliced luncheon meats, and salami), dairy products (soft cheeses made with unpasteurized milk, ice cream, butter, etc.), or minimally processed fruits (apple cider, strawberries, cantaloupe, etc.) and vegetables (sprouts, lettuce, spinach, etc.).

Annual economic losses resulting from deaths, illnesses, loss of work, loss of manpower, and product loss account for about $8.4–23 billion. Besides acute gastroenteritis, the sequelae of the foodborne infections results in chronic rheumatoid conditions, ankylosing spondylitis–autoimmune disease (HLA), hemolytic–uremic syndrome (HUS due to Shiga-like toxin from EHEC), atherosclerosis due to lipid deposition in arteries, Guillain-Barré syndrome from *Campylobacter* infections, autoimmune disease such as allergic encephalitis, and autoimmune polyneuritis. Foodborne infections also vary between countries due to the eating habits of the population. In Japan, high *Vibrio parahemolyticus* cases are seen due to consumption of raw fish. Scandinavians and people from middle/eastern countries suffer from botulism due to increased consumption of fish, meat and vegetables.

Why High Incidence of Foodborne Outbreaks?

It is believed that new pathogens are emerging, which are responsible for increased incidences of foodborne diseases (Table 1.4). Some are recognized recently whose ancestors probably caused foodborne illnesses for many thousands of years. An example is *E. coli* O157:H7, a new strain first reported on 1982, which is responsible for numerous outbreaks in recent years. This bacterium has evolved relatively recently from an enteropathogenic *E. coli* (EPEC) progenitor. In addition, many of the older pathogens are reemerging and contributing to the overall foodborne outbreak statistics. Besides, human sufferings and fatalities, the high number of foodborne outbreaks in recent years has devastating economic impacts on food producers and processors. It has been a challenging task for scientists to figure out the reasons for the greater numbers of outbreaks in recent years. Some factors (Table 1.5) are discussed below which may explain the plausible reasons for increased incidence; (1) increased surveillance and reporting, (2) changes in the food manufacturing and agricultural practices, (3) changes in consumption habits, (4) increased at-risk populations, (5) improved detection methods, and (6) emerging pathogens with survivability in stressed conditions.

Table 1.4 Rate of incidence of nine foodborne diseases in the US during 1996–2005

Pathogen	Rate per 100,000 population									
	1996	1997	1998	1999	2000	2001	2002	2003	2004	2005
Campylobacter spp.	23.5	25.2	21.4	17.3	15.7	13.8	13.37	12.6	12.9	12.72
Salmonella serovars	14.5	13.6	12.3	14.8	14.4	15.1	16.10	14.5	14.7	14.55
Shigella spp.	8.9	7.5	8.5	5.0	7.9	6.4	10.34	7.3	5.1	4.67
E. coli O157	2.7	2.3	2.8	2.1	2.1	1.6	1.73	1.1	0.9	1.06
L. monocytogenes	0.5	0.5	0.6	0.5	0.3	0.3	0.27	3.3	2.7	0.3
Y. enterocolitica	1.0	0.9	1.0	0.9	0.4	0.4	0.44	4.0	3.9	0.36
Vibrio spp.	0.2	0.3	0.3	0.2	0.2	0.2	0.27	3.0	2.8	0.27
Cryptosporidium	NR	3.0	3.4	2.9	1.5	1.5	1.42	10.9	13.2	2.95
Cyclospora	NR	0.3	<0.1	<0.1	0.1	0.1	0.11	0.3	0.3	0.15

NR not recorded; Data compiled from FoodNet (http://www.cdc.gov/foodnet)

Table 1.5 Factors affecting the emergence of increased foodborne illnesses from food

---

1. Increased surveillance and reporting

2. Changes in the food manufacturing and agricultural practices
    - Centralized production facility
    - Distribution to multiple states/other countries
    - Intense agricultural practices
    - Minimal processing (produce and fruits)
    - Increased importation of fresh produce

3. Changes in consumer habits
    - Eating more meals outside the home
    - Increased popularity of fresh fruits and vegetables

4. Increased at-risk populations (immunocompromised, elderly)

5. Improved detection methods and tracking of pathogens

6. Emerging pathogens with improved survivability in stressed conditions

---

Surveillance and Reporting

In the past, foodborne incidence reporting was poor or under reported. Sometimes, persons suffering from illness also did not always consult their doctors and sporadic cases were not reported routinely. In addition, the causative agents were not always identified because of lack of better methodologies. Sensitive detection methods are now available and the epidemiological survey has been improved. Computer-based databases such as FoodNet and PulseNet in the USA; EC Enter-Net for *Salmonella* species and *E. coli* O157:H7 in Europe; and WHO-Global Salm-Surv are now available to assess the trends, changes in expected numbers, and types from historical data (Table 1.6). These databases are used as an alert mechanism for future outbreaks.

Changes in the Agricultural Practices and the Food Manufacturing

Agricultural practices affect the incidence of microorganisms in the intestinal tract and on the surface of the animal. High ambient temperature encourages salmonellae growth in animal feed. Use of antibiotics in feed generally kill one type of pathogen but encourage others to grow. Transportation to the slaughterhouse creates stress and weakens immune system, thus, favoring microbial growth. High speed slaughter and evisceration also can result in product contamination. Intensive farming allows faster growth of pathogens and recycling of animal waste products and animal byproducts results in increased cases. Feeding animal products to another species allowed one pathogen (a prion) to adapt and transmit to another host (see Chap. 2). Sine 1997, the US government and several European countries (governments) have banned the use of animal-products (MBM; meat bone meal) as feed to control the spread of the prion that causes bovine spongiform encephalopathy (BSE) (see Chap. 2).

Changes in food manufacture and consumption practices are also contributing factors in increased foodborne diseases. Consumption of preprepared foods at home and outside the home, consumption of chilled and frozen foods, and increased consumption of poultry and fish (health reason) resulted in

Table 1.6 List of surveillance programs currently used in the US and other countries

US surveillance and monitoring programs	
Program	**Purpose**
FoodNet	Foodborne Diseases Active Surveillance Network (FoodNet): routine surveillance of select foodborne pathogens in 10 states in US (see Table 1.3)
PulseNet	DNA fingerprints of pathogens based on pulsed-field gel electrophoresis
CalciNet	Fingerprints of calciviruses
EHS-Net	Environmental health and cause of foodborne diseases
CAERS	CFSAN Adverse Events Reporting System for foods, cosmetics, dietary supplements
EFORS	Electronic Foodborne Outbreak Reporting Systems of CDC (Center for Disease Control and Prevention)
eLEXNET	Electronic Laboratory Exchange Network (data from FDA, USDA, DOD) from all 50 states
Global food industries and air travel	
EC Enter-Net	European Commission Enter Networks for *Salmonella* and *E. coli* O157:H7
OzFoodNet	Australian foodborne disease information for risk assessment and policy, training for food borne disease investigation
WHO-Global Salm-Surv	Surveillance resources and training in food borne disease for participating countries

increased incidence of salmonellosis and *Campylobacter*-induced enteritis and *Vibrio parahemolyticus*-induced gastroenteritis. Cross-contamination of raw foods with cooked/processed foods, and undercooking of meat and holding at higher temperature also were contributing factors. Reduced use of salt, less use of food preservatives (sorbate and benzoate) for health reasons and demand for more natural, fresher, healthier, and convenient meal may serve as contributing factors. These types of food require greater care during production; storage, and distribution, otherwise, pathogens can grow rather easily in these products. Tightening or enforcing strict hygienic measures in the food manufacturing facilities is needed to reduce the incidence of pathogens. Improved sanitary practices with rotation of sanitizers can avoid the emergence of resistant microbes.

The large centralized food processing facilities are thought to be a major contributing factor in recent years. The products produced by such processors are distributed to multiple states or even global markets. Thus, contamination in such products affects large populations with devastating consequences. For example, in 1985 in Chicago, a processing equipment breakdown in a milk pasteurizing unit resulted in an estimated 15,000 cases of illnesses from salmonellosis. In January 1993, in a fast-food restaurant, undercooking of hamburger meat resulted in *E. coli* O157:H7 outbreaks where more than 600 were infected including children and many were hospitalized with 35 showed hemolytic syndrome and 3 died. In 1997, *E. coli* O157:H7-tainted ground beef resulted in 25,000 lb ground beef recall. In 1998–1999, an outbreak of *L. monocytogenes* occurred due to consumption of hotdogs/lunchmeats resulted in 79 cases with 16 deaths and 3 miscarriages. In 2000–2001, consumption Mexican-style soft cheese resulted in 12 cases of listeriosis in North Carolina. In 2002, consumption of sliced turkey meat resulted in a multistate outbreak of *L. monocytogenes* with 50 cases, 7 deaths, and 3 abortions. In 2003, raw milk cheese was responsible for an outbreak in Texas, and in 2005, a multistate outbreak involving consumption of turkey deli meat affected nine states and caused 12 illnesses. In 2006, spinach outbreak with *Escherichia coli* O157:H7 resulted in 199 illnesses with 3 fatalities in 26 states in the US. In 2006–2007, *Salmonella* serotype Tennessee outbreak with peanut butter resulted in 628 cases in 47 states in the US.

Changes in Consumer Habits

Consumption of food outside the home also increased the chance of foodborne illnesses. Meal prepared and eaten at home also could cause disease because of poor hygienic practices during storage and preparation of food. Persons with underlying conditions also serve as contributing factors: liver disease patients are susceptible to *Vibrio vulnificus* infection; therefore, these patients are advised not to eat raw oysters; and immunocompromised and pregnant women should avoid RTE meats, pate, deli or prepared meals and cheeses made with unpasteurized milk because of possible *L. monocytogenes* infection. If outbreaks occur in public institutions such as restaurants, hotel, hospitals, cruise-lines, manufacturer, institutions, etc., the consequences are severe because large numbers are at risk. This happens because of ignorance, poor management, sloppy practices, and lack of education or understanding of safe handling of foods.

Increased At-Risk Populations

Populations that are susceptible to infections are increasing and moreover, people are living longer. Young, old, pregnant, and immunocompromised (YOPI) people are susceptible to various foodborne diseases. Also, cancer patients receiving chemotherapy, people with organ transplants, and AIDS (acquired immune deficiency syndrome) patients are vulnerable to foodborne illnesses. Pathogen-free food may lead to increased susceptibility to diseases because subclinical infection may strengthen immunity. "Delhi-belly," "Montezuma's revenge," or "Travelers Diarrhea" affects only travelers and not the indigenous populations. Also, Scandinavians are very susceptible to salmonellosis when they are abroad but not when they are at home. Nutritional factors, physiological status, and concurrent or recent infection of intestinal tract also can favor increased infection.

Improved Detection Methods and Tracking of Pathogens

Improved detection methods are now capable of detecting low numbers of pathogens in products. Biosensors, PCR, and immunoassays are very sensitive and now pathogens are detected from products, which would normally give negative results with earlier detection technologies.

Emerging Pathogens with Improved Survivability in Stressed Conditions

Newly identified and emerging pathogens are also a major contributing factor. Clinical, epidemiological and laboratory studies have identified a number of so called emerging and new pathogens (Table 1.1). Emergence due to unsafe food handling practices, under cooking of hamburger, consumption of raw milk, ice cream contaminated with raw liquid eggs, consumer demands for fresher foods, and increased consumption of chilled stored chicken resulted in high numbers of enteritis cases. Emergence of antibiotic or preservative-resistant organisms also is a contributing factor. Antibiotic and acid-resistant *Salmonella enterica* and *E. coli*, and also heat-resistant *Salmonella* and *Listeria* can survive processing conditions and persist in the product. Microorganisms acquiring virulent genes through vertical or horizontal transfer may generate a new pathogen with highly virulent gene sets. Viable but nonculturable organisms are also problematic such as Norovirus, *Campylobacter jejuni*, and *Vibrio* species that are difficult to detect.

Summary

Food microbiology is broadly classified into three focus areas: beneficial microorganisms, spoilage microorganisms, and pathogenic microorganisms. Beneficial microorganisms are those used for making traditional or ethnic fermented products and these microorganisms also are used as probiotics which are gaining increased popularity because of their health promoting effects. Spoilage organisms on the other hand are responsible for product spoilage and place an economic burden on the producers, processors and retail store owners for product losses. This is a serious issue in developing countries because of inadequate processing and refrigeration facilities. Foodborne pathogens present a serious challenge since contamination may result in severe diseases such as food intoxication, toxicoinfection, and infection. Mortality, morbidity, and product recalls are serious consequences of foodborne pathogens outbreaks. Foodborne pathogens are generally psychrophilic and mesophilic and their growth do not alter the aesthetic quality of foods. The pathogenic traits are sometimes acquired through plasmids, bacteriophages, or through pathogenicity islands. Foodborne pathogens can be zoonotic, genotic, or human origin and consumption of contaminated foods results in foodborne diseases. In order for a foodborne pathogen to cause disease they must be able to survive in the food and when transferred to human hosts, they must be able to find niches, multiply and express virulence factors to cause host cell damage. Worldwide, foodborne pathogens are responsible for large numbers of outbreaks, illnesses, and mortalities. However, recent foodborne statistics show there is a decline in foodborne outbreaks for several pathogens. Nevertheless, foodborne pathogens are still a serious public health concern and outbreaks are attributed to the emergence of new pathogens and reemergence of some old pathogens. The routine epidemiological and food product surveys are introduced by many countries in order to provide an accurate picture of global distribution and occurrence of foodborne diseases. The reason for emergence of increased foodborne diseases has been investigated and several factors are thought to be responsible: improved survey system and creation of database for various pathogens; changes in agricultural and food manufacturing practices; consumers habits of food consumption and preparation; increased population of susceptible group; improved survival and adaptation of pathogens in harsh food environments; and improved detection methods.

Further Readings

1. Baird Parker, A.C. 1994. Foods and microbiological risks. Microbiology 140: 687–695.
2. Brandl, M.T. 2006. Fitness of human enteric pathogens on plants and implications for food safety. Annu. Rev. Phytopathol. 44:367–392.
3. CDC. 2006. Preliminary FoodNet data on the incidence of infection with pathogens transmitted commonly through food – 10 states, United States, 2005. Morbid. Mortal. Weekly Rep. 55:392–395.
4. Flint, J.A., Van Duynhoven, Y.T., Angulo, F.J., DeLong, S.M., Braun, P., Kirk, M. et al. 2005. Estimating the burden of acute gastroenteritis, foodborne disease, and pathogens commonly transmitted by food: An international review. Clin. Infect. Dis. 41:698–704.
5. Falkow, S. 1997. What is a pathogen? ASM News 63:359–365.
6. Kramer, A., Schwebke, I., and Kampf, G. 2006. How long do nosocomial pathogens persist on inanimate surfaces? A systematic review. BMC Infect. Dis. 6:130.
7. Lynch, M., Painter, J., Woodruff, R., and Braden, C. 2006. Surveillance for foodborne-disease outbreaks – United States, 1998–2002. Morbid. Mortal. Weekly Rep. 55:1–42.
8. Mead, P.S., Slutsker, L., Dietz, V., McCaig, L.F., Bresee, J.S., Shapiro, C., Griffin, P.M., and Tauxe, R.V. 1999. Food-related illness and death in the United States. Emerg. Infect. Dis. 5:607–625.
9. Ray, B. and Bhunia, A.K. 2008. Normal microbiological quality of foods and its significance, Chapter 4, Fundamental Food Microbiology, CRC, Boca Raton, pp 35–41.
10. Reij, M.W., Den Aantrekker, E.D., and ILSI Europe Risk Analysis in Microbiology Task Force. 2004. Recontamination as a source of pathogens in processed foods. Int. J. Food Microbiol. 91:1–11.
11. Trevejo, R.T., Barr, M.C., and Robinson, R.A. 2005. Important emerging bacterial zoonotic infections affecting immunocompromised. Vet. Res. 36:493–506.

2

Biology of Microbes Associated with Food

Introduction

General properties of microorganisms including Gram-positive and Gram-negative bacteria, viruses, molds and mycotoxins, parasites, and toxins of seafood origin that are responsible for foodborne diseases are reviewed in this chapter (Table 2.1). In addition, the morphological and structural characteristics of microorganisms in relation to pathogenesis are reviewed so that this background knowledge will aid in understanding the mechanism of pathogenesis and host response to microbes in subsequent chapters.

Bacteria

Morphologically, bacterial cells are short or long rods (1–3 μm in length and 0.5–1 μm in diameter) or spherical or curved or spiral (Fig. 2.1). Some cells may form long chains depending on the growth environments or physiological conditions or some may exist in singlet or doublet. Some cells are motile and may display very unique motility such as tumbling, cork-screw rotation, and swimming.

Nutrients, temperature, and gaseous composition of the environment influence bacterial growth and metabolism. Aerobic bacteria require oxygen for respiration and energy while anaerobic bacteria grow in absence of oxygen. Obligate or strict anaerobes cannot withstand traces of oxygen while aerotolerant anaerobes have certain tolerance for oxygen. Bacteria are also classified based on their temperature requirements. Psychrophiles grow in subfreezing (1 °C) to above freezing temperatures (4–25 °C), mesophiles (25–37 °C), and thermophiles (above 40 °C). Most of the pathogenic bacteria are mesophilic and only some are psychrophiles. Thermophiles are rarely pathogenic. Bacteria are divided into two groups based on their cell wall structure; Gram-positive and Gram-negative. Gram-positive cell envelope retain crystal violet iodine complex, appearing purple to blue during Gram-staining, while Gram-negative cell wall is porous and does not retain the stain. Counter staining of Gram-negative cells with basic fuschin make those cells pink to red appearance.

Table 2.1 Microorganisms or toxins associated with foodborne infections or food intoxication

Gram-positive	Gram-negative	Virus	Molds (mycotoxins)	Parasites	Toxins (seafoods)
Staphylococcus	*Aeromonas*	Enterovirus	*Aspergillus* spp. (Aflatoxin, Ochratoxin)	(Protozoa)	Ciguater-atoxin
Clostridium	*Salmonella*	Hepatitis A		Giardia	Scombroid
Listeria	*Escherichia*	Rotavirus	*Penicillium* spp. (citrinin, patulin, penicillic acid)	Entamoeba	Saxitoxins
Bacillus	*Campylobacter*	Norwalk	*Fusarium* spp. (Fumonosin, zearalenone)	Toxoplasma	
Enterobacter	*Yersinia*	Bird-flu		Cryptosporidium	
	Shigella	Bovine spongiform encephalopathy (BSE)		(Flatworm)	
	Vibrio			Taenia	
	Arcobacter		*Claviceps purpura* (Ergot)	(Roundworm)	
				Trichinella	
				Ascaris	

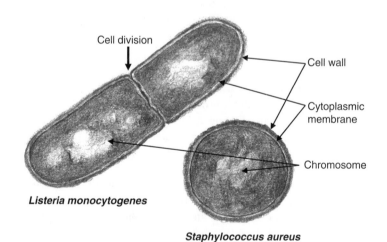

Fig. 2.1 Bacteria: transmission electron microscopic cross section of Gram-positive *Listeria monocytogenes* and *Staphylococcus aureus* cells

The cellular structure of Gram-positive (*Listeria, Staphylococcus, Streptococcus, Clostridium*) and Gram-negative (*Salmonella, Escherichia, Campylobacter*) bacteria is distinct. Outer most layers of Gram-positive bacteria contain a thick rigid cell wall or peptidoglycan (PGN) structure. Cell wall also contains protein and teichoic acid (TA), teichuronic acid, lipoteichoic acids (LTA), lipoglycan, and polysaccharides. The inner layer is a porous cytoplasmic membrane (CM) which consists of lipid bilayer (Fig. 2.2).

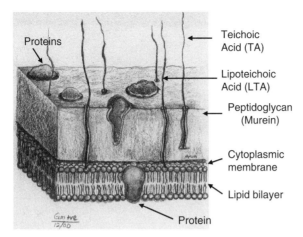

Fig. 2.2 Cross section of Gram-positive bacterial cell wall

Gram-negative bacteria, on the other hand, have outer membrane (OM) layer, a thin peptidoglycan layer and an inner cytoplasmic membrane. The OM consists of lipid bilayer with lipopolysaccharide (LPS) being located on the outer leaflet of the bilayer. The major components of LPS is lipid A, which is a glycophospholipid consisting of β-1, 6-D-glucosamine disaccharide. Phosphate and carboxylate groups of the lipid A provide strong negative charge to the outer surface.

Gram-Positive Bacteria

Cell Wall and Peptidoglycan

Cell wall carries a large numbers of molecules that have multitude of functions. In addition, cell wall protects cells from mechanical damage or osmotic lysis. The major component of the cell wall is peptidoglycan (PGN) and is also known as murein, which consists of peptide with sugar moieties (Fig. 2.2). PGN is highly complex and dynamic structure, which contains a disaccharide N-acetyl-D-glucosamine (NAG) and N-acetylmuramic acid (NAM) and linked by β 1, 4-glycosidic linkage (GlcNAc-(β1–4)-MurNAc) and pentapeptide. The enzyme transpeptidase helps in the formation of peptide crosslink with disaccharide GlcNAc-(β1–4)-MurNAc molecules and provides stability to the peptidoglycan backbone. In *Staphylococcus aureus* the tetrapeptide consisting of Ala, Glu, Lys, and Ala and forms a bridge with pentaglycine (Gly$_5$). In *L. monocytogenes*, pentaglycine is absent but the crossbridge is formed by amide bond between the ε-amino group of a *meso*-diaminopimelic acid (m-Dpm) and the D-Ala of the adjacent cell wall (Fig. 2.3). Peptidoglycan with low degree of crosslinking is much more susceptible to degradation by cell wall hydrolases than the one with high degree of crosslinkers. Penicillin or other β-lactam antibiotics inhibits transpeptidase, hence affect bacterial growth. In addition, β-lactam can activate autolysin, which degrades peptidoglycan. A carbohydrate hydrolyzing enzyme, lysozyme (14.4 kDa) breaks down peptidoglycan. Lipoteichoic acid, teichoic acid, and surface proteins also can prevent lysozyme action. In addition, other enzymes that hydrolyze PGN are; glucosaminidase, endopeptidase (ex, lysostaphin from *S. aureus*),

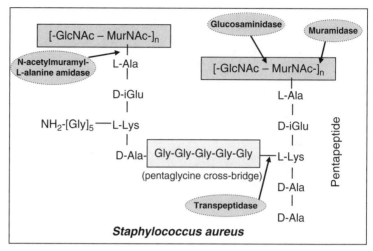

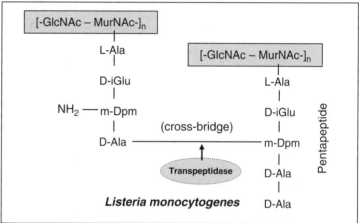

Fig. 2.3 Structure of peptidoglycan from *Staphylococcus aureus* and *Listeria monocytogenes* is shown. Enzymes are marked by oval circle showing their site of action. m-Dpm, *meso*-diaminopimelic acid (redrawn from Navarre, W.W. and Schneewind, O. 1999. Microbiol. Mol. Biol. Rev. 63:174–229)

muramidase, amidase, and carboxypeptidase (Fig. 2.3). Structural integrity and the shape of cells are largely maintained by the presence of intact PGN. When the cell wall is removed, it forms a *protoplast* while the bacterium cell with remnant of cell wall is called *spheroplast*.

Cell wall also carries several surface proteins containing L (Leu) P (Pro) X (any) T (Thr) G (Gly) motif in the C-terminal end, where X could be any amino acid. This LPXTG motif helps bacterial surface proteins to anchor to the peptidoglycan backbone. Examples of proteins that contain LPXTG motif are Internalin in *L. monocytogenes*, the M protein in *Streptococcus pyogenes*, Protein A in *Staphylococcus aureus*, and Fibrinogen binding protein in *S. aureus* and *Staphylococcus epidermidis*. These proteins serve as adhesion factors or binding molecules for these pathogens. Immune response against Gram-positive bacterial pathogens often target PGN. In addition, PGN performs multiple functions. PGN is a strong adjuvant and has been used as one of the components of some of vaccines. It also facilitates antibody production, activates macrophages, nitric oxide production, initiates complement activation through alternative pathway, stimulates cytokine production and suppresses appetite by inducing increased tumor necrosis factor (TNF-α) production (see Chap. 3). The PGN also can be cytotoxic to some cells. In innate immunity against bacteria, toll-like receptor (TLR) of immune cells such as macrophage binds to PGN for recognition. TLR-2 was thought to interact with PGN; however, recently it was determined that the nucleotide binding oligomerization domain protein (Nod)-1 and Nod-2 serve as mammalian pattern recognition receptor for PGN (see Chap. 3 for details).

Teichoic Acid and Lipoteichoic Acid

Teichoic acid is anionic polymer and has a polysaccharide backbone which consists of glycerol or ribitol linked by phosphodiester bonds, i.e., sugar–alcohol–phosphate and buried in the peptidoglycan backbone (Fig. 2.2). Cell wall TA is uniformly distributed over the entire peptidoglycan exoskeleton. The function of TA is not fully known but it is thought that the negatively charged TA captures divalent cations or provides a biophysical barrier that prevents the diffusion of substances and binds enzymes that hydrolyze peptidoglycan.

Lipoteichoic acid is a polyanionic polymer and is inserted into the lipid portion of the outer leaflet of cytoplasmic membrane (CM), travel through the peptidoglycan and is exposed outside the cell wall (Fig. 2.2). Both TA and LTA are unique for Gram-positive bacteria and are absent in Gram-negative bacteria. Function of LTA is unknown but it serves as species specific decorations of the peptidoglycan exoskeleton. LTA and the surface proteins provide unique serotype characteristics of a bacterium and are used for serological classification. LTA with different sugar molecules determines the serotype of the bacteria. This antigenic classification is called somatic antigen or O antigen.

Cytoplasmic Membrane

Cytoplasmic membrane consists of lipid bilayer and carries transport proteins, which binds specific substrate and transport unidirectionally toward the cytoplasm. The proteins are responsible for secretion of periplasmic extracellular proteins, energy metabolism (electron transport system, ATPase), and cell wall synthesis

Gram-Negative Bacteria

Outer Membrane

Cell envelop of Gram-negative bacteria is surrounded by an outer membrane (OM), which consists of phospholipids, proteins and lipopolysaccharide (Fig. 2.4). The OM has a lipid bilayer arrangement; the outer leaflet consists of LPS and the inner leaflet is phospholipid (Fig. 2.4). LPS is one of the most important molecules that contribute in bacterial pathogenesis and immune modulation in host cell. In Gram-negative bacterial pathogens when it is sloughed off, it is referred as endotoxin or pyrogen, and induces cytokines release and raises body temperature (fever). LPS consists of O side chain, which consists of repeating polysaccharide subunits of mannose, rhamnose and galactose, core oligosaccharide, and lipid A, which is very toxic (Fig. 2.5). O side chain is specific for individual strain. OM carries porin and receptor proteins. Porin is a

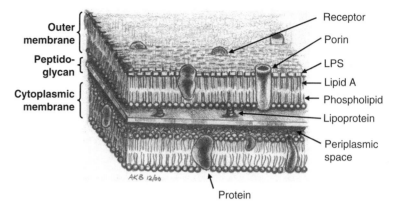

Fig. 2.4 Structural organization of Gram-negative bacterial cell wall and membranes

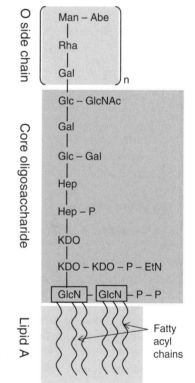

Fig. 2.5 Structure of lipopolysaccharide (LPS) from *Salmonella* Typhimurium. It has three regions; O side chain, core oligosaccharide, and the Lipid A. Abe; abequose; Man, Mannose; Rha, Rhamnose; Gal, galactose; Glc, glucose; GlcNAc, *N*-acetyl-glucosamine; Hep, heptulose; KDO, 2-keto-3-deoxyoctonate; GlcN, glucosamine; P, phosphate; EtN, ethanolamine

selective transport system for small molecules. It prevents access to lysozyme, therefore it cannot reach to the peptidoglycan thus lysozyme is ineffective against Gram-negative bacteria.

Peptidoglycan

Peptidoglycan structure is similar to the structure seen in Gram-positive bacteria but it is much thinner and less defined in Gram-negative bacteria. It is attached to OM by lipoprotein.

Periplasmic Space

Periplasmic space (PS) is the area specific to Gram-negative bacteria and located between peptidoglycan and the cytoplasmic membrane. PS serves as the storage space for numerous enzymes (example, phosphatases), binding proteins (binds nutrients and direct them to the cytoplasmic membrane for transport), oligosaccharides (prevents changes in osmolarity), toxins and some peptidoglycans which crosslinked into gel.

Protein Secretion Systems

Eight types of secretory channels also known as secretion system exist in Gram-negative bacteria that catalyze protein export across the outer membrane or insertion into the inner cytoplasmic membrane. Those systems, designated type I secretory system (T1SS), type III (T3SS or TTSS) and type IV (T4SS) can export proteins across both inner and outer bacterial membranes. Example of type I exporter is, those proteins belonging to ATP binding cassette (ABC). Type III pathway is related to pathogenicity, allowing transportation of cytoplasmically synthesized virulence proteins across the both outer membrane and cytoplasmic membrane in Gram-negative bacteria such as *Salmonella, Shigella*, and *Yersinia* species. It is also known as a "molecular syringe," which directly delivers virulence proteins across the cytoplasmic membrane of the host without exposing the proteins to extracellular milieu. Type IV are present in both Gram-positive and Gram-negative bacteria and exports proteins and DNA–protein complexes from the bacterial cell into the cytoplasm of the recipient cell. This system is very promiscuous and is capable of transporting DNA–protein complex to other bacteria, yeast, mold and plants. For example, VirB in *Agrobacterium tumefaciens* allows the transfer of DNA–protein complex in plant tissue causing cancerous growth. Other examples of Type IV system include CAG and ComB system in *Helicobacter pylori*, Ptl system in *Bordetella pertussis*, and Dot in *Legionella pneumophila*.

Accessory Structures in Gram-Positive and Gram-Negative Bacteria

The accessory structures including fimbriae, pili, flagella, and capsules provide structural integrity and facilitate bacterial colonization, motility, exchange of genetic materials and survival in in vitro and in vivo environments. These structures also serve as important virulence factors for pathogenic bacteria by promoting bacterial adhesion, colonization, motility, and evasion of host immune system (see Chap. 4). Mutations in the genes encoding these accessory structures provide antigenic shift thus allowing pathogens to overcome host immune system. Furthermore, genetic and phenotypic attributes of the accessory structures allow bacterial classification. The following key antigens are used as target for serotyping or serogrouping.

F Antigen or Fimbriae or Pili Antigen

Gram-negative bacteria carry cilia-like glycoprotein structures, which are used by bacteria to exchange genetic materials between organisms or used for attachment to different substrate including host cells and tissues. Fimbrial antigens are also called colonization factor antigens (CFA) and are important in bacterial pathogenesis (see Chap. 4 for details).

H Antigen or Flagellar Antigen

H stands for "Hauch" in German. Both Gram-positive and Gram-negative bacteria may carry one or more whip-like structures in their cell wall. Flagella consist of woven structure of flagellin proteins. Flagella aid in bacterial motility or locomotion and direct them to nutrient-rich environment. Flagella exist as either monotrichous (single polar flagellum) or peritrichous (multiple flagella present surrounding the bacterial cell). In some bacteria, flagella are considered as an important virulence factors aiding in adhesion and invasion of bacterial pathogens such as *Campylobacter jejunii, Legionella pneumophila, Clostridium difficile, Salmonella* Typhi, and *Vibrio cholera*. In some pathogens, flagella are also responsible for secretion of virulence factors.

O (Somatic) Antigen

In Gram-negative bacteria, O antigen represents O-polysaccharide chain of the LPS located in the outer membrane (Fig. 2.5). O-polysaccharide is highly conserved; however, it may be shared among different genera of Gram-negative bacteria. For example, *E. coli* O8 antigen is shared by *Klebsiella pneumoniae* and *Serratia marcescens; E. coli* O157 antigen is shared by *Citrobacter freundii* and *Salmonella enterica*. In addition, proteins in OM may contribute to O serotyping profile. In Gram-positive bacteria, O antigenic profile is shared by peptidoglycan, TA, LTA, and surface proteins.

K Antigen or Capsular Antigen

Capsule is composed of complex polysaccharides (i.e., sialic acid), it is a gel-like structure, takes negative stain (generally stained with India ink), and has antiphagocytic activity. Certain Gram-positive and Gram-negative bacteria produce capsules and examples are *Bacillus cereus, B. anthracis*, and *E. coli*.

Endospore Formation

Some bacterial species forms spores as a survival mechanism. Only a few bacterial genera, namely the Gram-positive *Bacillus, Alicyclobacillus, Clostridium, Sporolactobacillus*, and *Sporosarcina* and the Gram-negative *Desulfotomaculum* species are capable of forming spores. Among these, species of *Bacillus* (*B. cereus, B. anthracis*) and *Clostridium* (*C. perfringens, C. botulinum*) cause foodborne diseases in humans and are considered as agents for bioterrorism. Bacterial spores are called endospores since they are produced inside a cell and there is only one spore per cell. Endospores can be located terminal, central, or off-center, causing bulging of the cell. Under a phase-contrast microscope, spores appear as refractile spheroid or oval structures. The surface of a spore is negatively charged and hydrophobic. Spores are much more resistant to physical and chemical antimicrobial treatments than that of vegetative cells. This is because the specific structure of bacterial spores is quite different from that of vegetative cells from which they are formed. From inside to outside, a spore has the following structures (Fig. 2.6): a protoplasmic core containing important cellular components such as DNA, RNA, enzymes, dipicolinic acid (DPN), divalent cations, and very little water; an inner membrane, which is the predecessor of the cell cytoplasmic membrane; the germ cell wall, which surrounds this membrane, and is the predecessor of the cell wall in the emerging vegetative cell; the cortex, around the cell wall, composed of peptides, glycan, and an outer forespore membrane;

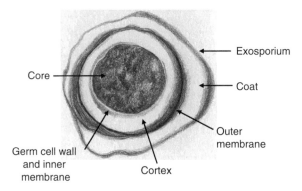

Fig. 2.6 Schematic drawing of a transmission electron microscopic cross section of a Clostridial bacterial spore showing different structures: (1) exosporium, (2) coat, (3) outer membrane, (4) cortex, (5) germ cell wall and inner membrane, and (6) core

and the spore coats, outside the cortex and membrane, composed of layers of proteins such as small acid soluble proteins (SASP) which protect spore DNA from thermal denaturation and other treatments. Spores of some species can have a structure called exosporium outside the coat. During germination and outgrowth, the cortex is hydrolyzed, and outer forespore membrane and spore coats are removed by the emerging vegetative cell.

The spores are metabolically inactive or dormant, and can survive for years, but are capable of emerging as vegetative cells (one cell per spore) in a suitable environment. The life cycle of spore forming bacteria has a vegetative cycle (by binary fission) and a spore cycle, which goes through several stages in sequence, during which a cell sporulates and a vegetative cell emerges from a spore. These stages are genetically controlled and influenced by different environmental parameters and biochemical events, which are briefly discussed here.

Sporulation in bacteria is triggered by the changes in the environmental factors such as reduction in nutrient availability (particularly carbon, nitrogen, and phosphorous sources) and changes in the optimum growth temperature and pH. Transition from cell division cycle to sporulation is genetically controlled, involving many genes. A cell initiates sporulation only at the end of completion of DNA replication and a triggering compound such as adenosine bis-triphosphate (Abt) synthesized by spore formers under carbon or phosphorous depletion may initiate the sporulation cycle (Fig. 2.7). Sporulation events have several stages (1) termination of DNA replication, alignment of chromosome in axial filament, and formation of mesosome; (2) invagination of cell membrane near one end and completion of septum; (3) engulfment of prespore or forespore; (4) formation of germ cell wall and cortex, accumulation of Ca^{2+}, and synthesis of DPN; (5) deposition of spore coats; (6) maturation of spore: dehydration of protoplast, resistance to heat, and refractile appearance; and (7) enzymatic lysis of wall and liberation of spore.

Activation of spores is necessary before spore germination. Spores can be activated in different ways, such as sublethal heat treatment, radiation, high pressure treatment with oxidizing or reducing agents, exposure to extreme pH, and sonication. These treatments probably accelerate the germination process by increasing the permeability of spore structures to germinating agents for macromolecular reorganization. This process is reversible, i.e., a spore does not have to germinate after activation if the environment is

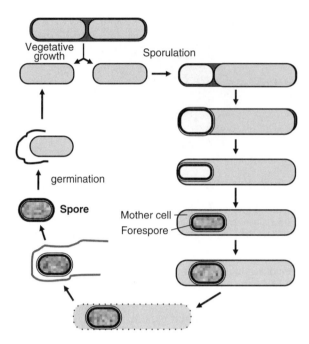

Fig. 2.7 Schematic presentations of the cycles of cell division and endospore forma-tion, germination, and outgrowth of spore forming bacteria (redrawn from Foster, S.J. 1994. J. Appl. Bacteriol. 76:25S–39S)

not suitable. During germination, several structural and functional events take place. Structural changes involve hydration of core, excretion of Ca^{2+} and DPN, and loss of resistance and refractile property. Functional changes include initiation of metabolic activity, activation of specific proteases and cortex-lytic enzymes, and release of cortex-lytic products. Generally, germi-nation is a metabolically degradative process. Germination can be initiated by low pH, high sublethal temperature shock, high pressure, lysozyme, nutrients (amino acids, carbohydrates), calcium-DPN, and other factors. The process can be inhibited by D-alanine, ethanol, EDTA, high concentrations of NaCl, NO_2, and sorbate.

Diseases Caused by Viruses

Introduction

Virus means "poison" in Latin. Virus contamination resembles food poisoning in humans. Sometimes it is also referred as "stomach-flu" characterized by watery diarrhea with nausea and vomiting. Enteric viruses are highly infec-tious and generally a low dose of viruses consisting of 10–100 particles is required to cause foodborne infection (Table 2.2). Viruses are obligate intrac-ellular parasites, i.e., they require a live host for replication and cannot grow outside their specific host and not even the food. They are "host-specific," i.e., a virus can be plant, animal, human, or bacteria-specific and generally do not show cross-species infection. However, the zoonotic viruses are able to infect humans and some viruses undergo genetic modifications to adapt themselves to different hosts. Viruses are metabolically dependent on the host. Animal

Table 2.2 Foodborne viruses

Name	Virus family (genus)	Foodborne
Polio, Coxsackie, echo, enterovirus	Picornaviridae (enterovirus)	Yes, mainly water, present in shellfish
Astrovirus	Astroviridae	Yes, shellfish
Hepatitis A virus	Picornaviridae (hepatovirus)	Yes
Hepatitis E virus	Unclassified	Mainly water
Rotavirus	Reoviridae	Rare often water
Saporovirus	Caliciviridae (saporovirus)	Yes (rare), mainly shellfish
Adenovirus group F, types 40 and 41	Adenoviridae	Shellfish?
Norovirus	Caliciviridae (norovirus)	Yes

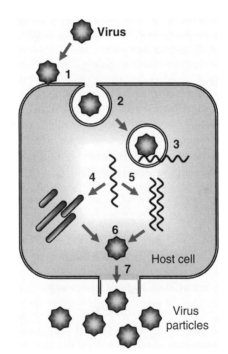

Fig. 2.8 Schematic diagram showing seven steps during viral replication in a host cell: (1) attachment to receptor; (2) penetration, (3) uncoating, (4) transcription and/or translation, (5) replication, (6) assembly of viral proteins and nucleic acids, and (7) release of matured virus particles

cell culture or chick embryos are used for viral growth and isolation. Enteric viruses are generally nonenveloped RNA viruses and they can survive in stomach acid. Environmental survival is critical. Viruses replicate rapidly to yield prodigious titers in cells.

Virus life cycle consists of seven steps (1) attachment to receptor, (2) penetration, (3) uncoating, (4) transcription and/or translation, (5) replication, (6) assembly, and (7) release of matured virus particles (Fig. 2.8). During infection, they destroy mature enterocytes causing decreased absorption/reabsorption resulting in gastroenteritis. Even though the damaged enterocytes are replaced by immature enterocytes, gastroenteritis continues since these immature cells are unable to perform normal physiological function. Viruses are shed in high numbers and are stable in outside environments. Norovirus and rotaviruses are true gut inhabitants while Polio and Hepatitis A virus (HAV) are found in nonenteric locations. Viruses generally do not respond well to conventional antibiotics, hence prevention is only possible through appropriate vaccination or through employing proper hygienic practices.

Significance of Viral Infection

There are four acute gastroenteritis-causing viruses: Calicivirus, Rotavirus, Astrovirus, and Adenovirus. Norovirus (Formerly known as Norwalk-like virus) belong to Calicivirus family, causes about 23 million cases per year. Rotavirus and Astrovirus affect primarily the infants and each is responsible for 3.9 million cases per year. Adenovirus affects children younger than 2. Hepatitis A virus causes gastroenteritis and is responsible for 83,000 cases per year. In general infants are most susceptible and more than 3.5 million of them are affected each year; resulting in 500,000 office visits; 55,000 hospitalizations and 30 deaths. In the US, the foodborne/waterborne cases are about 9.2 million (out of 13.8 million total from all causes). In general, foodborne viruses cause 221,000 cases annually with economic losses totaling around $337 million per year.

Sources and Transmission

Two ways a food can transmit virus. Primary contamination occurs in a food before harvesting such as shellfish, oyster (which concentrate virus particles), crustaceans, clams, vegetables when irrigated with contaminated or polluted water (example, Hepatitis A in strawberry). Secondary transmission occurs during processing or handling of products such as person-to-person transmission (fecal–oral route). Secondary transmission can also occur such as vomitus is aerosolized with virus particles and virus is transmitted to other persons. Asymptomatic carrier such as a food handler without wearing gloves can transmit virus. Other sources are foods that receive no heat treatment, salads, bakery products, raw shell fish, and ice.

Virus Classification/Taxonomy

Viruses are classified based on the size, shape, structure, and nucleic acid content. Viruses are either DNA or RNA virus with single or double stranded nucleic acid molecules. DNA virus has double stranded DNA and RNA virus have generally single stranded RNA molecule. Enteric viruses are found in the gastrointestinal tract and small round structured viruses are called "SRSVs," which are found in feces. Based on the genetic elements, Dr. David Baltimore has classified viruses into seven types (1) dsDNA, (2) ssDNA, (3) dsRNA, (4) (+) sense ssRNA, (5) (−) sense ssRNA, (6) RNA reverse transcribing viruses, and (7) DNA reverse transcribing viruses.

Structure

Common shape of virus is cubical or icosahedral, i.e., they are polyhedron with 20 triangular faces, 12 corners, and spherical or helical shape. The size of the virus is determined by electron microscopy or their ability to pass through a defined porous membrane (filtration unit). They are small particles ranging from 10 to 300 nm. The smallest virus is picorna virus, a causative agent for foot and mouth disease (FMD), and the largest one is the poxvirus

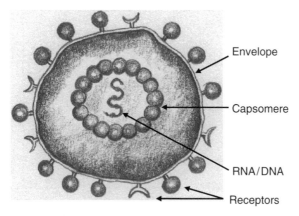

Envelope

Capsomere

RNA/DNA

Receptors

Fig. 2.9 Schematic drawing of a model virus showing various structural components

or vaccinia virus (~300 nm). Virus structure could be symmetrical. Picorna virus has an icosahedral symmetry. Tobacco mosaic virus (TMV) is helical. Herpes, Vaccinia, Polio are spherical. The nucleic acid core is surrounded by a "protein coat" called nucleocapsid, which consists of capsomere. In some cases protein coat is surrounded by an envelope made of lipid bilayer and accessory protein molecules. The envelope is sensitive to solvent. Envelope carries specific receptors that aid viral interaction with host cell receptors. For example, hemagglutinin (HA) and neuraminidase (NA) in human influenza virus or bird-flu virus binds to host epithelial cells (Fig. 2.9).

Adenovirus

It has a large icosahedral structure containing a double stranded DNA. There are 51 serotypes. It causes infection mostly in the upper respiratory tract and also grows well in gut. Only serotypes 40 and 41 induce gastroenteritis and are shed in feces in large numbers. It is found in shellfish. The incubation period is about 3–10 days and the illness lasts for about a week. Adults are immuned and serotypes 40 and 41 are responsible for 5–20% watery diarrhea in children below 2 years of age.

Astrovirus

It is a RNA virus with smooth round structure of 28 nm diameter and sometimes may carry surface projections. Astrovirus has eight serotypes and serotypes 1 and 2 are predominant in children. It causes diarrhea in children and the illness is generally mild. The incubation period is 2–3 days and the disease lasts for about 3–4 days. A major foodborne outbreak was reported in Japan in 1994 affecting 1,500 school children and the teachers. Serotype 4 was responsible for that outbreak. The virus can be cultured using mammalian cells. Electron microscopic diagnosis is generally inconclusive. Enzyme-linked immunosorbent assay (ELISA) is available for astrovirus diagnosis.

Rotavirus

It is a large RNA virus belonging to Reoviridae family. The virus looks like a wheel (rota means a "wheel") with capsid proteins are arranged like spokes of a wheel (Fig. 2.10). Rotavirus is grouped in five (A–E) serogroups and the

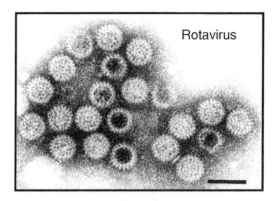

Fig. 2.10 Immunoelectron microscopic picture of rotaviruses (70 nm particles). Bar = 100 nm (adapted from Parashar, U. et al. 1998. Emerg. Infect. Dis. 4:561–570 with permission)

group A, B, and C are known to infect humans. It primarily infects children of less than 4 years of age and causes gastroenteritis. It is responsible for 3.5 million cases per year in the US with a mortality rate of 120 year^{-1}. Worldwide, rotavirus causes 611,000 deaths. It provides a long lasting immunity. A cell culture model (Ma104) is available for group A rotavirus, allowing the study of the pathogenesis of this organisms and augmented our understanding of this pathogen. The incubation period is about 4–7 days and the illness is manifested by diarrhea and vomiting lasting for a week. Virus first attaches and enters enterocytes at the tip of the small intestinal villi and induces structural changes such as villous atrophy and infiltration of mononuclear inflammatory cells in the lamina propria. Rotaviruses are released from infected epithelial cells without destroying them or causing cell death. Nonstructural viral protein called NSP4 has been shown to act as an enterotoxin. It promotes chloride secretion and fluid loss. Chloride secretion response is regulated by phospholipase C-dependent calcium signaling pathway. Viruses are released in the stool in high numbers (~10^9 particles per gram of stool) and can contribute to the fecal–oral transmission. Diagnosis is relatively simple which is accomplished by electron microscopy, agglutination assay or ELISA assay.

Polio

Polio belongs to picornaviridae family which infects man and monkey and the virus is shed in feces. It is a small RNA virus containing surface projections similar to the rotavirus. The virus can survive in stomach acid. It attaches and replicates in the throat and intestine and causes diarrhea. It resides in tonsils, lymph nodes, and blood circulation. One to two percent of infected children develop crippling disease.

Hepatitis A Virus

Hepatitis viruses can be grouped into hepatitis A, C, E, D, and B (cause serum hepatitis). Hepatitis A virus (HAV) and hepatitis E virus (HEV) are known as foodborne pathogens and they are major public health concerns. HAV has seven genotypes and four of them (IA, IB, II, III) are associated with human infection and three (IV, V, and VI) are associated with nonhuman primates.

Although hepatitis is an ancient disease, it was not identified until 1972 when immunoelectron microscopy was used for diagnosis. The most common vehicles are person-to-person (fecal–oral route) contact in a household or in homosexual men, intravenous drug users sharing the same needles, international travelers, contaminated food (feces or vomitus), and water. Foods including clams, mussels, raw oysters, lettuce, ice slush beverages, frozen strawberries, blueberries, raspberries, and green onions have been reported to cause outbreaks. Waterborne outbreak appeared to be less common with HAV. The largest outbreak occurred in 1988 in Shanghai (China) due to consumption of contaminated airy clams and about 300,000 people showed the symptom of acute hepatitis with 47 deaths. Generally, the HAV infection is asymptomatic in children under 6 years of age, while it is symptomatic in older children and adults with jaundice occurring in greater than 70% patients.

Hepatitis A Virus Pathogenesis
Hepatitis viruses are nonenveloped single stranded RNA viruses and belong to picornaviridae family. HAV virus is 27–32 nm in diameter and has icosahedral symmetry. The incubation period of HAV is about 28–30 days, after which symptoms appear. HEV has even longer incubation period of about 60 days. From intestine, HAV moves to liver and replicates inside hepatocytes, cause viremia and then released from gall bladder with bile and is shed in the stool. Infection results in inflammation of liver cells. Viruses impair liver function and as a result, bilirubin accumulates in blood and jaundice develops. Two to three weeks after infection, immune response to virus develops. Consequently, activated immune cells attack virus infected hepatocytes to eliminate the virally infected cells. As a result, hepatocytes are severely damaged manifesting characteristic viral pathogenesis. The major symptom of hepatitis is jaundice manifested by yellow discoloration of skin and the white part of the eye. In jaundice patient, the feces are pale colored and the urine becomes dark. Anorexia, vomiting, malaise, and fever are manifested in the hepatitis patients and virus particles are shed in large numbers in feces (10^9 particles per gram). Viruses are also shed through saliva. Liver failure may occur in patients with underlying chronic liver disease. Children shed viruses longer than the adults.

It is difficult to trace the food source because of long incubation period of the disease. HAV is active for 1 month in the environment and it is resistant to chlorine and requires 1 min exposure to 1:100 dilutions of household bleach (sodium hypochlorite). Inactivation by heating requires $>85\,°C$ for 1 min. Immunization has been effective strategy in reducing the hepatitis cases especially in children. The reduction in children hepatitis cases possibly affects the hepatitis infection cycle thus probably reduces the number of adult hepatitis cases in recent years. However, total numbers of sporadic hepatitis cases have not been reduced.

Detection of virus is achieved by immunological methods such as immunofluorescence assay, radioimmunoassay, dot blot assay, immunoblotting, and ELISA. In the cell culture using African green monkey kidney cells (Vero) and the fetal rhesus monkey kidney cell, virus replication could be detected in 2–4 weeks by immunoassays. Reverse transcriptase polymerase chain reaction (RT-PCR) and real-time PCR have been used to detect virus. More recently nucleic acid sequencing has been done on the PCR amplified products for confirmation and to determine the genetic relatedness among HAV isolates.

Norovirus

First outbreak of Norwalk virus occurred in children and adults in Norwalk, OH (thus the name Norwalk) in 1968 but the virus was not identified until 1972 by Albert Z. Kapikian. The Norwalk virus is now called Norwalk-like virus (NLV) or Norovirus (NoV). NoV has been responsible for numerous outbreaks in various establishments: the cruise ships, restaurants, swimming pool, school, nursing homes, and hospitals. Thirty-nine percent of reported foodborne illnesses in Minnesota from 1984 to 1991 were due to Norovirus. This virus is responsible for 60–80% of all gastroenteritis cases and there are 23 million cases per year (9.2 mill foodborne), with 50,000 hospitalization (20,000 foodborne), and 310 deaths (124 foodborne, i.e., 6.9% of all foodborne cases).

There are several serotypes exists and capsid protein is highly variable. Norovirus is a plus sense single stranded RNA virus, which codes for a RNA-dependent RNA polymerase. It is also known as small round structured viruses (SRSV) of 27 nm diameter (Fig. 2.11). It is nonenveloped and belongs to the family of *Caliciviridae*. Based on the gene sequence of PCR amplified products, virus is now grouped into three genogroups: Genogroup I (GI) infects humans and it has been isolated from Norwalk, Southampton, Desert Shield, and Cruise ships. Genogroup II (GII) infects humans and has been isolated from Mexico, White River, Bristol, Toronto and Hawaii. Genogroup III (GIII) has been isolated from cattle from Sapporo, Parkville, Manchester, Houston, and London.

Norovirus Pathogenesis

There is no cell culture system that can be used to study Norovirus pathogenesis. It has been demonstrated that recombinant virus like particles (rVLP) or rNV (baculovirus) binds to human histoblood group antigen (HBGA) – A, B, O, H, and Lewis. Human milk – oligosaccharide, milk-glycoprotein, and milk-glycolipid contain same epitope as HBGA and can block viral binding

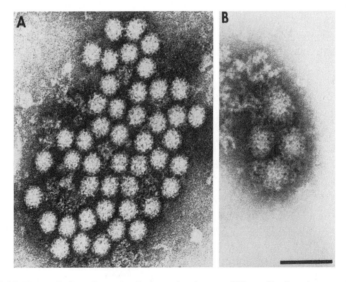

Fig. 2.11 Transmission electron microscopic picture of Norwalk virus (nonenveloped 27 mm sized virus) (adapted from Kapikian, A. 2000. J. Infect. Dis. 181:S295–S302 with permission)

to epithelial cells. Norovirus infects mature enterocytes covering the intestinal villi leading to massive cell damage and malabsorption. Damaged cells are rapidly replaced by undifferentiated immature enterocytes originated from Crypt, which are not susceptible to virus infection. These immature cells cannot function properly thus malabsorption continues until the cells mature. The virus causes gastroenteritis, characterized by explosive projectile vomiting, nausea, cramps, and diarrhea. Adults are more susceptible to Norovirus than children and the incubation period is 24–48 h. Symptoms appear within 12–24 h after ingestion and last for 1–2 days. The disease is self-limiting and the viruses are excreted in the feces of infected persons.

The virus is routinely detected by ELISA and by the RT-PCR assay targeting the RNA polymerase gene. Electron microscopy is used as a confirmatory test.

Bird-Flu Virus

Bird-flu is also known as avian influenza virus and the predominant strain is H5N1. Avian influenza virus infects birds including poultry (panzootic); however, it is capable of adapting and causing infection in humans, especially the poultry handlers and is thus considered a zoonotic organism (Fig. 2.12). Avian influenza viruses have been isolated from wild birds such as geese, ducks, and water fowls and mammals including pig, horse, dog, and sea mammals. The migratory birds can readily transmit the virus to different continents. Though human-to-human transmission has not been confirmed, scientists predict that avian influenza can possibly cause pandemic which is now responsible for epidemic in Asia and Eastern Europe. According to the World Health Organization (WHO), between 2003 and February 2007, avian influenza virus strain H5N1 infection resulted in 271 cases and 165 fatalities worldwide. In the past human history, three major pandemics of Influenza A virus have been reported: 1918 Spanish-flu, 1957 Asian-flu, and 1968 Hong Kong-flu. These three pandemics killed millions of people.

Influenza virus belongs to the family, Orthomyxoviridae. There are three types of influenza viruses; A, B, and C. Influenza virus type A is responsible for human epidemic every year and it was also responsible for past pandemics. Influenza virus A is an enveloped virus and the size is about 80–120 nm in diameter. It is a negative stranded RNA virus with eight different segments and encodes ten proteins including hemagglutinins (HA), neuraminidases (NA),

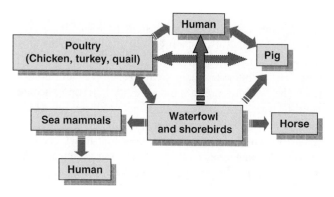

Fig. 2.12 Interspecies transmission pathways for bird-flu virus (avian influenza H5N1 virus)

matrix proteins M2 and M1, nonstructural (NS) proteins NS1 and NS2, the nucleocapsid, and the three polymerase enzymes; PB1 (polymerase basic 1), PB2 (polymerase basic 2), and PA (polymerase acidic).

HA and NA are surface antigens and binds the host epithelial cell receptors to initiate infection. These proteins bring antigenic variation resulting in antigenic shift and drift and allow the virus to evade the host immune system. These antigens allow sustained human-to-human transmission. There are 16 subtypes of HA (H1–H16) and 9 subtypes of NA (N1–N9) and many of these are associated with various animals including humans, dogs, pigs, horses, and birds. The subtypes H1, H2, and H3 and N1 and N2 are associated with human infection; H1 and H3 and N1 and N2 are associated with pigs, and H3 and N8 in dogs. Several different avian influenza virus types have been isolated from different regions or countries: H1N1 (Asia), H4N6 (Canada), H9N2 (China), and H5N1 (Asia). Bird-flu virus is currently named by type/place isolated/culture number/year of isolation. For example the strain isolated from China is designated as B/Shanghai/361/2002 (H5N1).

Bird-flu Virus Pathogenesis

Typically, influenza viruses affect the respiratory tract, resulting in typical flu-like symptoms. The HA of influenza virus binds to epithelial sialic acid containing receptor before initiating infection. In human, this receptor consists of sialic acid-galactose and the crosslink between these two molecules consists of α-2,6 linkage (SA α-2,6) while in birds it is α-2,3 linkage (SA α-2,3). This difference possibly prevents the avian influenza virus from readily infecting humans. Recent studies indicate that receptors with α-2,3 linkages are distributed in the lower part of respiratory tract of humans. Moreover, human isolates of avian H5N1 have shown to display antigenic variation to show binding to human cellular receptor containing α-2,6 linkage. The virus infects and multiplies in nasopharyngeal and alveolar epithelial cells. Virus also exhibits tropism for liver, renal system and other tissues showing signs of diarrhea, renal dysfunction, and lymphopenia.

The symptoms of H5N1 infection in humans appear 2–4 days after exposure and clinical signs include fever, cough, shortness of breath and pneumonia affecting primarily the lower respiratory tract. The patient requires mechanical ventilation and dies within 9 days form the onset of symptoms.

Virus can be cultured using Mardin–Darby canine kidney (MDCK) cell line or embryonated eggs. Viral antigens can be detected from clinical specimen by enzyme immunoassay or immunofluorescence assay, and viral RNA can be detected by using a reverse transcriptase PCR (RT-PCR) assay that targets genes for HA and NA. Antiviral drugs; amantadine, rimantadine, and NA inhibitors (oseltamivir and zanamivir) are found to show serotype specific effectiveness. NA inhibitors (oseltamivir and zanamivir) are shown to be effective against H5N1 in vitro and in a mouse model. Several vaccines based on killed subunit vaccines are under development but those must be able to protect against different strains currently causing infections in humans globally.

Can the bird-flu virus be a food safety concern? It has been demonstrated that conventional cooking temperature (70 °C or more) can readily inactivate the H5N1 virus; however, the virus may not be killed by refrigeration or freezing. If the poultry eggs are properly cooked it would eliminate the virus. The greatest risk of exposure is through handling and slaughter of live infected poultry.

Bovine Spongiform Encephalopathy

A group of infective agents capable of transmission to various hosts are termed transmissible spongiform encephalopathy (TSE). Several TSE agents are described in the literature: Bovine spongiform encephalopathy (BSE) or "mad cow disease" in cattle, "scrapie" in sheep, and Creutzfeldt-Jakob (CJD) or variant CJD (vCJD) or Gerstmann–Straussler syndrome, and "Kuru" in humans. In the early 1950s, in the eastern highlands of Papua New Guinea, Kuru was prevalent among the islanders due to a cannibalistic practice of consumption of infected brain tissues of relatives. The disease was characterized by degenerative brain with spongy appearance and the victims suffered from rapid physical and mental abnormalities, culminating in paralysis, coma, and death. It was called slow virus because the incubation period is about 2–10 years. It became a major concern in early 1990s when the disease was detected in cattle and the wasting of brain tissue resulted in abnormal behavior in cattle and was thus called "mad cow disease." Cattle over 24–30 months of age are susceptible to this infection. Though there is no human case directly linking the consumption of contaminated beef but finding the organism in late 1990s and early 2000 in Canada and the US caused a major beef embargo among developed countries with huge economic impacts. BSE is endemic in UK with reported 122 cases of vCJD. Before 1980, vCJD occurred due to the use of (a) cadaveric human growth hormone, (b) contaminated surgical instruments, (c) infected duramater graft, and (d) corneal transplant. In recent years, however, consumption of contaminated animal products with brain, lymph nodes or neurons are thought to be responsible for transmission. One suspected source of BSE in beef is thought to be due to the feeding of beef cattle with contaminated meat and bone meal (MBM) preparation, to promote fast growth and increased body weight. MBM is often prepared from sheep offal and/or condemned bovines which are not fit for human food. In 1997, FDA banned the use of proteins derived from mammalian tissues in feeding to ruminants in an effort to prevent transmission of TSE to food animals. Conversely, the UK delayed imposing such a ban and about 100 persons developed fatal cases of vCJD between roughly 1996 and 2005.

Infective Agent

BSE was originally thought to be caused by virus but later in 1982, Dr. Prusiner discovered that it is a proteinaceous infectious particle called Prion proteins (PrP), which is resistant to most treatments including heat and chemicals. It is found primarily in central nervous systems (CNS) including brain and neurons and also is found in lymphatic system in the gut. Amino acid sequences of PrP from normal and infected brains are identical but show differences in biochemical and biophysical behaviors. Monomeric form of the PrP protein contains 253 amino acids with a molecular mass of 22–36 kDa while abnormal or infective molecule is a macromolecular aggregates with molecular mass >400 kDa. The normal cellular version of PrP is called PrP^c and is encoded by a single chromosomal gene, *PRNP* located on chromosome 20 in humans. PrP^c is sensitive (PrP^{sen}) to proteases such as proteinase K and trypsin. PrP-mRNA is 2.1 kb long and is detected primarily in brain (neurons) and small amounts in lung, spleen and heart. The infective form is resistant to protease (PrP^{res}) and has a drastically different secondary structure: The α helical structures are predominant in PrP^C while the β structures are abundant in PrP^{res}; Prions are highly hydrophobic and form aggregates easily.

Transmission of prion through digestive tract has been the subject of much investigation in recent years. It has been suggested that prions pass through the M cells overlying the Peyre's patches and then are transported by dendritic cells to the central nervous system (CNS) and the brain. In a separate study, it is proposed that prion bypass the lymphoid system altogether and is directly transmitted via peripheral nervous system to reach to the CNS. PrP accumulates in the neural cells and disrupts normal neurological function, causing vacuolation (spongy appearance) and cell death.

Neurologic symptom is accompanied by cerebral ataxia (defective muscular coordination) and dementia. The psychiatric symptoms include; depression, withdrawal, anxiety, paranoid delusions, head and neck pain, progressive dementia. Mean duration of suffering is about 14 months. In the terminal stage patient becomes bed bound, akinetic and mute, a state in which a person is not able or will not move or make sounds.

Biophysical Properties
Prion is highly resistant to heat and certain chemicals. It can withstand dry heat treatments of 160 °C for 24 h, or at 360 °C for 1 h, and saturated steam autoclaving at 121 °C for 1 h. Prion is also resistant to chemical treatments such as 0.5% sodium hypochlorite for 1 h, 3% hydrogen peroxide for 1 h and ethanol. However, complete inactivation is possible by autoclaving at 132 °C for 1.5 h, and treatment with 1 M sodium hydroxide at 20 °C for 1 h, or sodium hypochlorite (2% chloride) for 1 h at 20 °C.

Detection
There is no test available that can be used in live animals for testing of abnormal PrP^{res}. Postmortem analysis of brain tissues show characteristics amyloid plaque (a waxy translucent substance composed of complex protein fibers and polysaccharides that is formed in body tissues in some degenerative diseases, for example, Alzheimer's disease) and spongy appearance. Immunoassays (Western blot or ELISA) are used to detect PrP antigens in cattle after slaughter.

Prevention and Control of Foodborne Viruses

Enteric viruses are shed in large numbers from host through feces and vomitus and they could be airborne or waterborne. Infectious dose is very low thus effective sanitization and control measures need to be employed to prevent contamination. Person-to-person transmission occurs readily when people are in close contacts, especially in cruise-ships, restaurants and in hospitals.

Depuration help remove the virus from shellfish however proper water temperature should be maintained. During depuration harvested shellfish are kept in clean fresh water for 24–48 h where viruses are escaped into the water. Food preservatives (chlorine compounds, detergents, etc.); freeze-drying; ultraviolet light; freezing; and heating at 100 °C can inactivate foodborne viruses. Viruses are highly stable. Most enteric viruses are RNA virus, which is labile and can be readily hydrolyzed by acid, alkali or radiation. DNA viruses are more stable than RNA viruses. Virus coat protein most likely provides protection against processing treatments. Viruses are remarkably stable at high temperatures such 90–100 °C. Virus particles may remain aggregated or protected by food particles. UV treatment can inactivate the virus. For fruits and salad vegetables, clean virus free-water should be used for irrigation. Food handlers serve as a source thus workers health and hygienic practices should receive greatest attention. Shellfish are potential source of Norovirus and hepatitis A virus and these animals should not be harvested from polluted water.

Bird-flu virus has the potential for causing pandemic thus precaution should be taken to prevent the spread. Contact with infected domestic or wild birds should be avoided. Routine surveillance of migratory birds (dead or alive) or birds in a poultry farm for the presence of influenza virus should be performed. Vaccination of human populations may be needed to control the spread; however, concerns of antigenic variation may challenge its efficacy and effectiveness.

BSE can be prevented by several ways (1) prevent entry of BSE agent in cattle population; (2) stop feeding beef cattle with animal proteins derived from other animals. Currently European Union and the US have banned such practices. (3) Identify and condemn the infected cattle before entering to human food chain.

Diseases Caused by Parasites

Foodborne parasites of interest include protozoa, flatworms, and roundworms. Some are unicellular and some are multicellular. Contamination of products with parasites often indicates inadequate sanitary hygienic practices employed during production and processing. Water, fruits and vegetables are known to be the major source for these groups of pathogens.

Characteristics

Parasites are significantly larger than the bacteria and vary in their sizes and many of them are intracellular. The general, life cycle consists of a stage when they are maintained as oocyst and contaminate products. Oocysts are then converted into larvae and then mature into adults. Parasites do not proliferate in food and are detected by direct means using genetic or antibody-based assays. They cannot be cultured in a growth media but require a live host. They require more than one animal host (intermediate host) to complete their life cycle. Generally, protozoa complete their life cycle in a single host and are known as homoxenous while other parasites require more than one host and are known as heteroxenous.

Protozoa

Giardia
Giardia duodenalis also known as *Giardia lamblia* or *G. intestinalis* is responsible for giardiasis. It is characterized by watery diarrhea referred as Traveler's diarrhea, which occurs in people during travel to countries where sanitary practices are inadequate. Giardia is also sometimes responsible for chronic illnesses. The common source for Giardiasis is contaminated water and vegetables. Cysts are also found in sewage affluent. Wild animals, such as beavers, can carry the protozoa and contaminate water.

Giardia life cycle has two stages: trophozoite and cyst. Infective dose is 10–100 cysts and the incubation period is 1–2 weeks. Once ingested, the cyst (ovoid; 9–12 µm long) arrives in duodenum, is dissolved by digestive enzymes and two trophozoites (9–21 µm long and 5–15 µm wide) are released.

Trophozoites are teardrop shaped and contain two nuclei and four flagella and they display tumbling motility (Fig. 2.13). They multiply rapidly through asexual reproduction and cause damage to the mucus membrane resulting in severe watery diarrhea, bloating and flatulence. Some trophozoites mature into cyst and are released into the feces completing its life cycle. Cysts are infective while trophozoites are not. Treatment with metronidazole or tinidazole can eliminate the infection. *Giardia* is resistant to chlorine treatment that is applied to municipal water supplies.

Entamoeba histolytica

The genus *Entamoeba* consists of several species: *Entamoeba histolytica, E. dispar, E. moshkovski, E. coli, E. hartmanni,* and *E. polecki. Entamoeba histolytica* is pathogenic and the most invasive member of the genus and is responsible for amebic dysentery. Remainders exist as commensal in human intestine. Worldwide, *E. histolytica* causes 50 million cases and 100,000 deaths and the most amebiasis cases are associated with the developing countries including Africa, Central and South America and Indian subcontinent. Amebic cases are also predominant in tropical and subtropical areas. Humans serve as the major reservoir of the organism and can contaminate water and foods. Unsanitary living conditions and unhygienic food preparation practices are major risk factors for the spread of amebic infection. Transmission may occur through contaminated water, fresh foods, sewage, insects, fecal–oral route, or in homosexuals through oral–anal contacts. *E. histolytica* produces cysts (10–15 μm in diameter), which are infective. Ingested organism reach to the lower small intestine (terminal ileum) and upper large intestine, where each cyst gives rise to eight daughter trophozoites. The trophozoites (10–60 μm in diameter) are motile and adhere and invade intestinal epithelial cells. They produce several virulence factors, which promote adhesion and tissue damage. The parasite uses Gal/GalNAc-inhibitable lectin to interact with host glycoprotein for adherence, to block complement activation, and to promote cytotoxicity of neutrophils and macrophages. It produces proteolytic enzymes (collagenase and neutral proteases) and cysteine proteases which promote parasite invasion and tissue damage. It produces amoebapore that forms ion channels in the host cell membrane causing cytolysis. Growth of trophozoites

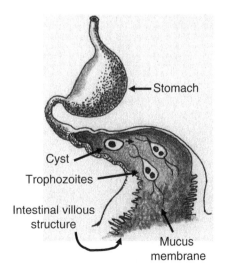

Fig. 2.13 *Giardia duodenalis* life cycle and tissue damage in the intestine

results in inflammation and ulceration in the mucosal layer. Each trophozoite carries single nucleus and can convert into precyst and matures into tetranucleated cysts which are released with feces. Cysts can survive outside for several weeks to months but are sensitive to temperature under −5 °C or over 40 °C.

The incubation period of disease is 1–4 weeks. The symptoms include abdominal cramp and diarrhea which is watery, mucoid or bloody and resembles bacillary dysentery caused by *Shigella dysenteriae* Type 1 and *S. flexneri* (see Chap. 15), and enteroinvasive *Escherichia coli* (see Chap. 10). Diarrhea may occur with ten or more bowel movements per day and some patient show high fever. The patients show anorexia and continue to loose body weights. The disease becomes chronic in some patients, characterized by intermittent diarrhea, flatulence and ulcerative colitis.

Toxoplasma gondii

The disease is called toxoplasmosis which normally is transmitted by domestic cat. It is considered a zoonotic infection. Drinking water contaminated with cat feces was thought to be responsible for several outbreaks worldwide including a recent outbreak in Brazil in 2002. Unwashed hands after contact with pet cats, vegetables washed in contaminated water and eating of undercooked pork are considered as potential source for this parasite.

T. gondii is an obligate intracellular parasite and it has very low host specificity. Its life cycle has two phases: intestinal (enteroepithelial) phase, which is seen in primary host, cats; and extraintestinal phase which is seen in secondary host, i.e., all infected animals. Cat can be infected by eating infected prey (Fig. 2.14). Cysts or oocysts in the meat are dissolved in the lumen of the digestive tract and the bradyzoites are released, which penetrate epithelial cells and undergo asexual multiplication. Later, sexual multiplication follows. Male and female gamets form zygotes, which develop into oocysts and released in the feces. Oocysts mature in the feces and infect warm blooded animals (secondary host) including rodents, farm animals and humans. Oocyst dissolves in the gut and sporozoites are released which then penetrate the intestinal epithelial cells, reach to blood circulation and invade muscle tissues. Sporozoites bind to host cell receptor (sialic acid) and exhibit gliding motility which helps parasite entry in enterocytes. Invasion is an active process for parasites and host cell does not play active role in invasion, i.e., invasion does not alter host cell actin cytoskeleton or phosphorylation of tyrosine residues. Inside the host cell, *Toxoplasma* is trapped in a specially modified vacuole primarily derived from host cell plasma membrane and avoids fusion with normal host endocytic or exocytic vesicle and replicates. *Toxoplasma* then migrates to subepithelial region in the basement membrane and penetrates deep into the submucosa and also disseminates into central nervous system (CNS), retina and placenta.

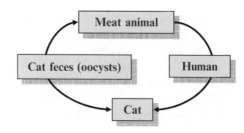

Fig. 2.14 *Toxoplasma gondii* transmission pathway

Human can also acquire by eating raw or undercooked meats, water or food contaminated with cat feces. In healthy individuals, the symptoms appear as flu-like, swollen lymph nodes, fatigue, joint and muscle pain, rash, headache and the disease is usually self-limiting. In immunocompromised host the disease could be fatal. In pregnant human host, there is a significant risk of transmission of parasite to the fetus, resulting in spontaneous abortion. Infection during later stage of pregnancy may be less severe and may not affect the fetus.

An experimental infection induced in mice shares morphologic and histologic characteristics with human inflammatory bowel disease (IBD) which was characterized by loss of intestinal epithelial architecture; shortened villi, influx of inflammatory cells (macrophages, monocytes, neutrophils, and lymphocytes), and scattered necrotic patches. The parasite translocated mostly to muscle and CNS. When the infection is unregulated, inflammatory process resulted in early mortality. Female animals died earlier than the males, suggesting that gender and sex hormones play important role for determining susceptibility of small intestine to *T. gondii*.

T. gondii infection exhibits significant increase in chemokine secretion and macrophage inflammatory proteins and recruitment of polymorphonuclear cells (PMNs) such as neutrophils, and macrophages, monocytes, dendritic cells (DCs), and T-lymphocytes. Further evidence suggests infiltration of neutrophils is essential to remove *T. gondii* during the first few days of infection. Neutrophil-depleted mice exhibit lesions and greater parasite burden. Activated CD4$^+$ T cells produce increased IFN-γ and TNF-α that are responsible for recruitment of inflammatory cells including macrophages for clearance of parasite.

Cryptosporidium parvum

The disease is called cryptosporiodiosis and the major source for this pathogen is water. Agricultural runoff as well as the sewage effluent is the major source. Vegetables or foods are exposed to contaminated water serve as the carrier for this pathogen. The first human case was reported in 1976. There are several species in the genus but *C. parvum* is the most common. Two types of *C. parvum* exist: Type 1 and Type 2. Type 1 is exclusively from humans and Type 2 from cattle or humans exposed to infected cattle. *C. parvum* completes its life cycle in one host (homoxenous). The oocyst is 4–6 μm in diameter and carries four sporozoites inside. The infective dose is thought to be very low ~10 oocysts and the incubation period is 2–10 days. After consumption of contaminated food or water, one oocyst releases four sporozoites, which undergo asexual follows by sexual reproduction and the zygotes are formed.

The sporozoites adhere to epithelial cells, and invade apical surface of the epithelial cells by gliding motility using parasite driven process. *Cryptosporidium* induces host cytoskeletal structure to create a platform made of host actin filaments (α-actinin, ezrin, talin, and vinculin) causing severe damage resulting in gastroenteritis. The disease is characterized by watery diarrhea or cholera-like illness with abdominal cramp, fever, and muscleache. Anorexia, weight loss, dehydration, and abdominal discomforts are generally associated with this disease. In immunocompromised patients (example, AIDS patients), the organism is highly invasive and can infect lungs and bile ducts which is often life threatening. Infected cattle or human shed a large numbers in the feces and the stool samples and used for diagnosis. Oocysts are resistant to chlorine treatment thus contaminated water supply become a threat for infection. A large outbreak of *Cryptosporidium* occurred among HIV patients in Milwaukee (Wisconsin) in mid 1990s due to the consumption of contaminated water from municipal water supply. It was determined that the water was contaminated with cattle manure runoff and the parasite survived the water treatment system.

Cyclospora cayetanensis

There are 17 species of *Cyclospora* exists. *Cyclospora cayetanensis* is the newest member of the coccidian family that is responsible for food and waterborne diseases. Consumption of oocyst contaminated food or water results in cyclosporiasis. The first human case was reported in 1977. Since then, outbreaks of cyclospora have been associated with raspberries, basil and lettuce. In late 1990s, contaminated raspberries from Guatemala caused widespread outbreaks in 20 states in the US and again in 2004, Guatemalan snowpeas caused outbreak in Pennsylvania (US). The oocysts are 8–10 μm in diameter oval shaped and each differentiates into sporocyst. Each sporocyst carries two infective sporozoites (crescent shaped). The oocysts are larger than *Cryptosporidium* oocysts. The infective dose is unknown but the incubation period is typically 1 week. The disease usually lasts for 2 weeks. Oocysts excyst in the small intestine usually in the jejunum and sporozoites infect epithelial cells. The detailed mechanism of infection is not clear; however, histopathological analysis revealed villous atrophy, crypt hyperplasia with inflammation in the small intestine. It causes self-limiting watery diarrhea, fatigue, nausea, vomiting, myalgia, anorexia, weight loss, and sometimes explosive diarrhea. Cyclosporiasis sometimes may lead to chronic sequelae like Guillain–Barre syndrome, reactive arthritis, and acalculous cholecystitis. Trimethoprim–sulfamethoxazole treatment is effective in controlling the infection.

Isospora belli

It causes isosporiasis which is manifested by diarrhea. It is distributed in tropical and subtropical countries including Latin America, Caribbean countries, Africa, Southeast Asia, and Australia. It is generally transmitted through vegetables and contaminated water. Isospora oocysts (10–19 × 20–30 μm) are much larger than *Cyclospora* or *Cryptosporidium*. Oocysts are elliptical, and one oocyst forms two sporocysts and each carry four sporozoites. Like other coccidian protozoa, the infective sporozoites invade intestinal epithelial cells, undergoes sexual and asexual reproduction, and cause tissue damage. Villous atrophy, hypertrophic crypts and inflammation and infiltration of epithelial cells with eosinophils are characteristic to isosporiasis infection. Watery diarrhea, abdominal pain, anorexia and weight loss are symptoms for this infection and the disease can persists for months to years. It could be fatal in immunocompromised hosts and the AIDS patients are very sensitive to this infection.

Flat Worm (Tapeworm)

Two species of flatworms are of major concern to humans; *Taenia saginata* known as "beef tape" is associated with beef, lamb, and fish. *T. solium* is associated with pork hence "pork tape." Length of an adult tapeworm varies between 4 and 12 m. Suckers in the scolex (head) help the worm to remain attached to the intestinal wall. The segments or proglotids could give rise to several new progeny. Gravid proglotids may carry up to 80,000 eggs and each egg represents a single unit. Before eggs are released, the scolex evaginates exposing the suckers outside. Upon ingestion, eggs can move to muscle and the heart. The tapeworms also can cause severe abdominal pain resembling colic-like sharp pain, headache and diarrhea.

Round Worm

Trichinella spiralis

It causes trichinellosis and mostly associated with the consumption of pork products. It completes its life cycle in one host. Larvae grow in the muscle and when the undercooked infected meat is ingested by man, swine, and horse, the disease propagates. Larvae penetrate epithelial lining of small intestine and the sexually matured adults circulate through blood. The cyst is primarily present in muscles but may be found in intestinal mucosa (Fig. 2.15). The disease is manifested by nausea, abdominal pain, muscle pain, diarrhea and vomiting. The cyst could be easily killed by freezing at −18 °C for 30 days or when the meat is stored below 10 °C.

Ascaris lumbricoids

Ascaris lumbricoids, a nematode is associated with food when prepared under poor sanitary conditions. Vegetables from contaminated soil may carry the worms. Children in tropical and subtropical areas carry this parasite for a long period. The disease is manifested by enlarged liver and malnutrition.

Diseases Caused by Mycotoxins

Characteristics

Fungi also called molds (microfungi) are ubiquitous in nature and cause diseases in humans, animals and plants. Fungi infection in mammals is referred as mycoses, which is seen in the form of skin infection (examples, ringworm and athlete's foot). Invasive mycoses acquired by inhalation of spores involve systemic spread of the infection in healthy as well as in immunocompromised hosts and infection can be fatal. Mycoses are caused by two categories of

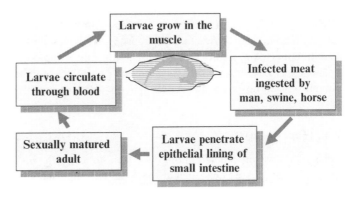

Fig. 2.15 Life cycle of *Trichinella spiralis*

pathogens: primary pathogens such as *Coccidioides immitis* and *Histoplasma capsulatum* and the opportunistic pathogens such as *Aspergillus fumigatus* and *Candida albicans*.

Molds are multicellular microorganisms and a typical mold possesses hyphae, conidiophore – consisting of stalk, vesicle, sterigmata, and conidia (spores) (Fig. 2.16). Often the identification of mold species is done by examining the morphological shape and size of the spores under a light microscope. The cell wall consists of polymer of hexose chitin and *N*-acetyl glucosamine (NAG). Mold spores or hyphae are allergenic to humans and frequent exposure may occur in damp buildings with high humidity with poor air circulations. Three major genera of molds; *Aspergillus, Penicillium*, and *Fusarium* are of significant interest in food safety for production of mycotoxins which are discussed below.

Mold contamination can occur in the field as well as during harvest, processing, transportation and storage. Mold infection in crops causes reduced yield and feeding of meat animals with contaminated feed results in poor animal health, death and huge economic losses. Molds also cause spoilage of foods. Toxigenic molds are major problems in agriculture products

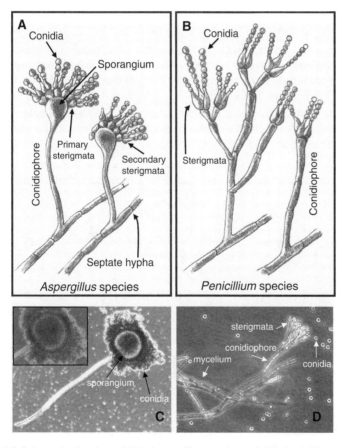

Fig. 2.16 Schematic drawing of (**A**) *Aspergillus* species and (**B**) *Penicillium* species. Panels (**C**) and (**D**) are the light microscopic photograph of *Aspergillus niger* (**C**) and *Penicillium citrinum* (**D**) (magnification 400×) (*See Color Plates*)

such as grains, cereals, nuts and fruits. Mycotoxins (Fig. 2.17) produced in these products can cause severe health problems in both humans and animals. Mycotoxins are produced as secondary metabolites that are low-molecular weight small molecules. Mycotoxins are highly stable and are difficult to destroy by traditional food processing conditions. Under UV light, some mycotoxins emit fluorescence thus can be used for screening of contaminated food/feed stuff. Mycotoxins can cause acute disease manifested by kidney or liver failure or chronic disease including carcinoma, birth defects, skin irritation, neurotoxicity, and death. Three general mechanisms of mycotoxin action are described as mutagenic, teratogenic, or carcinogenic. During the mutagenic action, toxin binds to DNA, especially the liver mitochondrial DNA resulting in point mutation or frame shift mutation due to deletion, addition or substitution in DNA and affect liver function (hence hepatotoxic). Teratogenic action leads to birth defects and the carcinogenic effect cause irreversible defects in cell physiology resulting in abnormal cell growth and metastasis. In recent years, the importance of mycotoxins has been highlighted for their potential use as weapon for bioterrorism.

Aflatoxin

Aflatoxin (*Aspergillus flavus* toxin) is produced by *Aspergillus flavus* and *A. parasiticus* (Fig. 2.17). Aflatoxins occur in different chemical forms; B_1, B_2, G_1, G_2, and M_1. The B and G stand for blue fluorescence and green fluorescence, respectively, which is evident when the samples containing those mycotoxins are exposed to UV light. Aflatoxins are found in nuts, spices, and figs and produced during storage under hot and humid conditions. The allowable toxin limits are 20 ppb in nuts (example, Brazil-nuts, peanuts, pistachio). Aflatoxin contaminated feed causes high mortality in farm animals. Cows fed with aflatoxin contaminated feed convert aflatoxin B_1 into hydroxylated form called aflatoxin M_1 and the toxin is released through milk and the allowable

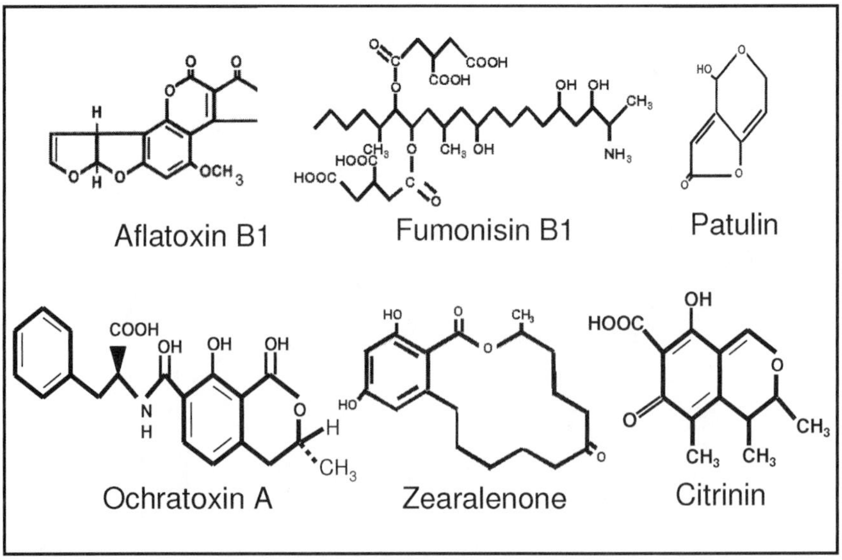

Fig. 2.17 Chemical structures of selected mycotoxins

limit in milk is 0.5 ppb. Allowable limit in meats, corn, and wheat is also 0.5 ppb. The 50% lethal dose (LD_{50}) for rat is 1.2 mg kg^{-1}. The acute lethal dose for adult human is thought to be 10–20 mg. The primary target organ for aflatoxin is the liver. Mitochondrial cytochrome P450 enzyme converts aflatoxin into reactive 8,9-epoxide form which binds to DNA and results in GC to TA transversions, leading to carcinogenesis. Aflatoxin causes gross liver damage, resulting in liver cancer (hepatocarcinogen). It can also cause colon and lung cancer. The International Agency for Research on Cancer (IARC) has classified aflatoxin B_1 as a group I carcinogen.

Ochratoxin

Aspergillus ochraceus and several other species including *Penicillium* spp. produce seven structurally related secondary metabolites called ochratoxin (Fig. 2.17). Ochratoxin is found in a large variety of foods including wheat, corn, soybeans, oats, barley, coffee beans, meats and cheese. Barley is thought to be the predominant source. The toxin is analyzed by using high performance liquid chromatography (HPLC) technique and mass spectrometry. Ochratoxin is hepatotoxic and nephrotoxic and a potent teratogen and carcinogen. Nephropathy and renal pathology are predominant consequences of ochratoxin poisoning. It inhibits cellular function by inhibiting the synthesis of phenylalanine–tRNA complex, and ATP production. It also stimulates lipid peroxidation. The LD_{50} value in rats is 20–22 mg kg^{-1}. The IARC considers ochratoxin as category 2B carcinogen.

Fumonosins

Fumonosins are synthesized by the condensation of amino acid alanine into acetate derived precursor and the most abundant form is Fumonosin B1 (Fig. 2.17). These are produced by *Fusarium verticillioides, F. proliferatum,* and *F. nygamai. Fusarium verticillioides* under ideal conditions can infect corn causing seedling blight, stalk rot and ear rot, and are present virtually in all matured corns. Corns, tomatoes, asparagus, and garlic are the major source of fumonosins. Fumonosins are highly water soluble and they do not have any aromatic structure or unique chromophore for easy analytical detection; however, HPLC with fluorescence detector has been used for detection. Fumonosins are highly stable to a variety of heat and chemical processing treatments. In animals, fumonosins cause varieties of diseases including leukoencephalomalacia, pulmonary edema, and hydrothorax. The toxins are reported to cause esophageal cancers in humans.

Tricothecenes

Over 180 tricothecenes are reported and they produced by number of fungal genera including *Fusarium, Trichoderma, Myrothecium, Stachybotrys, Tricothecium,* and others. The most common tricothecenes are; DON (deoxynivalenol), 3-acetyl DON, and T-2 (Tricothecene-2) toxin. These toxins are associated with several different cereal products, meat and dairy products. They cause hemorrhage in gastrointestinal tract, and vomiting. They are cytotoxic, immunotoxic, and direct contact may cause dermatitis. Tricothecenes inhibit protein synthesis and results in cell death. The toxins can be detected by HPLC and thin layer chromatography.

Patulin

Patulin (Fig. 2.17) is produced by *Penicillium clariform, P. expansum, P. patulum* and by *Aspergillus* spp. Bread, sausage, fruits (apricots, grapes, peaches, pears, and apples), and apple juice are the major source for this toxin. However, patulin does not survive the apple cider making process. Patulin is needed in high dosage to show pathogenesis. The LD_{50} value in rats is 15–25 mg kg^{-1}. It is a carcinogenic toxin and is reported to be responsible for subcutaneous sarcoma. The allowable daily intake limit is 0.4 mg kg^{-1} body weight.

Penicillic Acid

Penicillic acid is produced by *Penicillium puberulum*, and *A. ochraceus*. The biological properties of penicillic acid are similar to Patulin. It is known to cause liver and gastric cancer. The LD_{50} value in rat is calculated to be 60–65 mg kg^{-1}. Wheat, oats, cheese, and coffee beans are reported as source for this toxin.

Zearalenone

Fusarium graminearum and other *Fusarium* species produce zearalenone (Fig. 2.17) and it has strong estrogenic properties and resembles 17β-estradiol, the principal hormone produced by human ovary. Zearalenone is classified as nonsteroidal estrogen or a mycoestrogen. Thus the toxin designation does not appear to be appropriate for zearalenone. Occasional outbreak of mycotoxin in livestock results in infertility. Zearalenone concentrations of 1.0 ppm can cause hyperestrogenic syndrome in pigs and even higher concentrations can cause abortion and other fertility related problems. Reproductive problems are also reported in cattle and sheep. The major concern is that zearalenone can disrupt sex-steroid function in humans. It has been used to treat postmenopausal problems in women and has been patented as oral contraceptives. Zearalenone promote estrus in mice and LD_{50} in rat is reported to be 10,000 mg kg^{-1}. Corn, wheat, oats, and barley are known source for this toxin. The safe allowable limit in human is 0.05 µg kg^{-1} day^{-1}.

Citrinin

Citrinin (Fig. 2.17) is produced by *Penicillium citrinum*, and *P. viridicatum* and it has been also produced by several species of *Aspergillus*. The major source of this toxin is rice, moldy bread, ham, wheat, oates, rye and barley. Citrinin is a nephrotoxin and causes nephropathy in animals. The LD_{50} for citrinin in chicken is 95 mg kg^{-1}; in rabbits, 134 mg kg^{-1} and its significance in human health is unknown.

Alternaria Toxin

Alternaria toxin is produced by several species; *Alternaria citri, A. solani*, and *A. tenuissima*. This toxin is generally associated with apples, tomatoes, and blueberries.

Ergot Alkaloids

Claviceps purpurea produces a toxic cocktail of alkaloid, which is not considered a typical mycotoxin. Ergots grow on the heads of grasses such as wheat and

ryes and the disease is known as St Anthony's Fire because of severe burning sensations in the limbs and extremities of the victim. Two forms of ergotism are reported: gangrenous and convulsive. In the gangrenous form, the blood supply is affected causing tissue damage. In the convulsive form, the toxin affects the central nervous system. The ergotism is a serious problem in animals including cattle, sheep, pigs and chicken resulting in gangrene, convulsions, abortion, hypersensitivity and ataxia. In cattle, ergotism spreads around the hooves and animal may loose hooves and are unable to walk and die by starvation.

Prevention and Control of Mycotoxins

Good agricultural practices (GAP) and good manufacturing practices (GMP) to control molds in preharvest and postharvest crops should be employed. Those include soil testing, crop rotation, irrigation, antifungal treatments, appropriate harvesting conditions, drying, and storage. Traditionally mold control was achieved by controlling the temperature, pH and moister levels of the stored grains, cereals and fruits. In modern day HACCP is employed to reduce mold and mycotoxins in products. Implementation of HACCP is aided by improved analytical techniques for sensitive detection of mycotoxins and stringent regulatory standards to exclude products for human consumption that contain mycotoxin levels over the allowable limits. In addition, development of transgenic plants that are able to increase the insect and mold resistance may aid in reduced levels of mycotoxins in products.

Except supportive therapy, there is no treatment currently available for foodborne mycotoxin poisoning.

Diseases Caused by Seafood Toxins

The majority of seafood toxins are derived from single cell marine plant, algae. Examples of such toxins are ciguatera toxin and saxitoxin, while the scombroid toxin is produced by bacterial decarboxylation of fish proteins. Some of these toxins are acquired by fish and oysters living in the algae contaminated water.

Ciguatera Toxin

Ciguatera toxin is derived from single celled marine plant (macroalgae) – *Gambierdiscus toxicus* and is a neurotoxin cause paralysis. The symptoms appear after about 30 h of consumption with tingling sensation of lips, tongue and throat. It causes nausea, cramps, diarrhea, headache, muscle pain, and progressive weakness leading to paralysis. Fish such as snapper, grouper, barracuda, and seabass live in reefs or shallow water are potential source of this toxin. Toxin occurrence is very seasonal and found in fish in late spring through summer and found in coastal areas of the US; Florida, Hawaii, Puerto Rico, and Virgin Islands.

Scombroid Toxin

Scombroid toxin means mackerel-like. Toxin is produced due to the breakdown of flesh proteins (histidine) by bacterial decarboxylase within 3–4 h.

Bacterial decarboxylation of histidine produces small amines such as histamine, saurine, and cadaverine, which produces typical symptoms for the toxin. Microorganisms that are involved in the decarboxylation process are *Morganella, Hafnia, Citrobacter, Clostridium*, and *Vibrio* species. Symptoms are generally manifested by allergic reaction, which appears within hour of consumption of toxin. The typical symptoms are flush, body rash, headache, shortness of breath, dizziness, vomiting and diarrhea. This toxin is found in sardines, tuna, bonito, yellow tail (amberjack), and dolphin (mahi–mahi) fish.

Saxitoxin

Saxitoxin is found in mackerel, snapper, grouper, seabass, barracudas, mollusks, scallops, and oysters that feed on the alga called toxic dinoflagellates of genus *Gonyaulax catenella*. When dinoflagellates bloom it appears red and referred as "red tide." Saxitoxin is a heat-stable neurotoxin and is responsible for paralytic and neurotoxic shellfish poisoning. The symptoms are tingling, drowsiness, incoherent speech, numbness, rash, and paralysis.

Summary

Cell wall of Gram-positive bacteria contains a thick peptidoglycan (PGN), which is highly complex and dynamic structure containing a disaccharide *N*-acetyl-D-glucosamine (NAG) and *N*-acetylmuramic acid (NAM) and linked by β 1, 4-glycosidic linkage (GlcNAc-(β1–4)-MurNAc) and pentapeptide. PGN not only protects the cells against mechanical or physical damages but also hosts a numerous structural and functional proteins for rigid exoskeleton and for functional attributes such as bacterial pathogenesis and induction of host immune response. Gram-negative cell wall consists of outer membrane (OM) which carries LPS (an endotoxin), a thin peptidoglycan layer and a cytoplasmic membrane. Because of the presence of OM, the protein secretion system in Gram-negative is mediated by several secretory pathways designated as Type I–Type IV secretory pathways. Of which Type III secretion system is known as molecular syringe that delivers bacterial virulence proteins directly to the interior of host cell. Some bacteria produce endospores which are essentially a long term survival strategy for bacteria; however, in recent years endospores from pathogen such as *Bacillus anthracis* are considered as weapon of bioterrorism. Most foodborne viruses cause severe gastroenteritis and affect a large number of people every year. Foodborne enteric viruses are RNA virus and they are shed in large numbers (10^9 particles per gram) from infected patients through vomitus and feces. Person-to-person transmission is a common mechanism for viral infection. Since they are highly infectious, only a small dose of 10–100 particles is required to cause infection. Avian influenza virus (bird-flu) is transmitted primarily through contact or aerosol to the bird handlers, is not considered a foodborne pathogen, but it has the potential to cause human pandemics. Bird-flu infection is highly fatal and affects lower respiratory tract resulting in pneumonia. Transmissible spongiform encephalopathy (TSE) particles such as BSE, vCJD are caused by prion proteins, which are highly resistant to heat and protease enzymes and can be transmitted by consuming contaminated meat. Preventing the use of meat–bone–meal (MBM) can stop the spread of prions among meat-producing animals. Protozoan parasites are increasingly becoming concerns due to their spread through vegetables and fruits. Immunocompromised people are highly susceptible to the intracellular protozoan parasites. Tainted water and soil generally serve as the major source and presence of these pathogens with breaches in hygienic or sanitary practices during food production and harvest allowing them to infect man. Mycotoxins are carcinogenic, mutagenic and teratogenic and are acquired by consuming mold contaminated cereal foods, and meat. Some mycotoxins are hepatotoxic and nephrotoxic and are life threatening. Varieties of mycotoxins may be present in food and sensitive analytical tools are needed to monitor their presence in food. Seafood related toxins are generally derived from microalgae growing in the water from which fishes are harvested. The seafood related food poisoning is manifested by anaphylactic response and is mostly seen in people living in the costal areas because of their increased consumption of seafoods.

Further Readings

1. Bennett, J.W., Klich, M. 2003. Mycotoxins. Clin. Microbiol. Rev. 16:497–516.
2. Boneca, I.G. 2005. The role of peptidoglycan in pathogenesis. Curr. Opin. Microbiol. 8:46–53.
3. Carter, M.J. 2005. Enterically infecting viruses: pathogenicity, transmission and significance for food and waterborne infection. J. Appl. Microbiol. 98:1354–1380.
4. Cousin, M.A., Riley, R.T., and Pestka, J.J. 2005. Foodborne mycotoxins: chemistry, biology, ecology, and toxicology. In Foodborne Pathogen: Microbiology and Molecular Biology. Edited by Fratamico, P.M., Bhunia, A.K., and Smith, J.L., Caister Academic, Norfolk, UK, p 164–226.
5. Dawson, D. 2005. Foodborne protozoan parasites. Int. J. Food Microbiol. 103:207–227.
6. Dormont, D. 2002. Prions, BSE and food. Int. J. Food Microbiol. 78:181–189.
7. Foster, S.J. 1994. The role and regulation of cell wall structural dynamics during differentiation of endospores-forming bacteria. J. Appl. Bacteriol. 76:25S–39S.
8. Kapikian, A. 2000. The discovery of the 27-nm Norwalk virus: an historic perspective. J. Infect. Dis. 181:S295–S302.
9. Lorrot, M. and Vasseur, M. 2007. How do the rotavirus NSP4 and bacterial enterotoxins lead differently to diarrhea? Virol. J. 4:31.
10. Murphy, P.A., Hendrich, S., Landgren, C., and Bryant, C.M. 2006. Food mycotoxins: an update. J. Food Sci. 71:R51–R65.
11. Nainan, O.V., Xia, G., Vaughan, G., and Margolis, H.S. 2006. Diagnosis of Hepatitis A virus infection: a molecular approach. Clin. Microbiol. Rev. 19:63–79.
12. Navarre, W.W. and Schneewind, O. 1999. Surface proteins of Gram-positive bacteria and mechanisms of their targeting to the cell wall envelope. Microbiol. Mol. Biol. Rev. 63:174–229.
13. Ortega, Y. 2005. Foodborne and waterborne protozoan parasites. In Foodborne Pathogens: Microbiology and Molecular Biology. Edited by Fratamico, P.M., Bhunia, A.K., and Smith, J.L., Caister Academic, Norfolk, UK, p 145–162.
14. Parashar, U., Bresee, J.S., Gentsch, J.R., and Glass, R.I. 1998. Rotavirus. Emerg. Infect. Dis. 4:561–570.
15. Peiris, J.S.M., de Jong, M.D., and Guan, Y. 2007. Avian influenza virus (H5N1): a threat to human health. Clin. Microbiol. Rev. 20:243–267.
16. Ray, B. and Bhunia, A.K. 2007. Fundamental Food Microbiology, 4th edition, CRC, Boca Raton, FL.
17. Saier, M.H. Jr. 2006. Protein secretion systems in Gram-negative bacteria. Microbe 1:414–419.
18. Tanyuksel, M. and Petri, W.A. Jr. 2003. Laboratory diagnosis of amebiasis. Clin. Microbiol. Rev. 16:713–729.
19. Sibley, L.D. 2004. Intracelluar parasite invasion strategies. Science 304:248–253.

3

Host Defense Against Foodborne Pathogens

Introduction

Interaction of pathogenic microbes with a host leads to two consequences: disease or recovery. Microbial dominance results in the disease while successful host response averts the full-blown infection. Immune system plays a key role in either of these outcomes. The ability of the human body to protect against the microbial infections is continuously evolving because of continued exposure to different elements. Ironically, microbes are also evolving in the same fashion so that they can adapt themselves in the host. Human body has a rich source of nutrients that supports the growth of both pathogens and commensal microbes. Though the pathogens are the primary disease producing organism, the commensal organisms can also cause disease only under a favorable environment. In most situations, immune system restricts the infection and the spread of disease; however, breach in the immune response or overt reaction to the pathogen presence results in the onset of symptoms. Therefore, to understand the disease process one has to understand the host immune system and the defense mechanism. The host immune system can be compared as "impenetrable fort" impervious to attack and is guarded by soldiers as immune cells, armed with deadly weapons like cytokines, complement proteins and antibodies to stay at an "alert" position to combat sudden or deliberate attack by unsuspected enemies as pathogens. Immune responses are of two types; innate and adaptive.

Innate Immune Response

Innate immune response is the first line of defense against pathogens. It is constitutive but not antigen-specific, and cannot be induced with subsequent exposure to invading agents. Since the foodborne pathogen's major site of action is gastrointestinal tract (Fig. 3.1) and the onset of clinical symptom starts as early as 1 h of ingestion, the enteric defense is the most crucial against intoxication or infection. Enteric host defense is multifaceted and several factors work in concert to achieve protection against pathogens in the intestinal tract such as intestinal epithelial barrier, mucus, complement proteins, antimicrobial peptides (AMPs), resident microbiota, and pattern recognition

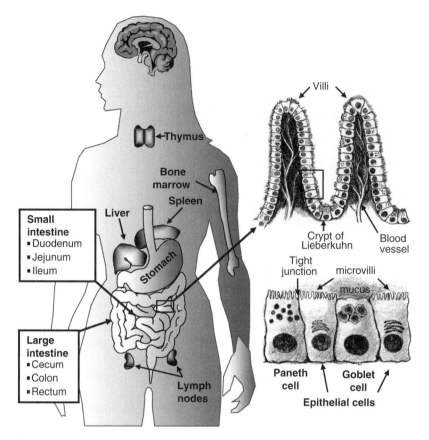

Fig. 3.1 Schematic drawing showing the gastrointestinal tract, various immunologic organs and tissues, and the mucus membrane in human (*See Color Plates*)

receptor of pathogens such as toll-like receptor (TLR) and nucleotide-binding oligomerization domain protein (Nod), which initiate signaling events for enhanced phagocytosis to eliminate pathogens from a host.

Adaptive Immune Response

Adaptive also referred as acquired immune response and is induced or stimulated by the pathogens and the response is generally pathogen-specific. The response increases in magnitude with successive exposure to antigen and the response is mediated by specific antibodies, cytotoxic T-lymphocytes (CTL), cytokines and chemokines. The adaptive immune response is slow to develop and typically requires 4–7 days. During foodborne pathogen infection, adaptive immune response is important for only a limited numbers of pathogens that require a long incubation period and are known to cause systemic infection. These include *Listeria monocytogenes, Shigella* species, *Salmonella enterica, Yersinia enterocolitica,* Hepatitis A virus, *Toxoplasma gondii,* and *Trichinella* species.

Innate Immunity of Intestinal Tract

Intestinal environment is a complex ecosystem and consisting of gastrointestinal epithelium, immune cells, and resident microbes. Pathogens overcome these barriers to cause infection. Pathogens first adhere and colonize in the gut and then modulate the signaling pathway. They use host cell cytoskeleton as target to gain entry. Intestinal environment is also conducive for pathogen survival and growth.

Skin

Skin is the first line of defense against foodborne pathogens since food handlers may easily be infected with bacteria when present in products. Skin is made up of epidermis and dermis layers. Epidermis consists of stratum corneum, stratum lucidum, stratum granulosum, stratum mucosum, and stratum germinativum. The dermis contains glands, adipose tissues, and lymphoid tissue. Skin has a slightly acidic pH (pH 5.0) and acts as a barrier for pathogen entry. Epidermis is dry thus prevent bacterial growth since bacteria require moisture for active replication. Also resident microflora in the skin prevents colonization of harmful bacteria. Lysozyme present in the sweat glands and in the hair follicle has antibacterial effect. Skin-associated lymphoid tissue (SALT) is located immediately underneath the skin and provides specific immune response against infection. Several phagocytic and antigen presenting cells (APC) are also located in the skin for elimination of pathogens; Langerhan's cells (localized phagocytic cells), dendritic cells, and keratinocytes. Keratinocytes in epidermis maintain acidic environments, produce cytokines, ingest and kill skin-associated bacteria like *Pseudomonas aeruginosa* and *Staphylococcus aureus*.

Mucus Membrane

Mucus membrane (Fig. 3.1) forms the inner layer of the gastrointestinal tract, respiratory tract, and urogenital tract. Mucus membrane consists of epithelial cells and depending on the location, the shape of epithelial cells vary, which may be stratified, squamous, cuboidal, or columnar. There are four epithelial cell lineages: epithelial cell, Paneth cells, goblet cell, and enteroendocrine cell.

Epithelial layer environment is moist, warm and suitable for bacterial colonization and growth. Epithelial cells are tightly bonded to form a barrier against pathogens or microbes in the intestine. They are highly selective in translocating molecules across the membrane. Endothelium cells (located in blood vessels) on the other hand, are not tightly bound and they allow movement of blood cells and bacterial cells. The junction between two epithelial cells has four areas: tight junctions, adherens junctions, desmosomes, and gap junctions. Tight junction (TJ) consists of 40 proteins including zona occludins (ZO-1, ZO-2, ZO-3), membrane-associated guanylate kinase family, occludin, claudins, cingulin, 7H6, and phosphoproteins. TJ proteins are regulated by classical signal transduction pathways such as heterotrimeric G proteins, Ca^{2+}, protein kinase C and raft-like membrane microdomain.

Adherens junction (AJ) consists predominantly of cadherin–catenin complex and PDZ proteins. TJ and AJ maintain cell polarization; however, certain pathogen or toxin such as *Clostridium perfringens* enterotoxin and *Vibrio cholerae* ZOT toxin target TJ domain to modulate intestinal permeability.

Epithelial cells also carry several receptor molecules for interaction with microorganisms. Those are called (1) PRR (Pattern Recognition Receptors), (2) Toll-Like Receptors, (3) Nucleotide-binding oligomerization domain proteins, and (4) PAMPS (Pathogen-Associated Molecular Patterns), which binds to bacterial lipopolysaccharides (LPS), peptidoglycan (PGN), and lipoteichoic acids (LTA).

Paneth cells located in the crypt of folded area of mucosal layer are phagocytic cells (Fig. 3.1) and they produce lysozyme and an antimicrobial peptide, cryptidin (a 35 amino acid long peptide), which is inhibitory to *L. monocytogenes, E. coli, S.* Typhimurium, etc. Goblet cells are located in between epithelial cells and secrete mucus and also play a major role in the intestinal natural immunity.

Goblet Cells and Mucus

Mucus is an important natural protective barrier against colonization and infection. Mucus is produce by Goblet cells, which are derived from stem cells located in the crypt and mature as they ascend upward to the villus (Fig. 3.1). They are polarized cells containing apical and basolateral domains. Brush borders (BB), also known as microvilli, are located in the apical side of the Goblet cells. BB consist of actin filament, and actin bundling protein (villin and fimbrin) and is responsible for vesicle trafficking. Mucin, after synthesis, is packaged in secretory granules. Goblet cells exfoliate (die) in lumen after 5–7 days. Proliferation and differentiation is regulated by Wnt and Hedgehog signaling pathways. Exfoliation (occurs in the tip of villi) initiated by a cell death program called "anoikis," which is mediated by focal adhesion kinase or β1-integrin related events, protein kinase signaling pathway (PI-3-kinase/Akt, mitogen-activated protein kinase, stress-activated protein kinase/Jun amino terminal kinase, and Bcl-2). Goblet cell function and composition of mucus are affected by numerous factors including changes in the normal microbial flora, harmful enteric pathogens, and nutritional or nutrients deficiencies.

Mucus consists of a protein called mucin and a polysaccharide, which makes mucus sticky. Mucin is a glycoprotein and the core protein moiety (apomucin) is crosslinked with carbohydrate chains that are attached to serine, and threonine residues by glycoside bond. Mucin forms a high order structure through polymerization of several molecules resulting in large molecule with molecular weights in millions of Daltons. There are three subfamilies of mucin: gel-forming, soluble, and membrane bound. Eighteen genes encode for mucin (MUC) synthesis and mucins are designated as MUC1, MUC2, MUC3A, MUC3B, MUC4, MUC5B, MUC5AC, MUC6–MUC13, MUC15–MUC17. The MUC2, MUC5AC, MUC5B, and MUC6 exist in secreted form and assemble via inter chain disulfide bridges. There are two types of mucin; membrane bound and the secreted form. Membrane bound mucin is highly adherent and remains associated tightly with the membrane while the secreted form is loosely associated. The secreted form remains associated with membrane bound by forming covalent/noncovalent bonds.

Mucin secretion occurs by two processes (1) immediate exocytosis, i.e., the mucin is produced and it is released immediately from the cells by exocytosis and (2) packaging and storage, i.e., mucin is first stored in large vesicles after synthesis and then is released. Release is regulated by specific stimuli – involving activation of signaling pathway. The activators may include acetylcholine, vasoactive peptides, neurotensin, IL-1, Nitric Oxide (NO), and cholera toxin.

In contrast, *Clostridium difficile* toxin A downregulates exocytosis thus favoring bacterial colonization and survival in the intestine.

The thickness of mucus layer varies throughout the gut. Mucus thickness is greater in large intestine than in the small intestine and the thickness increases progressively with descending section of intestine. Mucus contains sIgA, lactoferrin, lysozyme, and lactoperoxidase. Lactoferrin binds iron intimately thus it is unavailable for bacterial utilization. Lysozyme breaks down peptidoglycan of Gram-positive or Gram-negative bacteria; and lactoperoxidase produces superoxide radical "O," which is toxic to bacteria. Ciliated cells in the mucus membrane also propel or sweep away invaders. M (microfold) cells located in the intestine as part of gut-associated lymphoid tissue (GALT) or Peyer's patches deliver pathogens to local lymphoid tissue by transporting them across the mucosa. M cell is naturally phagocytic and allows translocation of intracellular bacteria such as *L. monocytogenes, Salmonella, Yersinia*, and *Shigella*. M cells constitute only less than 0.1% of the epithelial cells in the lining of intestine and are also present in the follicle-associated solitary lymphoid structure throughout the intestine. Mucus-associated lymphoid tissue (MALT) is present as secondary lymphoid tissues and are present in mucus to selectively respond to pathogens.

Mucin binds small molecules and proteins. The complex architecture of mucin performs several functions: it has hygroscopic effect, binds water; gel-filtration effect, allowing selective transport of molecules toward the membrane; ion-exchange effects; and sequestration effect. Membrane-associated mucin serves as receptor for selectins, lectins, adhesion molecules, and microorganisms.

Mucin serves as a barrier for pathogens. It protects and lubricates epithelial surface. It helps in the fetal development, epithelial renewal, carcinogenesis, and metastasis. Mucus gel serves as nutrient for resident flora, promote colonization, and a source of energy. Mucus also protects the cells against inflammation and injury from pathogens or toxins. For example, mucin inhibits the adhesion of enteropathogenic *E. coli* (EPEC), *Yersinia enterocolitica, Shigella*, and rotavirus. However, some pathogens regulate mucin secretion and favor bacterial colonization. For example, *Helicobacter pylori* colonizes mucus and reduces mucin exocytosis. *L. monocytogenes* induces exocytosis of mucin through listeriolysin O (LLO) and inhibits bacterial entry. *Lactobacillus plantarum* induces mucin production and prevents attachment of EPEC.

Antimicrobial Peptides

Antimicrobial peptides (AMP) are produced by fish, frogs, and humans. AMPs are found in airways, gingival epithelium, and cornea, and reproductive tract, urinary and GI tract and possess a broad-spectrum activity. They play a major role in innate immunity and help in pathogen elimination. AMPs are small peptides of 20–40 amino acids long and have α-helix and β-sheets. Intra disulfide bonds formed between the cystine residues stabilize the structure. AMP are produced as inactive prepropeptides and are activated after removal of the N-terminal signal sequence. Their antimicrobial activity is nonspecific. Peptides are inserted into the membrane to form pore by "barrel-stave," "carpet-like," or "toroidal-pore" mechanism. They induce leakage of ions and cause the collapse of proton motive force (PMF). In addition, the peptides may affect the cytoplasmic membrane septum, cell wall, nucleic acid, and protein and enzyme activity in microbes. They are effective against bacteria and enveloped viruses but not against nonenveloped viruses. Pathogens may develop resistance by expelling AMPs by energy-dependent pumps, altering membrane fluidity, and cleaving the AMPs with proteases.

AMP is produced by epithelial cells, Paneth cells, neutrophils, and macrophages. Examples of AMPs are lysozyme, phospholipase A2, α1-antitrypsin, defensin, cryptidin, angiogenin 4, and cathelicidins. Human β-defensin 1, 2, and 3 are cationic and are cysteine-rich peptides of M_r of 3.5–4 kDa and are produced by intestinal epithelial cells. Human α-defensin 5 and 6 are produced by Paneth cells.

Resident Microbiota

All mammals are born without any microorganisms in their system; however, they begin to acquire microbes immediately after birth and eventually every surface exposed to the environment is inhabited by commensal bacteria. Microbes are aerobic, facultative, and anaerobic and proportion of anaerobic bacteria increases from proximal to distal part of the intestine (Table 3.1). Facultative anaerobes are associated close to the epithelial layer for oxygen. The major genera present in the gut are *Escherichia, Clostridium, Bacteriodes, Eubacterium, Bifidobacterium, Ruminococcus*, and *Peptostreptococcus. Lactobacillus* and *Bifidobacterium* remain stable through out life. More than 10^{13-14} bacteria per gram of colonic contents are harbored in the lower gastrointestinal tract of mammals representing about 300–1,000 different species.

Interestingly, host requires the colonization by commensal microbiota for its development and health. Indigenous (autochthonous) bacteria provide essential nutrients and helps in the metabolism of indigestible compounds by synthesizing necessary enzymes. Microbiota also influence the release of biologically active gastro intestinal (GI) peptides, regulate GI endocrine cells, and contribute toward the development of intestinal architecture. They also influence the nutrient absorption, mucosal barrier fortification, xenobiotic metabolism, angiogenesis, and postnatal intestinal maturation. They also prevent colonization by opportunistic pathogens and participate in innate and the adaptive immune system. Pathogens induce proinflammatory response by activating NF-κB and by producing IL-8, IL-1β, and TNF-α. Natural microbiota (example, *Lactobacillus acidophilus*) suppress unnecessary inflammatory response (suppress IL-8 expression) and help maintain immune homeostasis.

Microflora or their components such as LPS, PGN, and FMLP (formylated chemotactic oligopeptide) translocate actively through mucosal barrier (mucus and epithelium) and are found in lamina propria and activate macrophages, dendritic cells, neutrophils, natural killer (NK) cells, and T-cells (Fig. 3.2).

Table 3.1 Natural microbial loads in the gastrointestinal tract

Intestinal segment	Microbial loads
Stomach, duodenum	10^1–10^3 cfu g^{-1}
Proximal small bowel	10^4–10^8 cfu g^{-1}
Terminal ileum to colon	10^{10} cfu g^{-1}
Colon to rectum	10^{13} cfu ml^{-1}

Microflora together with immune system regulate intestinal motility, secretion, proliferation, villous length and crypt depth. Furthermore, the amount of IgA produced is directly dependent on the number of gut flora. For example, the germ-free animals have low levels of plasma cells and IgA, and decreased number of immune cells. Bottom line, the natural microflora help in the development of immune system. Sometimes, immunity against natural microflora may lead to the onset Crohn's disease and ulcerative colitis.

Natural microflora in the gut acts as probiotics and prevent invasion of enteric pathogens. For example, microcin produced by *E. coli* prevents invasion of *Salmonella, Shigella*, and *Listeria. Lactobacillus* and *Bifidobacterium* inhibit invasion by *Salmonella. Lactobacillus* reduces the pathogen-induced drop in transepithelial electrical resistance. Bacteriocins produced by some natural flora can kill certain pathogens in the gut. In summary, AMPs and the microbiota provide frontline defense by inducing mucus secretion and protecting the host against invading pathogens.

Mouth/Throat/Respiratory Tract

Saliva in mouth contains lysozyme, lactoferrin, and sIgA and provides protection against pathogens. Resident bacteria compete for nutrients and the colonization sites with invading pathogens. Ciliated cells sweep away pathogens and also mucus ball help remove pathogens. Epithelial cells are sloughed off periodically to dispose of colonized bacterial cells on them as a strategy to eliminate pathogen. Large particles are sometimes expelled by coughing. Alveolar macrophages in the respiratory tract or nasopharynx also remove pathogens.

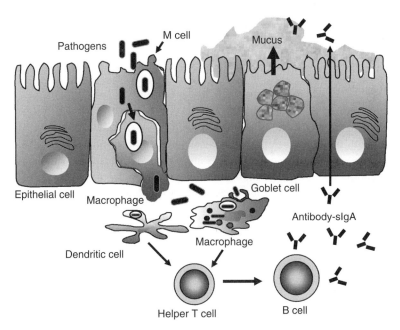

Fig. 3.2 Specific immunity in the mucosal surface. Bacteria are engulfed by M (microfold) cells and translocated across the mucosal barrier. Resident macrophages or dendritic cells can engulf the bacteria, process and present to the T-cells for production of antibody (sIgA) by B-cells. sIgA (secretory IgA) travels through the epithelial cell and are secreted into the lumen

Stomach

Stomach has a low pH (2–2.5) due to the secretion of HCl and most food-borne pathogens are inhibited by acid. However, food can neutralize the pH of stomach to pH 4–pH 5 thus allowing acid-resistant pathogens to pass through the stomach. Also liquid food like milk, soup, and beverages can rapidly transfer bacteria through the stomach thus preventing longer exposure to acid. Psychosomatic conditions or physiological abnormalities, achlorhydria may also raise stomach pH when an individual becomes susceptible to a foodborne infection. Stomach also secretes some proteolytic enzymes that can destroy pathogens. Natural microflora such as *Lactobacillus, Streptococcus*, and some yeast are present at 10^1–10^3 cfu ml^{-1} in the stomach and may contribute toward immunity.

Small Intestine

Small intestine has three regions; duodenum, jejunum, and ileum. In the small intestine bile salts (sodium taurocholate, sodium glycocholate) are secreted from the gall bladder and their detergent-like action help destroy cell walls of Gram-positive bacteria. Also there is a fast flow of mucus and sloughing cells, and peristaltic movement of the intestine may eliminate pathogens from the gut. As mentioned earlier, Paneth cells located in the crypt of folded area of mucosal layer (Fig. 3.1) in small intestine also provide protection against enteric pathogens. These cells possess phagocytic activity and produce lysozyme, and cryptidin that inhibits foodborne pathogens. GALT also removes bacteria by inducing specific immune response. Natural microbiota present in jejunum and ileum are in the range of 10^4–10^8 cfu ml^{-1}. The major genera present in this part of intestine are *Lactobacillus, Enterobactericae, Streptococcus, Bacteriodes, Bifidobacterium*, and *Fusobacterium* and they provide protection against pathogen colonization.

Large Intestine

Large intestine has three segments; cecum, colon, and rectum and is home to an abundant resident microflora (10^{10-14} cfu g^{-1}), and 97% of which are anaerobic. The microbes present in large intestine include *Bifidobacterium, Lactobacillus, Escherichia, Clostridium, Bacteriodes, Enterobactericae, Streptococcus, Fusobacterium, Pseudomonas, Veillonella, Proteus*, yeast, and protozoa. These microbes prevent colonization of pathogens in the gut by physical hindrance and by producing metabolic byproducts such as acids or bacteriocins that are inhibitory to the transient pathogens. Colon also possesses sloughing cells and it secretes mucus which also can prevent colonization and attachment by the pathogens.

Toll-Like Receptors

Toll-like receptors (TLRs) are known as signaling molecules and were originally discovered in *Drosophila* (fruit fly). Several TLRs are now identified in mammalian immune cells and they play a major role in innate immunity. For example they are present in macrophages and are known to form a "microbial recognition" system. Toll-like molecules are analogous to IL-1 receptor. Different TLRs are activated by different types of microbial factors (Table 3.2). For example, TLR-4 is activated by bacterial lipoarabinomannan (LAM), bacterial lipoprotein (BLP), and yeast cell wall particle (zymosan). TLR 4/TLR 6 are activated by LPS and LTA; TLR-5 by bacterial flagella; and TLR-9

Table 3.2 Toll-like receptors (TLRs) and the corresponding ligands

Receptor	Cell types that carry TLR	Ligand
TLR1	Ubiquitous	Triacyl lipoproteins
TLR2	PML, DC, monocytes	Lipoteichoic acid from G(+) bacteria
TLR3	DC, NK	dsRNA
TLR4	Macrophages, DC, endothelial cell	LPS
TLR5	Monocytes, immature DC, epithelial cells, NK, and T-cells	Flagellin
TLR6	High in B-cells, lower in monocytes and NK cells	Diacyl lipoproteins
TLR7	B-cells, DC	ssRNA recognition on endosomes (mouse)
TLR8	Monocytes, low in NK and T-cells	ssRNA recognition on endosomes (human)
TLR9	DC, monocyte, macrophages, PML, NK, microglial cells	CpG (unmethylated DNA)
Nod1		D-Glu-*meso*-DAP (diaminopimelate; G(−) peptidoglycan motif)
Nod2		Muramyl dipeptide from peptidoglycan

PML polymorphonuclear leukocytes, *DC* dendritic cells, *NK* natural killer cells, *Nod* nucleotide-binding oligomerization domain protein

by bacterial DNA. TLR-2 was thought to recognize peptidoglycan (PGN), but recently it was shown that it recognizes lipoproteins or lipoteichoic acid often present in the PGN preparations as a contaminant. It was shown that PGN interacts with Nod-1 and Nod-2 intracellular receptor in mammalian cells. TLR-7 and TLR-8 recognize single stranded RNA and small synthetic compounds. Following recognition of the microbe by TLR on macrophage or on immune cells, cascade of signaling pathways are activated to allow transcription of proinflammatory immune response genes for IL-1, TNF, IL-6, IL-12, etc. The transcription factor NF-kβ regulates the expression of these cytokines. IL-1 and TNF activate T-cells, IL-6 activates B-cells, and IL-12 activates NK cells, which eventually help eliminate the microbes through lyses of target cells or by inducing adaptive immune response.

Other Components of Innate Immunity

Transferrin, a glycoprotein is abundant in liver and it can prevent bacterial growth by sequestering iron. Complement proteins, especially those activated in the alternative pathway by bacterial LPS or LTA or viral proteins are present in the blood circulation. The complement activation products; "C3b" acts as an opsonin, and facilitates enhanced phagocytosis by macrophages; and "membrane attack complex" (MAC) directly destroys microbes by exerting in the cytoplasmic membrane by forming pores. Macrophages, neutrophils, and natural killer (NK) cells also try to eliminate pathogens nonspecifically during the early phase of infection. Macrophage-derived cytokines, α and β interferons (IFN), and TNF also are important during innate immune response. Interferons inhibit viral replication and TNF initiates inflammation to protect against bacteria-induced damage.

Adaptive Immunity

Adaptive immunity is characterized by specific response to an antigen resulting in the production of specific antibody or cytokines. The most important characteristics of adaptive immune response are: it remembers each encounter, i.e., a memory response and the responses are amplified with each successive encounter with the pathogen (Fig. 3.3). Specific immunity can be active or passive based on the form of immune response it generates. Active immunity is developed when host immune system is stimulated by an antigen. For example, immunization with tetanus toxoid (inactive form of tetanus toxin) develops antibody and it is capable of neutralizing tetanus toxin in subsequent exposure. Passive immunity is achieved when serum or cells from a previously immunized host is introduced to a recipient to confer immediate protection against the pathogen. Examples of passive immunity are antibotulinum serum against botulism and antivenom serum against snake bite.

Adaptive immunity is also classified based on the components of immune system that mediate the response, i.e., humoral or cell mediated. Humoral immunity involves production of antigen-specific antibodies by B-lymphocytes. Generally, extracellular antigens such as toxins or soluble protein antigens induce humoral immunity. Cell-mediated immunity (CMI) involves activation of T-lymphocyte subsets that are involved in the elimination of pathogens. Generally intracellular pathogens such as bacteria, virus, or parasites induce CMI.

Characteristics of Adaptive Immune Response

Adaptive immune response is highly specific, diverse, remembers first encounter (memory response), self-regulated, and is able to discriminate self from nonself antigens. (1) Specific immune response is generally raised against structural components of most complex protein or polysaccharides. Antigenic determinants or epitope-specific lymphocyte clones are selected and continue to produce specific antibodies or cytokines during their life span. (2) Specific

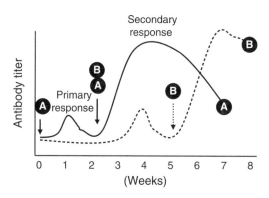

Fig. 3.3 A graph showing the primary and secondary immune response against hypothetical antigens A and B

immune response can respond to large numbers ($\sim 10^9$) of antigenic determinants due to the broad "lymphocyte repertoire" in a host. (3) Specific immune response remembers the first encounter with antigen and the response is enhanced during the subsequent exposure due to the presence of memory cells, which survive for prolonged period even in the absence of antigen. Memory cells respond to very low concentrations of antigen with increased immune response and with higher affinity. Memory response is larger, rapid and qualitatively different from the first. (4) Specific immune response is self-regulated and it does not produce antibodies or cytokines in absence of antigen. Lymphocyte performs the function for a brief period and then converts to memory cell. Feedback mechanism also controls the immune response such as blocking antibody. Antigen–antibody complex also regulates T-cell response and alters cytokine cascade to control immune response. (5) Specific immune response can discriminate self from nonself. It is an important feature of specific immune response to be able to distinguish self-antigen from a foreign antigen. T- or B-cells which are stimulated by self-antigen are generally eliminated from the body. Self-antigens are also known as tolerogens since no immune response is developed to them. Immunologic unresponsiveness is called "anergy." Foreign antigens may act as tolerogens or immunogens depending on the physicochemical state of the antigen, the dosage, and the route of administration used for antigens. Antigens administered orally or intravenously may not elicit response; however, if the same antigen is administered subcutaneously or intraperitoneally can induce immune response. If immune response is generated against a self-antigen, it leads to autoimmune disease. For example, systemic lupus erythematosus (SLE), where antibody is produced against cellular nucleic acids resulting in glomerular nephritis, and vasculitis. In myasthenia gravis, antibody is developed against acetylcholine receptor thus interfering with the transmission of nerve impulses causing muscular weakness.

Phases of Immune Response

Introduction of antigen initiates a cascade of events leading to the immunologic response and the elimination of the pathogen from a host (Fig. 3.4). Each antigen activates separate clone and each clone has the capacity to proliferate to produce specific effector response. In the cognitive phase, antigen binds to the membrane receptor of resting lymphocytes. Membrane bound IgM and IgD serve as receptors. Cells are activated and enlarged in the activation phase and begin to proliferate and differentiate into effector and memory cells. Memory cells remain dormant, circulate in the body, and are involved in immune surveillance while effector cells produce antibodies or cytokines to initiate effector functions through complement activation or activation of phagocytes to neutralize antigens.

Tissues and Cells of Immune System

Tissues

The primary lymphoid organs include bone marrow, thymus, and bursa (present in bird) and secondary lymphoid organs include lymph nodes, spleen, and lymphoid tissues such as gut-associated lymphoid tissue (GALT; Peyer's

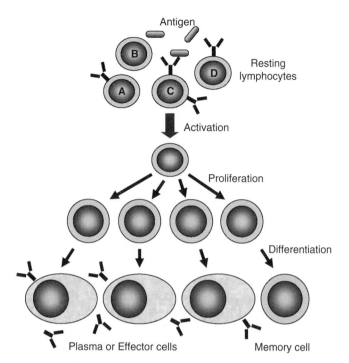

Fig. 3.4 Phases of immune response consist of activation, proliferation, and differentiation into antibody-secreting plasma cells or memory cells. Activation of a clone is antigen-specific

patch) (Figs. 3.1 and 3.2), mucus-associated lymphoid tissue (MALT), and skin-associated lymphoid tissue (SALT).

Bone Marrow

It contains stem cells, which are responsible for hematopoiesis, i.e., all blood cells are originated from stem cells. Growth factors or colony stimulating factors for B-cells (granulocyte–monocyte colony stimulating factor, GM-CSF) and macrophages (G-CSF, M-CSF, IL-1, IL-2, IL-7) are produced in the bone marrow.

Thymus

It is located adjacent to the larynx in the upper part of thoracic cavity (Fig. 3.1) and provides microenvironment for T-lymphocytes maturation and growth. It is a bilobed organ containing multiple lobules. Cortex part of the lobule is the resident site for thymocytes (T-lymphocytes) and cells with various stages of their maturation are found here. Immature T-cells do not carry any receptors. Mature T-cells migrate to medulla and encounter self-antigen for the first time. They differentiate into either CD4$^+$ or CD8$^+$ T-cells or express receptors molecules, i.e., CD4 or CD8 depending on the type of antigens they encounter. Epitheloid cells (giant macrophages) and macrophages located in the medulla present self-antigens to the T-cells. During development in thymus, only T-cells that cannot react to self-antigen are allowed to mature and enter blood circulation.

Lymph Node

It is a secondary lymphoid tissue. It has three regions – from outside in: cortex, paracortex, and medulla. Cortex area has lymphoid follicles containing a distinct germinal center. Germinal center is also called B-cell zone, where

B-cells proliferate and differentiate into the antibody-secreting plasma cells. Macrophage, dendritic and interdigitating reticular cells are also present in this site, which present antigen to B-cells. Paracortex has T-cell zone and it is primarily occupied by helper T-cells (T_H or CD4$^+$) cells. Cytotoxic T-cells (T_C or CD8$^+$) are also found in this area. Medulla is home to matured antibody-secreting B-cells (plasma cells), T_H cells, macrophages, and dendritic cells.

Spleen
It has two regions of interest, white pulp and red pulp. White pulp has periarteriolar lymphoid sheaths that contain lymphoid follicles. Lymphoid follicle has germinal center for B-cells. White pulp also contains CD4$^+$ and CD8$^+$ T-cells. Red pulp contains macrophages, dendritic cells, and the plasma cells, and the antigen-specific interaction takes place in this site.

Lymphoid Tissue
Mucosal immune system such as gut-associated lymphoid tissue (GALT) or Peyer's patch or solitary intestinal lymphoid tissue (SILT) is a regional lymphoid tissue. It contains "M" cells that engulf bacteria and transport them to subepithelial region in the lamina propria. Localized dendritic cells present antigens to T-cells, which in turn activate plasma cells for production of immunoglobulins. The primary immunoglobulin type found in the gut is secretory IgA (sIgA), which is transported to the lumen for neutralization of the antigen. Dendritic cells also transport the pathogens to the deeper tissues such as liver, spleen, and lymph nodes.

Cells of Immune System

Cells that are important in immune response are originated from pluripotent stem cells in the bone marrow and belonged to two distinct lineages; myeloid lineage, and the lymphoid lineage (Fig. 3.5). Cells originated from the myeloid lineage are granulocytes (neutrophil, basophil, eosinophil, and platelets), and phagocytes (monocyte and macrophage) while the cells of lymphoid lineage include, T-lymphocytes, B-lymphocytes, and natural killer (NK) cells.

Monocytes and Macrophages
Monocytes are produced in the bone marrow and circulate in the blood. Matured monocytes differentiate into macrophages and reside in various tissues. These tissue-specific macrophages are known as hitiocytes, epitheloid cells (skin), or multinucleated (polykaryon fused cells) giant cells. Macrophages are also present in various organs and differentiate into alveolar (lungs) macrophages, Kupffer cells (liver), spleenic macrophages (spleen), mesangial cells (kidney), synovial A cells (joints), microglial cells (brain), and Langerhan's cell (skin).

Cytoplasm of macrophage appears granular, which consists of phagocytic vacuoles (phagosomes) and lysosomes. Lysosomal contents have low pH and it contains proteolytic enzymes, reactive "O" radicals, prostaglandin, defensin, and lysozyme. After engulfment, the antigen is trapped inside the phagosome, which then fuses with the lysosome. Lysosomal antimicrobial components inactivate and subsequently degrade the pathogen. Macrophage also possesses surface receptors for antibody (FcR), complement, IL-2, and transferrin. During oxidative burst, NO (nitric oxide) is produced by macrophage and attacks metalloenzyme of bacteria to produce superoxide called peroxynitrite (OONO$^-$), which can cause collateral damage to the host. Macrophage has

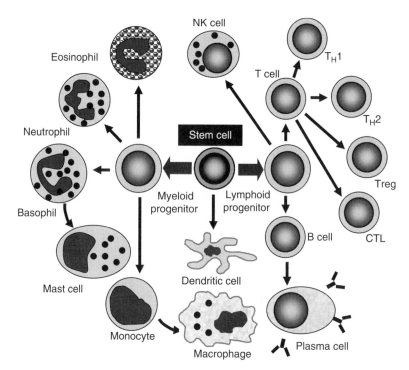

Fig. 3.5 Diagram showing the origin of various immune cells from the stem cells

three distinct functions: professional phagocytosis, antigen presentation, and production of cytokines:

1. *Professional phagocytosis.* As a first line of defense, macrophage engulfs pathogens and destroys them (Fig. 3.6). Macrophages can readily recognize antibody (act as an opsonin)-coated pathogen for elimination and this process is known as opsonization. They remove dead or injured self-cells, tissues, tumor cells, and apoptotic cells and prevent inflammation. The phagocytosis process involves several steps: First macrophage recognize an opsonin (antibody or complement or lectin-like molecules)-coated target using surface receptor or nonspecifically and then forms pseudopod to trap it inside the phagosome. Phagosome fuses with the lysosome which contains proteolytic enzymes, reactive oxygen radicals, and NO that aid in pathogen destruction and degradation. The vesicle containing degraded products are transported outside the cell by exocytosis.

2. *Antigen presentation.* Macrophage is rich in the major histocompatibility complex (MHC) class II antigens and serves as an antigen presenting cell (APC). Macrophage processes the antigen by lysosomal proteolytic enzymes and then presents the peptide fragments using MHC II molecules to T- or B-cells for effector function. The APCs are found in skin, lymph nodes, spleen, and thymus.

3. *Production of cytokines.* After encountering antigen, macrophage releases soluble effector molecules, called cytokines, which recruit inflammatory cells (neutrophils) and produce growth factors for fibroblast and vascular

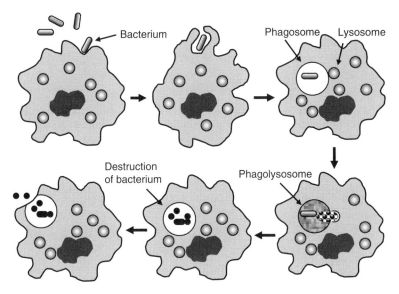

Fig. 3.6 Steps showing phagocytosis and destruction of a pathogen by a macrophage

endothelium to repair injured tissues resulting from inflammation. The cytokines produced by macrophages are IL-1α, IL-1β, TNF-α, IL-6, IL-10, and IL-12.

Granulocytes

Neutrophil
Neutrophils are polymorphonuclear (PMN) leukocytes and contain multilobed nuclei (Fig. 3.7). The granules known as lysosome contain myeloperoxidase, lysozyme, elastase, defensins, and acid hydrolases such as β-glucuronidase and cathepsin B. Neutrophils also contain receptor for antibody (FcR) and complement. They are involved in phagocytosis and chemotaxis, and are attracted to the site of infection during inflammation. Neutrophils are primarily effective against bacterial infection.

Eosinophil
Eosinophil, a polymorphonuclear leukocyte, responds during inflammation. The granules (lysosome) stain intensely with an acidic red dye, eosin, hence the name eosinophil (Fig. 3.7). It contains acid phosphatase and peroxidase. It also carries receptor for IgE (FcR$_\varepsilon$) and other immunoglobulins. It is also involved in phagocytosis, chemotaxis, and is effective against parasitic infection such as protozoa and helminthes, which are often resistant to the degradation by lysosomal enzymes of neutrophils and macrophages. Eosinophil counts are high in immediate hypersensitivity (allergic) reactions.

Basophil
The cytoplasmic granules stain intensely with basophilic dye – hematoxylin, hence the name basophile (Fig. 3.7). They are circulating counterparts of mast cells located in tissue and the granules contain vasoactive amines such as histamine and serotonin. They also carry receptor (FcR) for IgE. Binding of IgE to the receptor allows the release of vasoactive amines and cause immediate

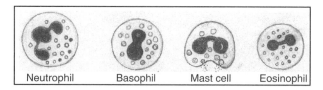

Fig. 3.7 Granulocytes have multilobulated nucleus and contain numerous granules. The granular contents are proteolytic enzymes, "O" radical, or amines, which aid in the destruction of microbes or induction of hypersensitivity reaction

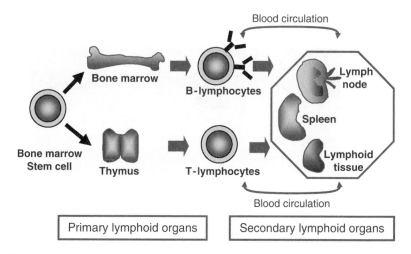

Fig. 3.8 T- and B-lymphocyte lineages: origin, differentiation, and maturation

hypersensitivity reaction. They are primarily effective in removing allergens from the system by increasing the membrane permeability and allowing other cells (NK, macrophages) to reach to the sites. They are nonphagocytic cells.

Lymphoid Tissue

Two populations of lymphocytes are most important: B-lymphocytes and T-lymphocytes (Fig. 3.8). B-lymphocytes are derived from bursa of bird and an equivalent organ, bone marrow of mammals, and they produce antibodies. T-lymphocytes are nurtured and educated in thymus and produce interleukins, interferons and cytokines. The T-cell subsets are designated as cytotoxic T-cells (CTL or Tc), helper T-cells (T_H), and regulatory T-cells (Treg).

T-Lymphocytes

They are originated from the bone marrow and migrate to thymus, where they differentiate into T_H and T_C. They produce cytokines. They also express MHC (class I and II) molecules on their surface and recognize peptide antigen presented in association with MHC molecules with the T-cell receptor (TCR). They do not respond to soluble antigen. They have receptors for immunoglobulins (Ig). They carry specific surface markers which are designated as CD (cluster differentiation) followed by a number. For example, CD2 molecules are expressed on all T-cells, which binds to sheep red blood cells (RBC) forming a rosette (appears like a rose). Numerous CD molecules are identified so far and T-cell subsets are classified based on the presence of different surface CD markers (Table 3.3).

Table 3.3 Selected surface markers for T-lymphocytes

Cluster designation	Distribution
CD1	Thymocytes
CD2	All T-cells-rosette formation
CD3	Mature T-cells
CD4	Helper T-cells
CD8	Cytotoxic T-cells, suppressor
CD4$^+$/CD25$^+$/Foxp3$^+$	Regulatory T-cells

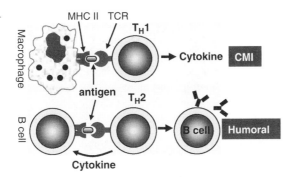

Fig. 3.9 Differences in antigen presentation by helper T$_H$1 and T$_H$2 cells and their respective effector functions. MHC, the Major Histocompatibility Complex; TCR, T-cell Receptor; CMI, Cell Mediated Immunity

Helper T Cells

Helper T-cells are also designated as T$_H$ or CD4$^+$ T-cells because of the presence of unique CD4 surface marker. T-cells are classified as T$_H$1, T$_H$2, T$_H$3, T$_H$0, and regulatory T-cells (Treg). Helper T-cells recognize antigens when presented in association with MHC class II molecules and secrete cytokines, which activate B-cells and macrophages.

T$_H$1. In general, the T$_H$1 cell induces inflammatory type of response. T$_H$1 responds to antigen when presented by macrophages and in turn activates macrophages (Fig. 3.9). They secrete IL-2, IL-3, interferon-γ (IFN-γ), lymphotoxin, granulocyte–monocyte colony stimulating factor (GM-CSF), TNF-β and lymphokines that promote B-cell proliferation. T$_H$1 helps B-cells to synthesize IgG2a. T$_H$1 cells are not stimulated by IL-1 (macrophage-derived). T$_H$1 helps in CMI and delayed hypersensitivity reactions and are preferentially developed during intracellular bacterial and viral infections. IL-12 and IFN-γ produced by macrophages and NK cells support the development of T$_H$1 cells.

T$_H$2. These cells respond to antigen presented by B-cells and play an important role during humoral immune response (Fig. 3.9). T$_H$2 expresses membrane CD30, a member of TNF receptor superfamily. T$_H$2 develops during helminth infection and common environmental allergens, and are responsible for removal of extracellular infective agents. IL-4 production by other cell type favors the development of T$_H$2 cells. T$_H$2 cells secrete IL-3, IL-4, IL-5 (eosinophil activation factor), IL-6, IL-10, IL-13, and GM-CSF. They stimulate B-cell proliferation and isotype switching for IgE, IgM, IgG1, and IgA. IL-4 promotes IgE and IgG1 production from B-cells.

$T_H 0$. These T-cells have less differentiated cytokine profiles than $T_H 1$ and $T_H 2$. They are predominant in the early stages of immune response and induce intermediate effector functions.

$T_H 3$. This subpopulation suppresses the immune response mediated by $T_H 1$ and $T_H 2$ and produce tumor growth factor beta (TGF-β). They may be related to regulatory T-cells (see below).

Regulatory T-lymphocytes. A subpopulation of T-lymphocytes is now recognized as the regulatory T-lymphocytes (Treg or RTL), which controls the immune response in a specific manner. RTL or the Treg cells are originated from thymus and are mostly CD4$^+$/CD25$^+$/Foxp3$^+$ T-lymphocytes. They inhibit proliferation and activity of CD4$^+$ and CD8$^+$ effector T-lymphocytes. They release IL-10 and TGF-β. Treg cell numbers increase in blood and other tissues in cancer patients. Chemotherapeutical drugs like decarbazine prevents tumor proliferation, and also have an immunosuppressive effect on T-lymphocyte populations, as well as on Treg.

Cytotoxic T-Lymphocytes

Cytotoxic T-cells are also called CD8$^+$ T-cells and provide immunity against viral and intracellular bacterial infections. They are responsible for graft rejection, and are the primary effector cells in tumor immunity (antitumor activity). They recognize antigen when presented in association with the MHC class I molecule and kill target cell with the help of cytotoxin (Fig. 3.10). Cytotoxin or perforin (pore-forming toxin) produced by CTL induces apoptosis in target cells. CTL also produces granulysin, which kills intracellular bacteria such as *Listeria monocytogenes*. CTL-mediated killing is antigen-specific, requires cell-to-cell contact; however, CTL itself is not injured during interaction. They secrete IFN-γ, lymphotoxin, and a small amount of IL-2.

Natural Killer Cells

Natural killer (NK) cells are called large granular lymphocytes (LGL) because they carry numerous large cytoplasmic granules. They are called non-T and non-B-cells because of lack of characteristic T-cell receptor like, CD3 or B-cell receptor like, surface immunoglobulin. They are also known as Null cells or lymphokine-activated killer (LAK) cells. They secrete IFN-γ and IL-2 and are effective against tumor cells or virus infected cells. The target cell killing is generally nonspecific, i.e., not restricted to MHC; however, specific target cell destruction is achieved by using immunoglobulin and is termed as antibody-dependent cell cytotoxicity (ADCC) (Fig. 3.11).

Cytokines

Cytokines are protein hormones produced by immune cells. Cytokines produced by mononuclear phagocyte (monocytes) are called monokines, by activated T-lymphocytes are called lymphokines and by leukocytes are called interleukins. Cytokines play important role in natural immunity and adaptive immunity and regulate lymphocyte activation, growth and differentiation. Cytokines also activate nonspecific inflammatory cells and stimulate immature leukocyte growth and differentiation.

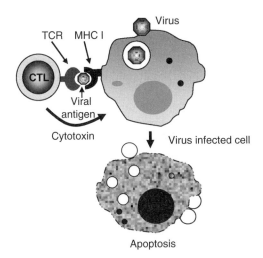

Fig. 3.10 Cytotoxic T-lymphocytes (CTL)-mediated killing of target cells

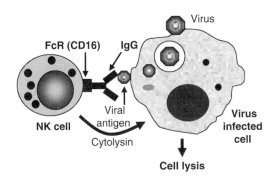

Fig. 3.11 Specific destruction of virus infected cell by the natural killer (NK) cell with the help of virus-specific antibody. This event is also called antibody-dependent cell cytotoxicity (ADCC)

Properties of cytokines are:

1. Cytokines are produced during the effector phase of natural and specific immunity. Example, LPS activates macrophage, which produces IL-1 during natural immunity. IL-1 stimulates thermoregulatory center that increases body temperature (fever) to inhibit bacterial growth.
2. Cytokine secretion is brief and self-limiting because the cytokine mRNA is unstable.
3. Many individual cytokines are produced by multiple diverse cell types. For example, IL-2 is produced by T, NK cells, while IL-6 is produced by B-cell, T-cell, macrophage, fibroblast, and endothelial cells.
4. Cytokines are produced by different cells and act upon different cells and cytokine action is redundant, and many cytokines have similar function.
5. Cytokines often influence the synthesis of other cytokines and can have antagonistic or synergistic action.

6. Cytokine action is mediated by binding to the receptor. Cytokine action can be autocrine, i.e., the action is on the same cell, it can be paracrine, i.e., the action is on the bystander cell or the endocrine, i.e., the action is on distantly located cells (secreted into the circulation).
7. Cytokine regulates cell division, for example, GM-CSF helps granulocyte–monocyte cell division and growth.

Cytokines in Natural Immunity

Two major cytokines are produced during natural immunity (1) interferon (IFN) and (2) tumor necrosis factor (TNF). Cytokines initiate inflammatory reactions that protect against bacterial and viral infections.

Interferon

Interferons are of two types; Type I and Type II. Type I interferon is again divided in two depending on the molecular weights and the type of cells that produce them: IFN-α and IFN-β. IFN-α is an 18-kDa polypeptide and is called leukocyte interferon since it is produced by mononuclear phagocytes. IFN-β is a 20-kDa polypeptide and is called fibroblast interferon since it is produced by fibroblast cells. Type I is involved in the natural immunity and inhibits the viral replication by establishing the antiviral state in infected cells. Type I IFN stimulates production of oligoadenylate synthetase, which acts on adenosine triphosphate (ATP) to produce adenine trinucleotide (AT). AT activates RNase L (endoribonuclease), which cleaves the viral mRNA thus inhibits viral protein synthesis. Interferon action is species-specific but not virus-specific.

Example of Type II interferon is IFN-γ, which is responsible for specific immunity. IFN-γ is 21–24 kDa polypeptide and is produced by CD4$^+$, CD8$^+$, and NK cells. IFN-γ activates mononuclear phagocytes, increases MHC Class II expression, aids in T- and B-cell differentiation.

Tumor Necrosis Factor

Tumor necrosis factor (TNF) is one of the major cytokines in innate immunity and it consists of three subunits of 17-kDa each with a total molecular mass of 51 kDa. It is produced by macrophages, T and NK cells in response to Gram-negative bacterial infections primarily due to the release of elevated levels of LPS from bacteria. TNF-α also stimulates macrophages to produce IL-1. TNF-α together with IL-1 raise body temperature (fever) through prostaglandin-mediated pathway to inhibit bacterial growth. TNF-α blocks lipoprotein lipase action, therefore the fatty acids are not released from the lipoprotein for use by the body cells as a result the patient looses muscle mass and body weight and leads to cachexia (wasting condition characterized by loss of body mass). The cachectic condition is also mediated by TNF-induced appetite suppression in host. Large amounts of TNF-α release affects heart muscle contraction and vascular smooth muscle tone resulting in lowered blood pressure and shock.

Interleukins

Interleukins (ILs) are produced by leukocytes in response to antigens and they serve as effector molecules in innate and adaptive immune response by activating other leukocytes. Several interleukins are identified and their source and functions are summarized in Table 3.4.

Table 3.4 Summary of interleukins and their target cells

Name	Molecular weight (kDa)	Produced by	Target cells
IL-1α	17	Macrophages	T, B
IL-1β	17	Macrophages	T, B
IL-2	15	T, NK	T (T_H, CTL), B, NK, macrophages
IL-3	28	T	Hematopoietic stem cells, B, mast, macrophages
IL-4	20	T_H2	Stem cell, T, B, macrophages
IL-5	18	T_H2	T, B, eosinophil
IL-6	28	T, B, macrophage, fibroblast, endothelial cells	B, T, stem cells, neurons, hepatocytes, macrophages
IL-7	25	Bone marrow stromal cells	Pre-B, pre-T, and T-cells (immature lymphocytes)
IL-8	8	Macrophages, lymphocytes, hepatocytes, endothelial cells	T, neutrophil, basophil (chemoattractant)
IL-9	32–39	T	T_H cells, none on CTL
IL-10	19	T_H2, B-cells	T_H1
IL-11	24	Stromal cell	B-cells, platelet production
IL-12	35	NK, macrophage	T_H2
IL-13	40	T_H2	
IL-15	13	Macrophage	NK, T-cells
IL-18	17	Macrophage	NK, T-cells

B-Lymphocytes

B-lymphocytes express surface receptors for antigen. The B-cell receptors are in fact membrane bound antibody. Antibody secreted from B-cells is present in the γ-globulin fraction of the serum and hence called immunoglobulin (Ig).

Structure of Immunoglobulin

A typical immunoglobulin (Fig. 3.12) consists of two identical heavy chains and two light chains. Two heavy chains (55 or 70 kDa) are joined by two disulphide bonds in the hinge region. Heavy chain has one variable domain called V_H and three constant domains designated C_H1, C_H2, and C_H3. Variable region (V_H) has hypervariable region, which is also called complementarity determining region (CDR). CDR again has three regions; CDR1, CDR2, CDR3, of which CDR3 is the most variable region and is located close to the N-terminal end. Two light chains, κ and λ (24 kDa each) contain one variable (V_L) and one constant (C_L) domain. Light chains are linked to heavy chain by disulphide bonds. Each structurally distinct unit or domain consists of 110 amino acids and is joined by disulphide bridges. Based on the heavy chain types, immunoglobulins are classified as IgA, IgG, IgM, IgE, and IgD (Fig. 3.13). IgA is made with α-chain, IgG with γ, IgE with ε, IgD with δ, and IgM with μ-chain (Table 3.5).

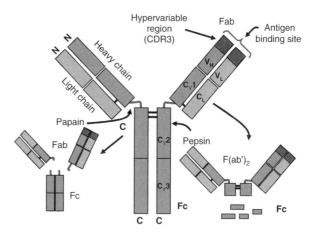

Fig. 3.12 Structure of immunoglobulin and the products of proteolytic (papain and pepsin) digestion

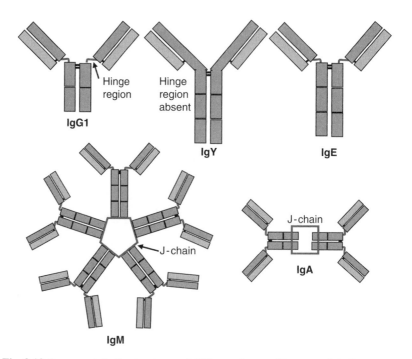

Fig. 3.13 Immunoglobulin structures of different classes of immunoglobulin

Table 3.5 Classes and subclasses of immunoglobulins.

Class	Heavy chain type	Subclass	Molecular weight (kDa)
IgA	α1, α2	IgA1, IgA2, sIgA	150, 300, or 400
IgD	δ	None	180
IgE	ε	None	190
IgG	γ	Mouse: IgG1, IgG2a, IgG2b, IgG3	150 each
		Human: IgG1, IgG2, IgG3, IgG4	
IgM	μ	None	950 (pentamer)

Antigen-binding end, which is also the N-terminal end is called Fab and C-terminal end is Fc (fragment crystalline), which binds Fc receptors present on various cell types. The Fc region contains carbohydrate molecules and complement-binding site. Proteolytic enzyme is used to cleave Fab from Fc region. Papain cleaves immunoglobulin molecule in the hinge region resulting in two separate Fab molecules and one Fc molecule, while pepsin cleaves the hinge region below the disulphide bond to produce an intact $F(ab)_2$ appears as the letter "V" and fragmented Fc.

Classes of Immunoglobulins

IgA. It consists of a heavy chain type α1 or α2, and exists either as a monomer or a dimer. In the dimeric form, a joining chain (15 kDa) holds two monomers together. IgA is found in serum at low concentrations (3 mg ml⁻¹). Dimeric IgA also exists as a secretory form (sIgA) where a secretory molecule of 70 kDa is attached, which protects sIgA from proteolytic degradation and helps in secretion. sIgA is about 400 kDa and is abundant in saliva, mucus and other body fluids.

IgE. It is about 190 kDa and it contains additional domain in Fc region. Generally IgE level is very low in blood (0.05 mg ml⁻¹), but its concentration increases during allergic response or during parasitic infection such as helminthes infection.

IgD. It is about 180 kDa and it is mostly remains membrane bound thus it is present in serum in trace amounts. IgD acts as a receptor for antigen when B-cells act as an APC.

IgM. It is present in pentameric form (950 kDa) where 5 units are joined by a joining chain (J-chain). This is the predominant antibody in blood (1.5 mg ml⁻¹) during primary immune response when challenged with an antigen. IgM is very unstable and rapidly looses its activity if it is subjected to temperature abuse.

IgG. Molecular mass of IgG is 150 kDa. Subclasses of IgG vary between mice and humans depending on the heavy chain type. In mouse, subclasses are IgG1, IgG2a, IgG2b, IgG3; in human – IgG1, IgG2, IgG3, IgG4. IgG concentration in blood is very high (IgG1, 9 mg ml⁻¹; IgG2, 3 mg ml⁻¹; IgG3, 1 mg ml⁻¹; IgG4, 0.05 mg ml⁻¹) and is the predominant immunoglobulin during secondary immune response. IgGs are very stable.

Chicken antibodies are consisted of three immunoglobulin subclasses; IgA, IgM, and IgY. IgA and IgM are similar to mammalian IgA and IgM, while IgY is equivalent to mammalian IgG. These antibodies are found in serum as well as in eggs. In egg, IgA and IgM are present in albumen in low concentrations (0.15 and 0.7 mg ml⁻¹, respectively) while IgY is found in yolk in large quantities (~25 mg ml⁻¹). Structurally IgY (180 kDa) is larger than the IgG (150 kDa) and it can be readily harvested in large quantities from eggs. The H-chain in IgY is 68 kDa and consists of four constant domains (Cv1–Cv4) instead of three for IgG. IgG constant domain Cγ1, Cγ2, and Cγ3 are equivalent to IgY Cv1, Cv3, and Cv4, respectively. IgY lacks a hinge region (Fig. 3.13).

Diversity of Antibodies

Total number of antibody specificities that an individual can produce is called "antibody repertoire." There are about 10^7–10^9 different antibody molecules with unique amino acid sequences in the antigen-binding site. This generates diversity in a host. Variable region has hypervariable region (CDR), and the unique determinant of the CDR varies from antibody to antibody and is called *idiotope* while the collection of idiotopes on a particular antibody molecule constitute *idiotype*.

Antibody Production

Sequences of events that take place for B-cells to produce active antibody molecules for a specific antigen begin in the stem cell (Fig. 3.14). B-cells originate from stem cell and give rise to pre-B, immature B, and mature B-cells. Immature or mature B-cells leave bone marrow and migrate to secondary lymphoid organs (lymph nodes and spleen) and transform into activated and antibody-secreting B-cells. Isotype switching (Fig. 3.15) takes place at this stage in the presence of antigen and B-cells can produce antibody subclasses of IgM, or IgA or IgG.

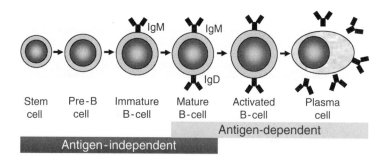

Fig. 3.14 B-cell maturation, growth phases, and synthesis of immunoglobulin

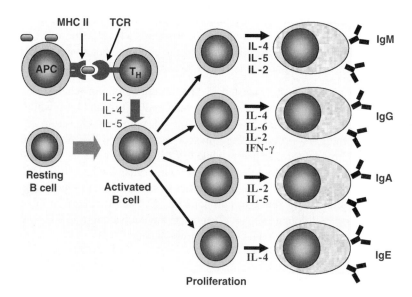

Fig. 3.15 Steps showing isotype switching of immunoglobulins during humoral immune response

Function of Antibody

Acts as B-cell receptor. Membrane bound antibody binds antigen which also leads to processing and presentation of the peptides to T-cells via MHC II molecules (Fig. 3.9).

Neutralization of antigen. Antibody neutralizes toxins, virus, or bacteria. It binds to the antigenic determinant and prevents the antigen from interacting with the host cell by "steric hindrance."

Activation of complement. IgG or IgM after forming a complex with the antigen initiates complement cascade in classical pathway to produce complement end products (C3b, membrane attack complex (MAC)) that facilitate lysis of the microorganisms.

Opsonization. Antibody also serves as an opsonin, a molecule that facilitates the recognition of target microorganisms by phagocytes (for example, macrophages) for elimination. Antibody first forms a complex with microbes and then the Fc part of the antibody binds to the Fc receptor on the phagocytic cells (macrophage or neutrophil) for engulfment and destruction.

Antibody-dependent cell cytotoxicity. NK, neutrophil, eosinophil recognize and specifically destroy target cells when coated with antibodies (IgG, IgE, IgA). Target cell coated with antibody binds to the Fc receptor (CD16) of phagocytic cell for specific recognition and destruction of virus or tumor infected target cell (Fig. 3.11).

Immediate hypersensitivity reaction by IgE. Basophils or mast cells have Fc receptor (FcεR1) for IgE. IgE forms a complex with antigen (allergen) and binds to the Fc receptor of mast cell and causes release of vasoactive amines (histamine, leukotrienes, or prostaglandins) that are responsible for immediate hypersensitivity reaction (Fig. 3.16). Examples are hay fever, asthma, and food allergy.

Mucosal immunity by IgA. Secretory IgA (sIgA) is abundant in mucus and other bodily fluids and provides specific or nonspecific immunity at the mucosal surface (Fig. 3.2).

Neonatal immunity-mediated by IgG. Colostrum milk produced by mother immediately after birth of fetus contains a high level of maternal IgG, which protects neonates against pathogens in the early part of its life.

Feedback inhibition of immune response is mediated by IgG. Binding of IgG to Fcγ receptor on B-cells inhibits activation of B-cells and this process is called antibody feedback inhibitions.

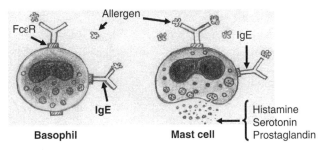

Fig. 3.16 Allergen-mediated activation of basophil and mast cell for secretion of vasoactive amines in the IgE-mediated hypersensitivity response

Antigen

A foreign molecule that is capable of stimulating immune system is called antigen. Most effective antigens are large, rigid, stable, and chemically complex. Factors that influence antigenicity are:

1. *Molecular size of the antigen.* Large molecules are more antigenic than the small molecules. For example, 14.4 kDa lysozyme is better antigen than 1 kDa angiotensin, the 69-kDa albumin is better antigen than lysozyme; γ-globulin (156 kDa) is better than albumin; fibrinogen (400 kDa) is better than γ globulin; and IgM (900 kDa) is better than the fibrinogen.
2. *Complexity of antigen.* LPS and proteins are complex, whereas lipid, carbohydrate, nucleic acids are polymers of repetitive units and are considered poor antigen.
3. *Structural stability of antigen.* Immune response recognizes shape; flexible molecules have no shape thus considered poor antigens. For example, gelatin is wobbly and is a poor antigen. Similarly flagella are structurally unstable thus are considered poor antigens.
4. *Degradability of antigen.* Antigen should be degraded relatively easily for processing and presentation by APC. Stainless steel pin, plastic joints are made of large inert organic or inorganic polymers, which immune system cannot degrade and are recognized as poor antigens. Proteins made of D-amino acids are poor antigen because immune cells cannot degrade them.
5. *Foreignness.* Immune system does not respond to self-antigen, the immune cells which respond to self-antigens are eliminated or turned off. Degree of foreignness also determines the antigenicity.

Types of Antigens

Bacterial antigens include cell wall or somatic antigen (O), capsule (K), pili or fimbriae (F), flagella (H), cell membrane, proteins, ribonucleoprotein, enzymes, and toxins (exotoxin).

Viral antigens include envelope (lipoprotein, glycoprotein), capsomere (capsid), and receptor proteins. Cell surface antigens of mammalian cells include (1) blood group antigens, ABO, Rh, MN, Kell, Duffy, Luthern, and Lewis; (2) histocompatibility antigens (MHC class I, class II); (3) cluster differentiation (CD) or lymphocyte surface antigens; and (4) autoantigens such as hormones, myelin, and DNA. Protein antigens are called T-dependent antigen, i.e., these antigens activates T-cells for effector function. Whereas, lipopolysaccharide (LPS) or peptidoglycans (PGN) consisting of carbohydrates is called T-independent antigen, which activates B-cells. However, some zwitterionic polysaccharides (that carry both positive and negative charges) such as polysaccharide A from *Bacteriodes fragilis* is able to activate T-cells (T_H1 cells).

Epitope or Antigenic Determinant

Antigenic determinant or epitope is the most immunodominant structure in an antigen and binds to the antibody. Number of epitopes in a molecule depends on the size of the antigen. Epitopes could be either linear, conformational, or neoantigenic. Linear antigenic determinant remains intact even after denaturation. Conformational epitopes are destroyed by denaturation by the action of physical or chemical treatments and are unable to bind to antibody. In this case, antibody only recognizes conformation of on an antigen. Neoantigenic determinants are available only after treatment with an enzyme or heat treatment.

Hapten

Hapten means grasp or fasten. Small molecules that do not induce immune response by themselves called haptens. Examples are penicillin, allergen, and peptides. They need carrier molecules for antibody production. Carrier molecules such as KLH (kehole limpet hemocyanin) and BSA (bovine serum albumen) are large and immunogenic and are used as carrier for experimental production of antibodies against haptens. Penicillin or other allergen though small, can interact with serum proteins in the blood forming a large complex that is capable of inducing immune response. Generally the immune response against penicillin is hypersensitive response or allergenic response.

Antigen–Antibody Reaction

Antigen–antibody reaction is specific and a reversible process. Different binding forces are involved, which include (1) Vander Waal force, where positive and negative electron clouds form bonding between antigen and antibody; (2) hydrogen bond [O]–H–[O] or [N]–H–[N] between molecules in antigen and antibody is formed; (3) electrostatic bond ([+]–[−]) formed by positive and negative charges contributed by both antigen and antibody; and (4) hydrophobic bond that is generated by removing water molecules from binding sites of antigen and antibody. Binding strength between an antigen and antibody molecule is expressed as *affinity*. That is the strength of binding between a single combining site of the antibody and an epitope in an antigen is called affinity. The *avidity* defines the strength of attachment of combining sites of all available epitopes, i.e., the overall strength of binding.

The Major Histocompatibility Complex

The major histocompatibility complex (MHC) molecules or MHC antigens are expressed on the surface of varieties of cells. They are essential for antigen presentation and also responsible for graft rejection. During organ or graft transplant if the host receives tissues or organs that are expressing same MHC, there is an acceptance and if different, rejection. There are two classes of MHC antigens; Class I and Class II. T-cell recognizes foreign antigen only when presented with MHC class I or class II. Exogenous peptide fragments always bind to class II (example, bacterial exotoxins or other surface proteins), whereas endogenously synthesized peptides are presented with class I (example, viral protein or tumor proteins).

Structure of MHC

MHC Class I
MHC class I consists of a single α-chain (also known as heavy chain) (Fig. 3.17). The molecular mass of α-chain in human is 44 kDa and in mouse 47 kDa. The α-chain has three domains: α1, α2, and α3, which appears as a constant region

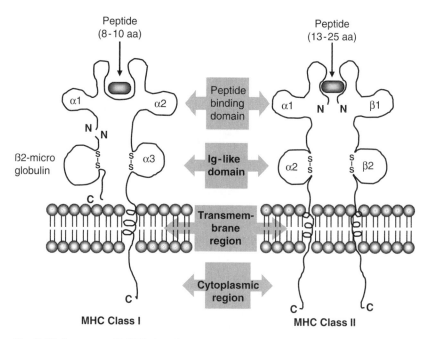

Fig. 3.17 Structure of MHC class I and class II molecules

of immunoglobulin. β-chain (12 kDa; a β2-microglobulin), a non-MHC coded gene product noncovalently interacts with the α-chain to provide stability in MHC class I molecule. Antigen (in the form of peptide) binding pocket is formed by the α1 and α2 domains which can accommodate a small peptide consisting of 8–10 amino acids. MHC I molecule is expressed by most nucleated cells in the body and present antigen to CD8$^+$ T-cells (CTL) for effector function (Fig. 3.18).

MHC Class II
MHC Class II antigen consists of two noncovalently associated polypeptide chains (Fig. 3.18). The α-chain is about 32–34 kDa, while the β-chain is 29–32 kDa. The peptide-binding pocket is made of both α1 and β1 domains and hold a peptide of 13–25 amino acids. Both α- and β-chains are encoded by two separate MHC genes and are expressed only in few cells; macrophage, dendritic cells, Langerhan's, B-cells, some T-cells in humans and rats but not in mouse, and some endothelial or epithelial cells after induction by IFN-γ. CD4$^+$ T-cells (T$_H$ cells) respond to the antigen when presented by Class II molecules (Fig. 3.18).

Antigen Presenting Cells

Antigen presenting cells (APC) are defined as those, which after internalization of antigen (either by phagocytosis or by active invasion of intracellular microbes), process and present antigen on the surface either in association with MHC class I or class II molecules for recognition by T- or B-cells. Only a limited numbers of specialized cells can express class II molecules and while most cells in the body when infected by intracellular pathogens can express class I molecule and present the antigen. (1) *Mononuclear phagocytes.*

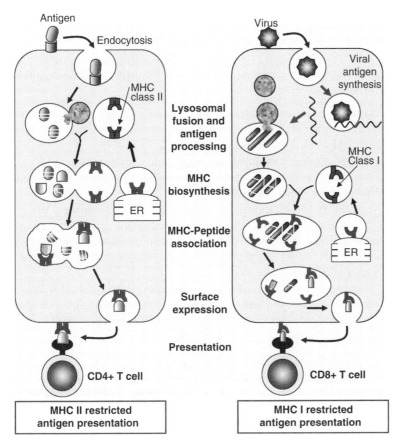

Fig. 3.18 MHC class I- and class II-restricted exogenous and endogenously synthesized antigen processing and presentation

Macrophages actively phagocytose large particles, infectious organisms, extracellular bacteria and parasites, and degrade them and the immunodominant peptide is presented in association with class II molecules to the T_H or B-cells. (2) *Dendritic cells* are originated in bone marrow and reside in spleen, lymph nodes and submucosal layer, and present antigen to T-cells. (3) *Langerhan's cells* are located in skin and contain characteristic Birbek granules in their cytoplasm and express class II molecules and serves as an APC in skin. (4) *Venular endothelial cells* can also express class II molecule and present antigen to T-cells. (5) *Epithelial cells* also can express class II molecules and serve as APC when activated by IFN-γ. Note: In general, IFN-γ can induce class II expression in any APC cells. (6) *B-lymphocytes* are considered an efficient antigen presenting cell because they carry surface immunoglobulin, which serves as a receptor for a protein antigen and present antigen to T-helper cells. B-cells bind antigen efficiently even at lower concentrations with higher affinity, and are important for T-cell dependent antibody production. (7) Most cells in the body also can express class I molecule during infection with virus, intracellular bacteria, protozoa and tumor and present antigen to CTL cells.

T-cells recognize peptides presented in context with MHC molecules while B-cells recognize proteins, carbohydrates, nucleic acid, lipid, small chemicals.

MHC-Restricted Antigen Processing and Presentation

Class II-Restricted Antigen Processing and Presentation
Class II-restricted antigen processing takes place with the exogenous protein antigens. The sequence of events (Fig. 3.18) for antigen processing involves several steps:

1. Initially, binding and internalization of protein antigens or native proteins occurs by phagocytosis, endocytosis, or pinocytosis and then the antigens are trapped in the endosome or phagosome.
2. In the processing step, fusion of phagosome with lysosome results in the proteolytic digestion of proteins in the acidic environment and the generation of immunogenic peptides.
3. Next, the fusion of endosome containing peptides with the MHC class II proteins containing vesicle takes place. In this stage, class II molecule binds to the appropriate immunodominant peptide fragments (13–25 amino acids). Unbound peptides are further degraded to amino acids and discarded from the cells by exocytosis.
4. Subsequently, the fusion of vesicles containing MHC class II with bound peptide and the cytoplasmic membrane occurs. Through an exocytosis (reverse phagocytosis) process, the MHC with bound peptide is displayed outside for recognition by CD4$^+$ T-cells.

Class I MHC-Restricted Antigen Presentation
Intracellular bacteria, virus, parasites, or tumor antigens are capable of infecting different types of host cells and are processed and presented by MHC class I molecule on the surface of infected cells also known as "target cells." After invasion, intracellular pathogens escape form the phagosome and multiply inside the host cytoplasm. The sequence of events are somewhat similar to class II-mediated pathway; however, in this case, the pathogen-specific proteins are synthesized inside the host cells through active gene transcription and translation process (Fig. 3.18). In the first step, (1) the bacterial DNA, or viral DNA or RNA are transcribed to produce proteins. In the same time, DNA for MHC class I is also transcribed to produce class I molecules. (2) Pathogen-specific proteins are synthesized in the cytoplasm and then are degraded by host cell proteosome into small peptides. (3) The peptides are transported into the lumen of rough endoplasmic reticulum. In the same time, MHC class I proteins are also transported to the rough endoplasmic reticulum and binds to the peptides (8–10 amino acids) and the peptide bound MHC complex is transported to the Golgi for further processing. (4) Vesicle containing MHC class I with bound peptides are fused with cytoplasmic membrane and are transported to the exterior of cells through exocytosis and presented to the CD8$^+$ T (CTL) cells.

Endogenously synthesized antigens are restricted to class I-mediated presentation while exogenously synthesized antigens are restricted to class II-mediated presentation.

Accessory Molecules Involved During MHC-Restricted T-Cell Activation
During antigen presentation by MHC class I or class II molecules to T-cells, several other molecules are also involved (Fig. 3.19). T-cell receptor (TCR) consisting of αβ- or γδ-chains is the primary component that recognizes the MHC–peptide complex on the surface of APC. The CD3, and CD4/CD8

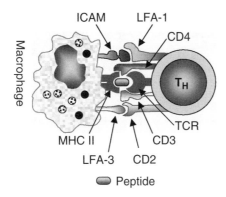

Fig. 3.19 Involvement of accessory molecules during antigen presentation to T-cells

present on T-cells also bind to the MHC–peptide complex and stabilize the structure. In addition, CD2 molecule on the surface of T-cell interacts with the LFA-3 (lymphocyte function-associated antigen) of APC, and LFA-1 of T-cell interacts with ICAM-1 (intracellular adhesion molecule) of APC. These accessory molecules increase the strength of the adhesion between T and APC or target cells, serve as surface markers, and transduce biochemical signal to the interior of the cell.

The Complement System

The complement proteins are synthesized in liver and are present in serum. They play important roles during innate and adaptive immune responses. Jules Border in his classical experiment demonstrated that if antibacterial antibody is mixed with a fresh serum at 37 °C caused lysis of bacteria; however, when the serum was heated to 56 °C or higher, lysis did not occur. Lysis was restored when fresh serum was added, suggesting the involvement of a heat-labile component which can complement the antibody function, hence the name complement. Complement consists of a series of proteins designated C1 to C9, which remains in serum in inactive form. Complement cascade can be activated sequentially by proteolytic enzymes by three mechanisms (1) classical pathway, (2) alternative pathway, or (3) lectin pathway (Fig. 3.20). The classical pathway is activated by antigen–antibody (IgG or IgM) complex in adaptive immunity while an infectious agent or a lectin-like glycoprotein activate alternative and lectin pathway, respectively. The primary and the most abundant component of the complement cascade is C3, which is converted to C3a and C3b by C3-convertase. The persons with complement deficiency are susceptible to many infective agents.

The Classical Pathway

The classical pathway is activated during the specific immune response (Fig. 3.21). Circulating IgG or IgM in the blood form a complex with specific pathogen or antigen and activates inactive form of C1. C1 is a large multimeric protein consisting of three subunits; C1q, C1r, and C1s. This active protein complex (C1qrs) is then catalyzes inactive C4 to produce active C4b and a

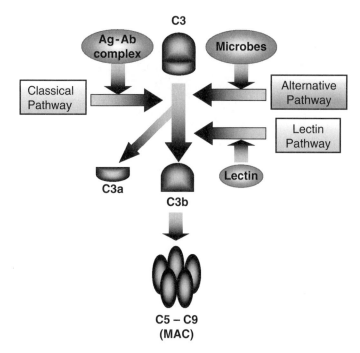

Fig. 3.20 Activation of complement by the classical, alternative, and lectin pathways

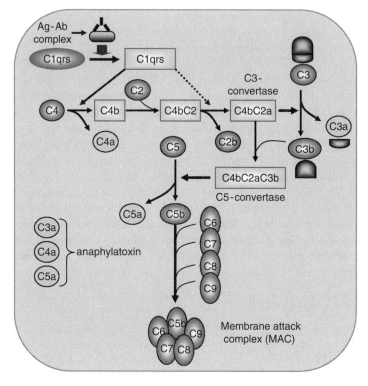

Fig. 3.21 Complement activation through classical pathway

soluble byproduct, C4a. C4 can also be activated by mannose-binding lectin (MBL), which is a pattern recognition receptor-specific to bacterial carbohydrate. C2 catalyzes the C4b to form C4b2. C1qrs complex can also activate C4b2 to form the C4b2a, which is also called C3-convertase. C3-convertase catalyzes C3 to form C3a and C3b, a major component of the complement activation pathway. C3b also forms a complex with C4b2a forming the C4b2a3b complex, which is known as C5-convertase. The C5-convertase catalyzes C5 to form C5a and C5b. C5b forms a complex through sequential reactions with C6, C7, C8, and C9 components and is called C56789 or a membrane attack complex (MAC). MAC makes donut shaped holes on the surface of target pathogens. The complement activation byproducts C3a, C4a, and C5a are called anaphylatoxin, which are known to induce hypersensitivity response.

Alternative Pathway

In the innate immunity, complement system plays a major role in the absence of antibodies. Viral proteins, bacterial LPS, peptidoglycan, LTA, and microbial surface polysaccharide can activate complement and the complement byproducts play important roles in the host natural immunity. In the alternative pathway (Fig. 3.22), several serum factors including B, D, H, I, and properdin systems are involved. Initially, microbial factors activate C3 to form C3b. In the presence of Factor B, Factor D catalyzes C3b to form C3bBb. The C3bBb complex is called C3-convertase, which is very unstable and Factor I rapidly degrade this

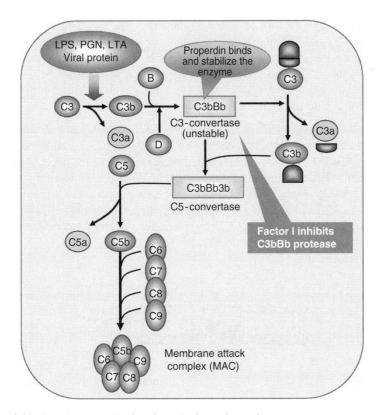

Fig. 3.22 Complement activation through alternative pathway

protease. Properdin proteins, however, bind and stabilize C3-convertase. C3-convertase catalyzes C3 to form C3b and a small molecule, C3a. C3b forms a complex with C3bBb to produce C3bBb3b, a C5-convertase, which converts C5 to form C5b and C5a. Sequential catalysis of C6, C7, C8, and C9 result in the formation of MAC.

Function of Complement

Complement system plays a major part in the defense against microbes in both innate and adaptive immune responses. (1) Complement-mediated bacterial lysis is accomplished by MAC, produced in both classical and alternative pathway. MAC also can attack host cells by inserting into plasma membrane (Fig. 3.23); however, most cells in the human body is protected by CD59, a membrane bound glycoprotein that inhibits MAC formation by blocking the aggregation of C9. (2) Complement protein C3b aids in the opsonization (phagocytosis) process by serving as an opsonin which binds to bacteria. Phagocytes (macrophage and neutrophil) express C3b receptor (CD35) which binds to opsonin coated microbes for engulfment. In combination with antibody and C3b, opsonization can be enhanced several-fold. (3) Complement protein byproducts C3a, C4a, and C5a also serve as anaphylatoxins. They induce the release of soluble inflammatory mediators such as histamine to increase membrane permeability, to increase smooth muscle contraction. C5a acts as chemoattractant for neutrophils at the site of inflammation thus facilitating increased clearance of pathogens from the site of infection. (4) Complement proteins also aid in the phagocytic clearance of immune complex. Antigen–antibody complex is formed during immune response may dislodge in variety of organs/tissues including kidney resulting in pathological consequences such as glomerulonephritis. Since the complement system enhances phagocytic activity, the antigen–antibody complex is rapidly removed from the body.

Control of Complement Activation

Activation of complement and resulting byproducts are able to eliminate pathogens from the host; however, activation can also cause damage to the nearby host cells and tissues. If C3b binds to host cells, those cells become the target of phagocytic cells causing cell/tissue damage. Also MAC can cause host cell damage. As a protective measure, host cells carry surface proteins, which prevent C3b binding. For example, the C3bH formed during the alternative pathway of activation allows degradation of C3b by protein I that in turn discourage formation

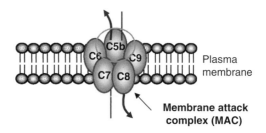

Fig. 3.23 Schematic diagram showing the insertion of MAC (membrane attack complex) into the plasma membrane

of C3bBb. In addition, host cells contain sialic acids and C3b does not bind to sialic acid and therefore prevent host cell damage by phagocytes. However, some bacteria take advantage of the situation. Some bacterial capsule contains sialic acid, thus prevent the binding of C3b thus are protected from the phagocytic killing. Mannose-binding protein also prevents activation of complement. Some bacteria carry mannose on their surface thus prevent complement activation. Macrophage produces IL-6, which can activate liver cells to produce mannose-binding protein, which binds to mannose on the bacterial surface. This alters configuration and can activate complement.

Immunity to Microbes

The principal physiologic function of the immune system is to protect the host against pathogenic microbes such as extracellular bacteria, intracellular bacteria, virus or parasites. Important features of immunity to microbes are (a) defense is mediated by both innate and adaptive immunity, (b) different types of microbes stimulate distinct population and subpopulations of lymphocytes, (c) survival of microbes or production of disease depends on their ability to evade immune system, and (d) the tissue injury and disease consequent to infection caused by host response rather than the microbes itself.

Extracellular Bacteria

These organisms replicate outside the cells, i.e., in circulation, interstitial space, lumen of respiratory tract, and intestinal tract. Examples of extracellular bacteria include *Clostridium, Staphylococcus, Bacillus, Streptococcus, Escherichia coli*, and *Vibrio* species. They cause disease either by inducing inflammation or by direct action of toxins or enzymes they produce.

Inflammation
Bacterial cells provoke inflammation that result in tissue destruction and influx of neutrophils and macrophages at the site of infection. During inflammation, complement activation takes place and cytokines are released from macrophages. Release of enzymes and toxins from neutrophils and macrophages causes cell injury and damage which is also known as suppurative infection (characterized by pus formation). Sequence of events in inflammation include (1) increased blood supply to the site of infection characterized by the development of "redness," (2) increased capillary permeability and fluid accumulation results in "swelling," and (3) the recruitment of neutrophils by chemotaxis (C5a) and macrophages result in tissue injury that invoke "pain."

Toxins
Exotoxins cause diverse pathological effects resulting in cell injury or cell death (see Chap. 4 for details). For example, Shiga and diphtheria toxins block protein synthesis; botulinum toxin blocks neurotransmitter release; and cholera toxin stimulate cAMP synthesis resulting in Cl^- secretion and H_2O loss. Endotoxin (LPS and PGN) also called pyrogen, stimulates the release of cytokines IL-1 and TNF-α; activate B-cells; and act as an adjuvant to stimulate macrophage to produce more cytokines. These cytokines, induce fever, decrease smooth muscle contraction, increase membrane permeability, lower blood pressure, and induce shock.

Innate Immunity

Innate immunity against extracellular bacteria include (1) phagocytosis by neutrophils, monocytes, and tissue macrophages, (2) activation of complement in the alternative pathway or mannose lectin pathway by bacterial peptidoglycan or mannose-binding lectin. The complement activation byproduct C3b enhances opsonization while MAC induces cell lysis. (3) LPS activates macrophages to produce cytokines TNF-α, IL-1, IL-6 which activate neutrophils and macrophages to remove bacteria. These cytokines also could induce fever to retard bacterial growth or may induce "septic shock" or "endotoxin shock" leading to fatal consequences.

Adaptive Immune Response

1. In the specific immune response, cell components, i.e., LPS or PGN act as T-independent antigens and activate B-cells to produce specific IgM. Antigen-mediated specific T-cell activation results in specific cytokines release which are responsible for isotype switching to produce specific antibodies.
2. Exotoxins are T-dependent antigens and are processed by APC (macrophages, B-cells) and presented via MHC class II molecules to activate CD4$^+$ T-cells and B-cells for cytokine and antibody production, respectively. Certain toxins act as superantigens, which induce T-cells to produce increased cytokines leading to toxic shock syndrome (TSS). For example, staphylococcal enterotoxins (SEA, SEB, SEC, SED, SEF, etc.) bind to MHC class II on macrophages which directly interacts with TCR of T-cells and activate CD4$^+$ T-cells to produce IFN-γ, which in turn induce enhanced MHC expression, antigen presentation, and cytokine production in macrophages leading to toxic shock syndrome.
3. Antibodies (IgG, IgM) help in the opsonization of pathogens, neutralization of toxins, and activation of complement via the classical pathway. The activation byproducts, C3b, and MAC cause increased phagocytosis and cell lysis, respectively.

Evasion of Immune System by Extracellular Bacteria

In order to cause a successful infection, bacteria must be able to evade the immune system. Indeed, they have developed various strategies to achieve that. (1) Antigenic variation is an important strategy for some microorganisms (*Salmonella enterica, Haemophilus influenzae, E. coli*). Bacteria use surface molecules for adhesion and colonization; however, genetic variation in bacterial surface molecules such as pili, fimbriae, flagella or surface proteins will evade antibodies antibodies developed against previous form of molecules. (2) Some bacteria (*Bacillus anthracis, B. cereus*, and *Pneumococcus* spp.) exhibit antiphagocytic mechanism. They possess capsules, which are made of sialic acid and hyaluronic acid (both are also present in host cell membrane) and prevent binding of C3b. As a result these bacterial cells are not recognized by macrophages for destruction. (3) Sialic acid component of capsule also inhibits complement activation. (4) Some intracellular bacteria (*Listeria monocytogenes, Salmonella*) are capable of scavenging reactive oxygen using superoxide dismutase (SOD) thus avert toxic effects of oxygen radicals. (5) Some bacteria (*Staphylococcus aureus, Streptococcus pyogenes*) cover themselves with host antibody thus immune system is unable to recognize them as foreign. *S. aureus* produces Protein A and *S. pyogenes* produces Protein G, which bind IgG in Fc region thus macrophages (carry receptor for Fc) are unable to recognize these bacteria. (6) Some bacteria produce immunoglobulin-specific proteases which degrade antibodies.

Intracellular Bacteria

Several foodborne pathogens maintain intracellular life cycle as part of their infection strategy. After binding to specific host cell receptor, they modulate signaling events resulting in cytoskeletal rearrangement and facilitating their own entry. Bacteria trapped inside a phagosome is either escaped by lysing the vacuolar membrane or produce virulence factors that sustain their intracellular life style. Bacteria either spread from cell-to-cell, or induce apoptosis, or cause host cell lysis as part of their pathogenic mechanism. The examples of intracellular foodborne bacterial pathogens are *Listeria monocytogenes, Yersinia enterocolitica, Shigella* spp., and *Salmonella enterica*.

Innate Immunity

Natural immunity including gastric acid, bile salts, antimicrobial peptides, mucins, and natural microflora provide some protection; however, many pathogens are resistant to those thus natural immunity is less effective. Furthermore, phagocytes (macrophages and neutrophils) are also less effective since bacteria are resistant to degradation by lysosomal contents.

Adaptive Immunity

Humoral immune response does not contribute toward the protection since bacteria is mostly localized inside the cell thus the cell-mediated immunity provides the most effective protection. However, it is important to point out that despite the lack of protection by humoral immunity, antibody against bacterial antigens are found in the serum and thus indicating that the antigens are processed and presented by MHC class II pathways for response by CD4+ T-cells and B-cells.

(1) Cell-mediated immunity (CMI) is most effective against intracellular bacteria, where CD8+ T-cell act as the primary effector cell, which recognizes the infected target cell expressing surface antigen in association with MHC class I molecule. (2) Macrophages activated by T-cell-derived cytokines (IFN-γ) also play important role by actively phagocytosing bacteria. (3) In some cases, the host induces granuloma (nodule) formation to contain the infective agent from spreading, which is seen in chronic infections caused by *Mycobacterium, Histoplasma*, and some mold species. There is an onset of delayed type of hypersensitivity reaction (DTH) to these infections. Both CD4+ and CD8+ cells respond to the soluble protein antigens, and intracellular bacterial antigens, respectively. Following activation these cells secrete cytokines, TNF, IFN-γ, which activate vascular endothelial cells, which in turn recruits neutrophils, lymphocytes, and monocytes. IFN-γ also activates

macrophages, which convert them into epitheloid or giant cells and try to eliminate the bacteria. If the antigen stimulation persists; macrophages are chronically activated to secrete additional cytokines and growth factors, which recruit fibroblast cells to encase the bacteria containing cells to form a protective nodule called granuloma.

Evasion of Immune System

Intracellular bacteria utilize various strategies that allow them to maintain intracellular life style. (1) Some bacteria (example, *Mycobacterium*) inhibit phagolysosomal fusion, thus protects the bacteria from toxic lysosomal contents. (2) Escape from phagosome: *Listeria monocytogenes* produces hemolysin that forms pores in the phagosome and allow the bacteria to escape into the cytoplasm preventing the degradation by lysosomal contents. Furthermore, hemolysin also blocks the antigen processing thus *Listeria* antigen is not presented on the surface by MHC class I molecule for recognition by T-cells. In addition, *Listeria* and *Salmonella* produce superoxide dismutase, which inactivates toxic "O" radical and bacterial catalase breaks down H_2O_2. (3) Antiphagocytic activity: *Yersinia enterocolitica* outer membrane proteins such as YopE has antitoxin activity, YopH has tyrosine phosphatase activity and YpkA blocks signaling pathway that is required for phagocytosis. (4) Intracellular localization also prevents the bacteria to be seen by the immune system.

Immunity to Virus

Viruses are obligatory intracellular pathogen which upon entry replicate within the cell. Viral protein binds to host cell receptor, enter the cell and the viral genome is released into the host cell. Following vial protein synthesis, nucleic acid is packaged and matured virus particles are released and proceed to infect neighboring cells (see Chap. 2). Viral proteins also block host cell protein synthesis thus induce host cell death showing characteristic cytopathic effects. Many viruses infect immune cells as their primary target. They first binds to surface molecules, for example, HIV I binds to CD4 molecule on the helper T-cells and then enter the cells; Epstein Barr virus (EBV) binds to Type 2 complement receptor (CR2) on B-cells, and Rhinovirus binds to intercellular adhesion molecules (ICAM). SV40 (simian virus) binds to MHC class I to enter cells. HIV causes AIDS; EBV causes infectious mononucleosis and associated with Burkitt's lymphoma and other carcinomas; Rhino virus causes inflammation of nasal passage; and SV causes cancer in monkey.

Innate Immunity

1. Virus infection stimulates production of type I interferon (both IFN-α and IFN-β) from infected cells, which inhibit viral replication called "antiviral state," discussed earlier.
2. NK cells can kill or lyse a variety of virally infected cells. Moreover, the Type I interferon enhances NK activity.
3. Viral protein can activate complement cascade in the alternative pathway and the complement byproduct, C3b, can enhance phagocytosis.

Adaptive Immunity

Both humoral and cell-mediated immunity (CMI) are important in protecting host against viral infection.

Humoral Immunity

During humoral immune response, antibody is developed against envelope or capsid proteins, which are responsible for binding to host cell receptor. Circulating antibodies bind viral surface proteins and block the viral binding and entry in host cells. Antibodies are generally very effective in the initial stage of viral infection. sIgA are found to be most important since they are abundant in the mucus of respiratory and the digestive tract which serve as primary portal for virus entry. (1) Opsonization: Antibody-forming complex with the viral particles can help macrophages for enhanced phagocytic clearance from blood circulation. (2) Antiviral antibody forming a complex with viral antigen can activate complement through the classical pathway and promote increased phagocytosis. (3) Humoral immunity against viral antigen can be used as a prophylactic vaccine to protect the host from future infections. Attenuated or killed virus generates circulating antibodies in serum that prevent viral binding to host cells. Current vaccination strategy include induction of mucosal immunity since many foodborne viruses (Hepatitis A, Norovirus, Rota virus) use intestinal mucosa as the site of entry for initiation of infection.

Cell-Mediated Immunity

Antibodies are important protective components of immunity to viruses but they may not be effective since virus hide inside the cells thus CMI is most important. The principal mechanism of specific immunity against viral infection is the generation of cytotoxic T-cells (CD8$^+$ T-cells). These cells recognize endogenously synthesized viral proteins in association with MHC class I molecules on the surface of target cells and destroy them by producing pore-forming cytolysins.

Evasion of Immune System by Viruses

Viruses use several strategies to overcome immune system and the most important of all is their intracellular life cycle and utilization of host cell machinery for replication and spread, in which immune effector cells are unable to find them. Antigenic variation is an important characteristic. There is a large numbers of serologically distinct strains are present which show huge antigenic variations in their proteins. Therefore, protective antibody or vaccine is ineffective. This characteristic is most important with influenza virus (flu-virus) in which genetic variation is very common. In addition, some viruses directly infect immune cells and cause immune suppression. For example, influenza and rhinovirus suppress immune system, HIV-I suppress CD4$^+$ cell populations, and EBV inhibits production of IL-2 and IFN-γ.

Immunity to Parasites

Many of the foodborne protozoan species cause chronic disease where the natural immunity is weak and less effective. Protozoan species such as *Giardia, Entamoeba, Cryptosporidium*, and *Cyclospora* invade intestinal cells and cause massive damage to the site of infection and are responsible for diarrhea. Helminths such as *Trichinella* and *Taenia* species (tapeworms) invade liver and muscle tissues and cause inflammation resulting in chronic infection. During chronic infection, the clinicopathological consequences are mostly due to the host response to the parasites.

Innate Immunity

Besides acid, mucus, bile salts, and antimicrobial inhibitors in the gastrointestinal tract, several other molecules play important role in immunity. Those include complement and phagocytic cells. Complement is activated through alternative pathway and results in the formation of MAC, which causes lysis of parasites. Many parasites are resistant to phagocytic killing since they are capable of multiplying inside the macrophages. Parasites also induce inflammation and recruit polymorphonuclear cells, macrophages, dendritic cells, and neutrophils.

Adaptive Immunity

Adaptive (specific) immunity to parasites is effective which is mediated by both humoral and CMI. In humoral response, IgE plays important role. Antigens from parasites are presented on the surface of APC and are recognized by CD4$^+$ T-cell population, which produces IL-4 and IL-5. IL-4 promote IgE production from B-cells and IL-5 recruit eosinophil. IgE bound parasite complex is then recognized by eosinophil via receptor, FcεR and the parasite is destroyed by toxins produced by eosinophil. Parasites also induce granuloma formation which is seen in infection caused by *Trichinella*. In response to antigen both CD4$^+$, CD8$^+$ T-cells are activated and produce cytokines that recruit macrophages and ultimately the fibroblasts to encase the infective agents. CMI also induces DTH response. *Trichinella* eggs cause liver cirrhosis when deposited in the liver. Some parasites like *Schistosoma mansoni* form polyps during infection. Cytotoxic T-cells (CD8$^+$) cells destroy host cells that carry intracellular protozoa and serve as the primary effector component of the CMI.

Evasion of Immune System

Parasites also use numerous strategies to evade immune system. (1) Since protozoa grow intracellularly, immune cells cannot find them. This strategy is called anatomic concealment. Also, some parasites develop cyst (example, *Trichinella*), which is not detected by immune cells. (2) Some parasites mask themselves with a coat of host cell proteins on the surface, for example, *Schistosoma mansoni* larvae coat themselves with ABO blood group antigens and MHC molecules before they reach to lungs thus are not recognized by host immune cells. (3) Some parasites (example, *Entamoeba histolytica*) spontaneously sheds antigen coats after binding to antibody thus are not recognized by phagocytic cells.

Summary

To understand the pathogenic mechanism of infection one has to have a clear knowledge of immune response, both innate and adaptive or acquired that is in place to protect the host against the pathogens. Microbial dominance results in the disease while successful host response averts the full-blown infection. Since foodborne pathogens affect primarily the digestive system, the natural immunity of the gastrointestinal tract is very important. Moreover, the onset of symptoms is very fast appearing as quickly as in 1 h, thus protection by specific immune response would have little impact since it is slow to develop typically requiring 4–7 days. However, specific immune response is essential for food-borne pathogens that have a prolonged incubation time and are responsible for systemic infection such as *Listeria monocytogenes, Shigella* spp., *Salmonella enterica*, Hepatitis A virus, *Toxoplasma gondii*, and *Trichinella* species. In the innate immunity, gastric acids, bile salts, mucus, natural microflora, macrophages, neutrophils, interferons, and complements play critical role while in the specific immune response T- and B-lymphocytes produce cytokines and antibodies, respectively, are important. $CD4^+$ T-cells are most important for protection against extracellular bacterial infections and the exotoxins, while $CD8^+$ T-cells are involved in the elimination of intracellular bacterial, viral, and parasitic infective agents. Antibodies neutralize pathogens by preventing them from binding to host cell receptors or binding to antigens and become the target for elimination by macrophages and neutrophils. However, immune system sometimes fails to protect the host. It is because that the pathogens have developed strategies to overcome immune defense by producing virulence factors that ensure their invasion, survival, replication, and spread inside the tissues. It is important to recognize that the toxic action by foodborne toxin is very quick and the host has literally no time to mount any immune response, thus everyone suffers from this form of food poisoning (intoxication) irrespective of their health status.

Further Readings

1. Abbas, A.K., Lichtman, A.H., and Pober, J.S. 2000. Cellular and Molecular Immunology, 4th edition, WB Saunders, Philadelphia.
2. Boneca, I.G. 2005. The role of peptidoglycan in pathogenesis. Curr. Opin. Microbiol. 8:46–53.
3. Brandtzaeg, P. 2003. Role of secretory antibodies in the defense against infections. Int. J. Med. Microbiol. 293:3–15.
4. Gaskins, H.R. 2003. The commensal microbiota and development of mucosal defense in the mammalian intestine, Vol. 1. Proceedings of 9th International Symposium on Digestive Physiology in Pigs, Banff, AB, Canada, pp 57–71.
5. Heller, F. and Duchmann, R. 2003. Intestinal flora and mucosal immune response. Int. J. Med. Microbiol. 293:77–86.
6. Janssens, S. and Beyaert, R. 2003. Role of toll-like receptors in pathogen recognition. Clin. Microbiol. Rev. 16:637–646.
7. Kovacs-Nolan, J., Marshall, P., and Mine, Y. 2005. Advances in the value of egg and egg components for human health. J. Agr. Food Chem. 53:8421–8431.
8. Lievin-Le-Moal, V. and Servin, A.L. 2006. The front line of enteric host defense against unwelcome intrusion of harmful microorganisms: mucins, antimicrobial peptides, and microbiota. Clin. Microbiol. Rev. 19:315–337.
9. Shao, L., Serrano, D., and Mayer, L. 2001. The role of epithelial cells in immune regulation in the gut. Semin. Immunol. 13:163–175.

4

General Mechanism of Pathogenesis for Foodborne Pathogens

Introduction

The diseases caused by foodborne pathogens can be classified into three forms: *foodborne infection, foodborne intoxication*, and *foodborne toxicoinfection* and each are discussed below (Fig. 4.1). The principal route of infection for foodborne pathogens is oral and the primary site of action is the intestine. Most foodborne microorganisms cause localized infection and tissue damage but some spread to deeper tissues to induce systemic infection. For successful enteric infection, several factors must work cooperatively in a host. First of all, pathogens must gain access to the host in sufficient numbers to initiate infection. The primary vehicle of transmission is food and water. However, they can be acquired from direct contact with an animal or a human, such as a food handler, from environments (soil, air) or from an arthropod vector. Once inside the host, the pathogens must survive in the changing environment, multiply and propagate. Pathogens must find a suitable niche for colonization, which is facilitated by adhesion factors, invasion factors, and chemotaxis (for example, bacterial affinity for iron allows the organism to reach the liver which has a rich source of iron in the form of transferrin). The microbial cell envelope also helps bacteria to survive in the hostile environment, as the capsule protects the bacteria from being engulfed by phagocytes. In addition, bacterial toxins and enzymes protect cells from elimination by the host immune system. Presence of commensals in the site can assist the invading bacterium to find a niche. For example – in "wound botulism," aerobic organisms first grow and multiply in the wound, where they utilize oxygen to create an anaerobic microenvironment. *Clostridium botulinum*, transmitted to wound through sharp object or dust, now have the perfect niche for growth and botulinum toxin production to induce botulism. Pathogens also damage the host tissues and cells by using exotoxins, endotoxins, or enzymes that cause cell death by apoptosis or necrosis and promote bacterial survival and multiplication.

Foodborne Infection

Foodborne infection is committed by intact living microorganisms, which must enter the host to cause infection. Following ingestion with food or water, microorganisms pass through the acidic stomach environment and move to

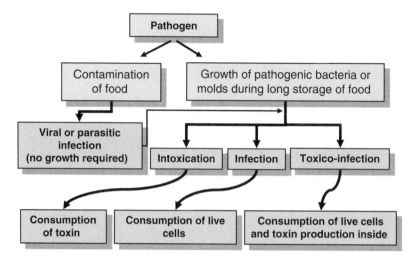

Fig. 4.1 Flow diagram showing the various forms of foodborne diseases

the intestine, where they colonize and cross the intestinal barrier using an active invasion process, or via translocation by phagocytic M cells. Some microorganisms cause local tissue damage and induce inflammation, while others spread to lymph nodes, liver, spleen, brain, or other extraintestinal sites. Foodborne infection can be acute or chronic. In acute infection, the onset of disease is quick and lasts only for a short duration due to a rapid immunological clearance of the microorganism. In chronic infections, the disease is prolonged and immune clearance is not effective. Often, the prolonged infection is perpetuated by the strong immune response mounted by the host rather than the infective agent itself, such as seen in chronic Shigellosis cases. Patients recovering from a foodborne infection may shed the organism for a while. Some foodborne infections may lead to chronic sequelae such as Reiter's syndrome, arthritis, and Guillain–Barre syndrome.

Infectious Dose

The infectious dose of pathogens or their toxins varies depending on the immunological status of the host and the natural infectivity of the organism (Table 4.1). The infectious dose decreases if consumed with liquid food that traverses stomach rapidly or food (milk, cheese, etc.) that neutralizes the stomach acid. Persons with high gastric pH or those undergoing antibiotic therapy for other ailments are also susceptible to foodborne infections because antibiotics reduce the natural microflora loads in the intestine, which renders the host more susceptible to foodborne infections.

Colonization and Adhesion Factors

The gastrointestinal tract, consisting of the mouth, esophagus, stomach, small intestine, and large intestine, is very dynamic and active as the epithelial cell surface is constantly washed by mucus and fluids. The GI tract uses peristaltic movements, mucus production, and epithelial ciliary sweeping action in an attempt to expel pathogens. In addition, bile salts, proteolytic enzymes, and

Table 4.1 Infectious dose and incubation periods of common foodborne pathogens

Pathogens	Infectious dose	Incubation period
Bacteria		
Escherichia coli O157:H7	50–100 cfu	3–9 days
Listeria monocytogenes	10^2–10^3 cfu	7–14 days or even longer
Salmonella enterica	>10^5 cfu	6–24 h
Shigella spp.	10–100 cfu	12 h–7 days, but generally in 1–3 days
Vibrio cholerae /hemolyticus/vulnificus	10^4–10^{10} cfu g^{-1}	6 h–5 days
Staphylococcus aureus cells	10^5–10^8 cfu g^{-1}	–
Staphylococcal enterotoxin	1 ng g^{-1} of food	1–6 h
Bacillus cereus	10^5–10^8 cfu or spores g^{-1}	1–6 h (vomiting); 8–12 h (diarrhea)
Bacillus anthracis (inhalation anthrax)	8×10^3–10^4 spores	2–5 days
Clostridium botulinum neurotoxin	0.9–0.15 µg (i.v. or i.m. route) and 70 µg (oral route)	12–36 h; 2 h when large quantities are ingested
Clostridium perfringens	10^7–10^9 cfu	8–12 h
Campylobacter jejuni	5×10^2–10^4 cfu	1–7 days (24–48 h)
Yersinia enterocolitica	10^7–10^9 cfu	24–30 h and lasts for 2–3 days
Virus		
Norovirus	~10 particles	24–48 h
Hepatitis A	10–100 particles	15–45 days, average 28–30 days
Protozoa		
Giardia duodenalis	10–100 cysts	1–2 weeks
Cryptosporidium parvum	10 oocysts	2–10 days

resident microbiota can also prevent pathogens from colonizing. In order for a foodborne pathogen to cause foodborne toxicoinfection or infection, it must colonize the intestine or adhere to the epithelial surface before exerting its pathogenic action. Several adhesion or colonization factors are used and general descriptions of such factors are presented below.

Pili or Fimbriae

Pili, also known as fimbriae, are rod-like surface adhesin structures found in Gram-negative bacteria. Pili consist of pilin (20 kDa), and have a long helical–cylindrical structure. Pili are located on the surface of bacteria, and have a specialized protein tip that help the bacteria attach to host cell receptors composed of glycolipids or glycoproteins (α-D-galactopyranosyl-β-D-galacto-pyranoside). Bacteria growing in the body constantly lose and reform pili.

The host sometimes produces antibodies against pili that prevent bacterial adhesion to the host cells.

There are four types of pili: Type I, Type IV, P pili, and bundle-forming pili (BFP) (Table 4.2). Type 1 pili are also known as colonization fimbriae antigen (CFA: CFA/II, CFA/IV) and are flexible. P pili (CFA/I) and Type IV are rigid, whereas bfp (CFA/III) is flexible. Type I and P pili are found in *E. coli, Haemophilus influenza*, and *Yersinia pestis*. Type IV pili is found in *Pseudomonas, Vibrio*, enteropathogenic *E. coli* (EPEC), and *Neisseria*. BFP is found in enteropathogenic *E. coli* (EPEC). Pili are encoded by the *pap* gene in *E. coli* and the *pap* operon consists of *papD, papC, papA, papH, papE, papF, papG*, and *papK* genes.

Some *E. coli*, including O157:H7, express curli, a thin coiled fimbriae-like structure composed of 15 kDa protein. It is typically produced under stress conditions such as suboptimal growth temperature, osmolarity and in nutrient limiting environments. It aids in bacterial adhesion to extracellular matrices such as fibronectin, laminin, and type 1 collagen on eukaryotic cells, as well as to inert substance like polystyrene.

Adhesion Proteins

Adhesion proteins promote tighter binding of pathogens to the host cells and are important for attachment and invasion. For example, *Yersinia* YadA adhesin binds to fibronectin, laminin, collagen, and β1-integrin while *Yersinia* invasin also binds to β1-integrin. In enteropathogenic *E. coli* (EPEC) and enterohemorrhagic *E. coli* (EHEC), the attachment and effacement (EAE) protein intimin binds to translocated intimin receptor (TIR) and aids in the formation of a pedestal. In *Listeria monocytogenes*, internalin A binds to E-cadherin located in the tight junction of epithelial cells; internalin B binds to globular complement protein (gC1q-R/p32), glucosaminoglycans and a tyrosine kinase Met-receptor (a hepatocyte growth factor); LAP (*Listeria* adhesion protein) binds to heat-shock protein 60 (Hsp60); and fibronectin-binding protein binds to host fibronectin to initiate contact. Other factors such as AMI (autolysin amidase) and LTA (lipoteichoic acid) are also involved in *Listeria* adhesion.

Biofilm Formation

Biofilm formation facilitates bacterial colonization in host tissues as well as in inert food processing or medical equipments. Some bacteria produce extracellular polymeric substances (EPS) consisting of polysaccharides, proteins,

Table 4.2 Pili/fimbriae of Gram-negative bacteria

Types of pili	Classification	Property	Present in
Type I	CFA/II, CFA/IV	Flexible	*E. coli, Haemophilus influenza; Yersinia pestis*
Type IV	–	Rigid	*Pseudomonas, Vibrio,* enteropathogenic *E. coli* (EPEC), and *Neisseria*
P pili	CFA/I	Rigid	*E. coli, Haemophilus influenza, Yersinia pestis*
BFP (bundle-forming pili)	CFA/III	Flexible	EPEC
Curli		Flexible	*Salmonella*

phospholipids, teichoic acids, nucleic acids, and other polymeric substances. The exopolysaccharide contains of poly-β-1,6-N-acetylglucosamine. Bacteria can form dense, multiorganism layers on food contact surfaces, in the mouth or intestinal lining. The first layer of the biofilm attaches directly to the host cell surface. Additional layers attach to the basal layer with a polysaccharide matrix. EPS provides protection against biocides, desiccation, antibiotics, and toxins. Biofilms are important in human diseases, including dental plaque caused by *Streptococcus mutans*; mastitis caused by *S. aureus*; intestinal colonization by *Salmonella*; endocarditis by *Enterococcus faecalis* and lung infections by *Pseudomonas aeruginosa* in cystic fibrosis patients. Type IV pili in *Vibrio* also promote biofilm formation.

In addition to exopolysaccharides, other proteins are found to play critical role in biofilm formation. A biofilm-associated protein (Bap) and the Bap family of proteins have been found in *S aureus, S. epidermidis, E. faecalis*, *Salmonella* Typhimurium, *E. coli*, and *Streptococcus pyogenes*, and are involved in biofilm formation.

Invasion and Intracellular Residence

Some bacteria are intracellular and can enter cells by two mechanisms (1) passive entry into host through uptake by natural phagocytic cells such as M cells and macrophages and dendritic cells or (2) active invasion via induced phagocytosis in nonprofessional phagocytic cells such as epithelial, endothelial, and fibroblast cells. Examples of passive entry are *Shigella* spp., *Listeria monocytogenes*, and *Salmonella enterica*. Examples of induced phagocytosis are *L. monocytogenes, Salmonella, Yersinia*, and *Shigella*, which use invasin proteins to promote bacterial uptake. Some pathogens use sphingolipid–cholesterol raft that forms invaginations during pinocytosis. Some pathogens also use caveolae or clathrin to gain entrance such as *Listeria* and *Yersinia*. Some intracellular parasites such as *Toxoplasma gondii* and *Cryptosporidium parvum* exhibit gliding motility (1–10 μm s^{-1} and counter clockwise), which helps parasite entry into enterocytes. The gliding motility is an active process for the parasite but the host cell does not play active role in invasion, as invasion does not alter host cell actin cytoskeleton or induce phosphorylation of tyrosine residues.

Phagocytosis

M (microfold) cells are naturally phagocytic cells and are present throughout the intestine in follicle-associated lymphoid tissue such as Peyer's patches. Their primary function is to transport intact particles across the epithelial membrane without degrading or processing them. Some pathogens such as *Salmonella, Shigella*, and *Listeria* use M cells to reach to the subepithelial layer. In contrast, professional phagocytic cells such as macrophages, dendritic cells, and neutrophils spontaneously phagocytose bacteria. However, intracellular pathogens resist killing by using specialized virulence factors such as hemolysin, superoxide dismutase (SOD), and catalase. Hemolysin forms pores in the phagosomal membrane, SOD inactivates toxic oxygen radicals, and catalase breaks down hydrogen peroxide. Some pathogens such as *Salmonella, Legionella, Mycobacterium*, and *Brucella* block or alter the maturation of phagosomes which allows them to survive in the phagosome. Some pathogens such as *Listeria monocytogenes* and *Shigella* spp. escape the phagosome with the help of hemolysin, multiply in the cytoplasm, induce actin polymerization, and infect neighboring cells. *Shigella* also avoids autophagosomal lysis using a specialized virulence factor. Some pathogens such as *Shigella, Yersinia*, and *Salmonella* induce apoptosis in macrophages/dendritic cells and persist in subepithelial regions in the intestine.

Invasin-Mediated Induced Phagocytosis

Invasin molecules bind to their corresponding receptors (Table 4.3) to provoke phagocytic ingestion of the bacteria (example, *Shigella, Listeria*). Induced phagocytosis involves the coordinated interaction of sophisticated quorum sensing and signal transduction mechanisms. During this process, cytoskeletal proteins undergo rearrangement and accommodate bacterial entry. Induction of actin polymerization is a crucial event which aids in cytoskeletal rearrangements to accommodate bacterial entry. Two distinct mechanisms are identified for induced phagocytosis (1) zipper-like mechanism and (2) trigger mechanism. *Yersinia pseudotuberculosis* and *Listeria monocytogenes* induce the zipper mechanism while *Salmonella enterica* and *Shigella* spp. use the trigger mechanism for entry.

The zipper mechanism (Fig. 4.2) is a three-step process (1) contact and adherence; (2) phagocytic cup formation; and (3) phagocytic cup closure. Contact and adherence is independent of actin cytoskeletal rearrangement and involves ligands and receptors. For example, *L. monocytogenes* uses internalin A to interact with E-cadherin, located in the adherence junctions of polarized epithelial cells; *L. monocytogenes* also uses internalin B to interact with Met, a transmembrane receptor tyrosine kinase. *Yersinia* invasin protein interacts

Table 4.3 Adhesion factors for select foodborne bacterial pathogens

Bacteria	Adhesion factors	Receptor
Listeria monocytogenes	Internalin A (88 kDa)	E-cadherin
	Internalin B (65 kDa)	c-Met, gC1q-R/p32
	Vip (virulence protein) (43 kDa)	Gp96
	LAP (*Listeria* adhesion protein) (104 kDa)	Hsp60
Campylobacter species	CadF (37 kDa)	Fibronectin
Arcobacter	Hemagglutinin (20 kDa)	Glycan receptor
Enteropathogenic and enteroinvasive *E. coli*	Intimin (94 kDa)	Translocated intimin receptor (TIR)
Yersinia enterocolitica	YadA (160–240 kDa)	Collagen/fibronectin/ laminin/β1-integrin
	Invasin (92 kDa)	β1-integrin
Staphylococcus aureus	Fibronectin-binding protein (FnBP)	Fibronectin
Vibrio cholerae	Toxin-coregulated pili (TCP)	Glycoprotein
	Mannose–fucose-resistant cell-associated hemagglutinin (MSHA), mannose-sensitive hemagglutinin (MSHA)	Glycoprotein?

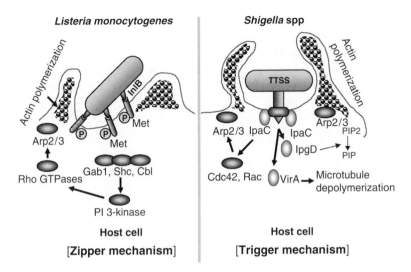

Fig. 4.2 Schematic drawing showing *Listeria monocytogenes*-induced bacterial entry into cell by zipper mechanism and *Shigella* entry by trigger mechanism. In zipper mechanism, listerial InlB interacts with Met receptor, which autophosphorylates and recruits protein adapters; Gab1, Cbl, and Shc. These proteins activate phosphatidylinositol 3-kinase (PI 3 kinase) and small Rho GTPase kinase, Rac, which in turn promotes actin polymerization via Arp2/3. In zipper mechanism, *Shigella* bypass the cell adhesion to receptor but inject effector proteins by Type III secretion (TTSS) apparatus directly into host cell cytosol and induces massive actin polymerization. Shigellar IpaC activates small GTPases, Cd42, and Rac and promotes actin polymerization by Arp2/3. Shigellar VirA inhibits microtubule formation, and IpgD hydrolyzes phosphatidylinositol (4,5) biphosphate (PIP2) to phosphatidylinositol (5) phosphate (PIP) and disconnects actin from the membrane (redrawn from Veiga, E. and Cossart, P. 2006. *Trends Cell Biol.* 16:499–504)

with β1-integrin located on the basolateral side of the polarized epithelial cells. Phagocytic cup formation is mediated by the initiation of signaling events and induction of actin polymerization through the Arp2/3 pathway. Cup formation occurs by membrane extension. Bacteria become enclosed in the phagocytic cup in a process mediated by membrane retraction and actin depolymerization. In the zipper mechanism, relatively modest cytoskeletal rearrangements and membrane extension occur following binding to a receptor. Interaction allows signaling events to take place, which in turn promote recruitment of adapter and effector molecules to form cup-like structure that accommodates the bacteria inside for phagocytosis.

The trigger mechanism (Fig. 4.2) consists of four steps; (1) preinteraction stage, (2) interaction stage, (3) formation of macropinocytic pocket, and (4) actin depolymerization and closing of the macropinocytic pocket. In the preinteraction stage, bacteria synthesize necessary proteins in preparation for initiating infection. In the interaction stage, bacteria such as *Salmonella* and *Shigella* inject dedicated bacterial effector proteins through the type III secretory system (TTSS) made of SipB/C in *Salmonella* and IpaB/C in *Shigella*. During formation of the macropinocytic pocket, bacteria initiate signaling events that trigger massive actin polymerization and extension to form entry foci, known as membrane ruffling. In the final stage, actin depolymerization occurs and invasion proceeds.

Once inside the cytoplasm, bacteria employ adaptation strategies to promote intracellular survival, effectively avoiding lysosomal enzymes, antibacterial peptides, low pH, reactive oxygen radicals, and low nutrient concentrations. Some bacteria also induce actin polymerization for inter- and intracellular movement. When they reach the plasma membrane of the adjacent cell a protrusion is created which is endocytosed by the neighboring cell. Bacteria are then released from the double membrane and perpetuate infection. Some bacteria use a sugar uptake strategy to garner energy from host cells. There are several advantages of intracellular residence; bacteria can avoid phagolysosomal degradation, there is an abundance of nutrients in the cytoplasm, and cells are protected from antibiotics, antibodies, and complement-mediated cell lysis.

Iron Acquisition

Iron is essential for bacterial growth and survival and bacteria must acquire iron from host. However, iron concentration in the human body is very low because iron is bound to proteins in the form of lactoferrin, transferrin, ferritin, and hemin. There are three different ways bacteria can acquire iron from host (1) use of siderophores, (2) direct binding to the host cells, and (3) killing of host cells. Siderophores are low molecular weight compounds (example, catechols, hydroxamates) that are excreted into the medium, and which chelate iron with very high affinity. The siderophore–iron complex is then taken up by the siderophore receptor located on the bacterial surface and is internalized. The complex is broken inside the cell to release iron. For example, *Salmonella* uses enterobactin (cyclic trimer of dihydroxy benzoic acid) to acquire iron while *Yersinia* uses yersiniabactin (catechol-type). During iron acquisition, mechanism, pathogenic bacteria can acquire iron by directly binding to the host cells that contain iron on their surface. Some bacteria produce exotoxins such as hemolysin when iron level is low. The toxins kill the cells containing ferritin or heme, allowing the bacteria to acquire the iron.

Motility and Chemotaxis

The intestine is constantly washed with fast moving fluids, making colonization by bacteria difficult. Motile bacteria have a complex sensory system that allows them to move in the direction of nutrients (sugars and amino acids). They move directionally toward the mucosal membrane, which provides a greater chance for colonization. For example, *Vibrio* produces long filamentous pili called TCP (toxin-coregulated pili) which promote bacterial motility and colonization of epithelial cells.

Evasion of Immune System

Certain bacteria (*Streptococcus pyogenes, S. pneumoniae, Bacillus* species, *Yersinia pestis*) possess capsule, which has antiphagocytic action. The capsule prevents serum protein B from binding to C3b but helps protein H to bind to C3b. Binding of H to C3b forms C3bH, which is easily degraded by protein I. As a result, no C3bBb (C3-convertase) is formed. Therefore no MAC (membrane attack complex) is formed and the bacteria will be protected from complement-mediated killing. Sialic acid and hyaluronic acid in the capsule also prevent binding of C3b, which also prevents phagocytosis. Binding of LPS (lipopolysaccharide) to C3b prevents formation of C3-convertase and prevents formation of MAC and complement-mediated killing. Intracellular bacteria survive phagocytosis by producing hemolysin, superoxide dismutase (destroys "O" radical) and catalase (breaks down H_2O_2).

Some pathogens evade the host antibody response by inducing antigenic variation, such as structural variations in pili as well as variations in other surface proteins. The antigenic variation disguises the pathogen and fools the immune system from recognition. In addition, some pathogens evade the immune system by shrouding themselves with host proteins. For example, *Streptococcus* species cover themselves with host fibronectin. Some bacteria express protein A (*S. aureus*) or protein G (*S. pyogenes*), which binds to the Fc part of the IgG therefore will prevent macrophages from recognizing the antigen–antibody complex. Synthesis of immunoglobulin-specific proteases such as sIgA-specific proteases can cleave IgA in the hinge region to make it ineffective for antibody-dependent removal of pathogens.

Intoxication

Ingestion of preformed toxins such as staphylococcal enterotoxin, botulinum toxin, *Bacillus cereus* toxin, and seafood toxins results in food poisoning or intoxication (Table 4.4). Microorganisms present in foods will grow under favorable conditions and produce toxins in the food. Following ingestion, toxins are absorbed through the gastrointestinal epithelial lining and cause local tissue damage and may induce inflammation resulting in diarrhea or vomiting. In some cases, toxins are translocated to distant organs or tissues such as liver, kidney, peripheral, or central nervous system where they can cause damage. In most cases, the microorganisms pass through the digestive system without causing any harm.

Toxicoinfection

Some bacteria cause toxicoinfection, which occurs when ingested bacteria first colonize the mucosal surface and then produce exotoxins in the intestine. Toxins can induce toxic effects on the local cells or tissues and in some cases toxins enter blood stream and induce disease. Examples include cholera toxin produced by *V. cholerae*; heat-labile (LT) and heat-stable (ST) toxin produced by enterotoxigenic *E. coli*; and enterotoxin from *Clostridium perfringens*. In some cases, toxins kill polymorphonuclear leukocytes and aids bacterial growth and spread as in gas gangrene that is caused by *Clostridium* spp. In addition, *Clostridium* produces hydrolytic enzymes such as lecithinase (breaks down lecithin), hyaluronidase and protease (break down extracellular matrix and disrupts tissue structure), and DNase (reduces the viscosity of debris from dead host cells and aid in the spread of the bacteria) which aid in bacterial propagation.

Toxins

There are two types of toxin: exotoxin and endotoxin. Exotoxins are usually cell associated or excreted and sometimes they are released after cell lysis. Examples are botulinum toxin and Shiga toxin. Endotoxins are part of the cell structure and are cell associated, such as LPS in Gram-negative bacteria or PGN and LTA in Gram-positive bacteria. Some toxins such as cholera toxin and *E. coli* LT toxin, are designated as enterotoxins because they are responsible for enteric diseases (or gastroenteritis). Cytotoxins are cell- or tissue-specific such as neurotoxins (affect nerve cells), leukotoxins (attack leukocytes), hepatotoxins (attack liver cells), and cardiotoxins (damage cardiac tissue). Toxin designation is sometimes based on the bacterial species that produced them; Cholera toxin is produced by *Vibrio cholerae*; Shiga toxin by *Shigella* species, Diphtheria toxin by *Corynebacterium diphtheriae*, tetanus toxin by *Clostridium tetani*, botulinum toxin by *Clostridium botulinum*. Sometimes the toxin is named on the basis of the action it exerts such as adenylate cyclase produced by *Bordetella pertussis*, lecithinase by each *Clostridium perfringens* and *Listeria monocytogenes*. Toxins may also be designated by a letter such as staphylococcal enterotoxin A (SEA), SEB, SEE, etc., by *Staphylococcus aureus*.

Structure and Function of Exotoxins

There are six types of exotoxins and they are grouped based on the structure and the mode of action (Table 4.4; Fig. 4.3) (1) A–B type toxin: These types of toxins have two subunits; A and B. B binds to cell receptor and A exerts enzymatic activity such as inhibition of protein synthesis or cleavage of target protein. Examples are, Shiga-like toxin, cholera toxin, and botulinum toxin. (2) Membrane-disrupting toxin: These toxins either insert into the membrane to cause pore formation (example, hemolysin) or remove lipid head groups to destabilize the lipid bilayer membrane (example, phospholipases). (3) Activation of secondary messenger pathway: Toxins like *E. coli* LT and ST interfere with signal transduction pathways. (4) Superantigens: These toxins activate immune system to produce excess amounts of cytokines which exert their deleterious effects on host cells, provoking toxic shock syndrome. The examples of superantigens are staphylococcal enterotoxin, exfoliative toxin, and toxic shock syndrome toxins (TSST). (5) Proteases: Some toxins (i.e., neurotoxins) inactivate metalloproteases (Zinc metalloprotease) action, thereby interfering with nerve impulses. (6) Toxin-induced apoptosis: Some toxins kill cells by inducing programmed cell death or apoptosis (see below).

A–B Toxins

A–B type toxins consist of A and B subunits and are linked by a disulphide bridge. The B subunit binds to the host cell receptor and the entire toxin is internalized. The A subunit possess enzymatic activity, which is physically separated from the B subunit after being internalized. The sequence of events for A–B type toxins are; binding to the receptor, internalization, membrane translocation, and enzymatic activity and target modifications (Fig. 4.3).

The B subunit binds to a specific receptor such as a glycoprotein or glycolipid on the cell surface, and binding is very specific. If the receptor is found only on neurons, B portion will bind to the neuron only. The A subunit is translocated by endocytosis, and has a nonspecific action such as inhibition of protein synthesis or proteolytic cleavage of target protein. If A is delivered experimentally into a cell it will exert its function independent of the B subunit. In some toxins, the A subunit has ADP ribosyltransferase activity.

Table 4.4 Characteristics of bacterial toxins

Toxin type	Toxins	Producing bacteria	Mode of action	Target
Membrane damaging toxin	Hemolysin	*E. coli*	Pore formation	Plasma membrane
	Listeriolysin O	*L. monocytogenes*	Pore formation	Cholesterol
	Perfringolysin O	*C. perfringens*	Pore formation	Cholesterol
	α-toxin	*S. aureus*	Pore formation	Plasma membrane
	Streptolysin O	*S. pyogenes*	Pore formation	Cholesterol
Inhibit protein synthesis (A–B type)	Shiga toxin or Shiga-like toxin	*Shigella* spp. *E. coli*	N-glycosidase	28S rRNA
	Diphtheria toxin	*C. diphtheriae*	ADP ribosylation	Elongation factor-2
	Edema factor	*B. anthracis*	Adenylate cyclase	ATP
Activate second messenger Pathways (A-B type)	Heat-labile toxin (LT)	*E. coli*	ADP ribosyltransferase	G-proteins
	Heat-stable toxin (ST)	*E. coli*	Stimulates Guanylate cyclase	Guanylate cyclase
	Cholera toxin	*V. cholerae*	ADP ribosyltransferase	G-protein
Activate immune response	Enterotoxins, toxic shock toxins	*S. aureus*	Superantigen	TCR and MHC II
Protease	Lethal factor	*B. anthracis*	Metalloprotease	MAPKK1/ MAPKK2
	Botulinum neurotoxin	*C. botulinum*	Zinc metalloprotease	Synaptobrevin, SNAP-25, syntaxin
Apoptosis	IpaB	*Shigella* spp.	Apoptosis	
	LLO	*L. monocytogenes*	Apoptosis	

Adapted from Schmitt, C. et al. 1999. Emerg. Infect. Dis. 5:224–234

This removes ADP ribosyl group from NAD^+ and attaches it to a host cell protein, such as elongation factor 2 (EF2) in a process called ADP ribosylation of EF2. Modified EF2 are unable to help in host cell protein synthesis. For example, diphtheria toxin (DT) and *Pseudomonas* exotoxin-A cause ADP ribosylation of EF2, and consequently lead to cell death. In contrast, cholera toxin and *E. coli* heat-labile toxin (LT) cause ADP ribosylation of the G-protein, a signal transduction protein which controls the adenylate cyclase activity which in turn increases the cAMP levels. The increased cAMP controls ion (Na^+, K^+, Cl^-) pumps and fluid movement in cells. Shiga-like toxin (Stx) displays a different type of action – it has adenine glycohydrolase activity. It removes a single adenine residue form 28S rRNA and blocks protein synthesis.

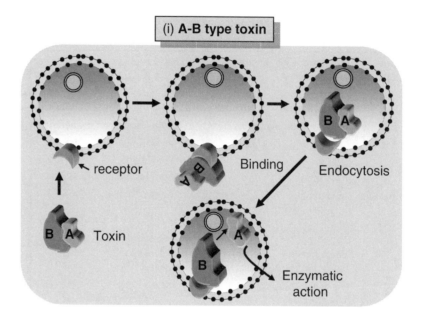

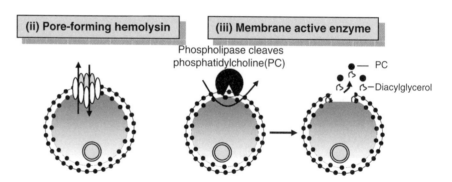

Fig. 4.3 (continued)

Membrane-Disrupting Toxins

Membrane-disrupting toxins such as hemolysins and phospholipases cause lysis of the cell membrane. Pore-forming hemolysins are sulfhydryl-activated toxins, and are produced by several pathogens such as *L. monocytogenes* (listeriolysin O), *Streptococcus pyogenes* (streptolysin O), *S. pneumoniae* (pneumolysin O), and *C. perfringens* (perfringolysin O). These hemolysins bind to the membrane receptor cholesterol, insert into the membrane and oligomerize to form a channel. As a result intracellular ions are released and as water enters, the cell swells and the membrane eventually ruptures. Phospholipases are also considered to be hemolysins and they remove the charged phosphate head group (PO_3^+) from the lipid (diacylglycerol) portion of the phospholipid, which destabilizes the membrane structure to cause cell lysis.

Induction of Second Messenger Pathways

Some toxins alter signal transduction pathways, which are required for various cellular functions. For example, *E. coli* CNF (cytotoxin necrotizing

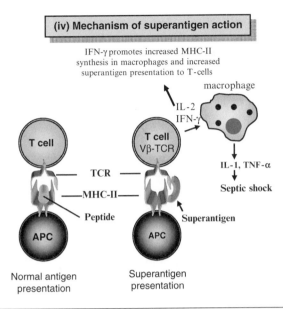

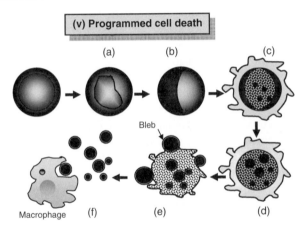

Fig. 4.3 Schematic drawing showing the mechanisms of actions for different toxins: (i) A–B type toxin; (ii) pore-forming toxins; (iii) membrane active toxins; (iv) superantigens; and (v) induction of programmed cell death

factor) 1 or 2, and heat-labile toxin (LT), Clostridial botulinum toxin and C3 toxin, and *Vibrio cholerae* cholera toxin inactivates or modify a small GTP-binding signal transduction protein, Rho, which regulates actin cytoskeleton formation. *E. coli* heat-stable toxin (ST) binds to the membrane guanylate cyclase that increases cellular cGMP. As a result, electrolyte balance in the bowel is disrupted which leads to fluid loss and diarrhea.

Superantigens
Staphylococcal enterotoxins (SEA, SEB, SEC, SEE), exfoliative toxins, and toxic shock syndrome toxins (TSST) are called superantigens. Unlike other protein antigens, they are not processed by proteolytic digestion by antigen presenting cells (APC). Instead these toxins bind directly to the

MHC II of APC and activate T-cells possessing a T-cell receptor (TCR) composed of the Vβ chain. Activated T-cells then produce large quantities of IL-2, which in turn activates macrophages for increased production of IL-1, IL-6, TNF-α, and IFN-γ. IFN-γ induces increased expression of class II molecules and subsequently enhances presentation of superantigen toxin. Production of large quantities of IL-2 and TNF-α induces nausea, vomiting, malaise, fever, erythematosus lesion, and toxic shock.

Protease

Botulinum toxin types A, B, C, D, E, F, and G, and tetanus toxin inactivate zinc metalloprotease, which cleave synaptobrevin and syntaxin. These proteins are responsible for acetylcholine release from synaptic vesicle. As a result, neurotransmitter release is impaired causing paralysis (see Chap. 8 for details).

Programmed Cells Death

Some toxins induce programmed cell death or apoptosis in host cells as a mechanism for infection (Fig. 4.3). The dead cells are then removed by macrophages, thus preventing inflammation. Shiga toxin and IpaB protein in *Shigella*, LLO in *L. monocytogenes*, cholera toxin from *V. cholerae*, and *Clostridium perfringens* enterotoxin (CPE) are some examples which induce programmed cell death. In contrast, some toxins such as LLO and CPE at high concentrations induce oncosis (necrosis) by causing physical trauma that leads to cell swelling, lysis, random DNA shearing, and release of toxic intracellular contents, which induce inflammation.

The typical signs of apoptosis include (a) cell shrinkage – condensation of cell cytoplasm and nucleus; (b) entrance of cells into zeiosis stage – cell morphology alters and shrinkage continues; (c) chromatin condensation/margination – crescent shaped nucleus localizes along the nuclear membrane; (d) DNA fragmentation – caspases (proteases) and nucleases are activated and cleave DNA in the factor of 168 bp fragments resulting in laddering of DNA bands, which can be detected by gel electrophoresis; (e) membrane blebbing of apoptotic bodies containing DNA fragments which can be seen under a microscope; and (f) clearance by phagocytic cells – macrophages engulf apoptotic cells/bodies to prevent inflammation. Phosphatidylserine (PS) is normally located in the inner leaflet of cytoplasmic membrane; however, during apoptosis, PS translocates to the outer leaflet and becomes the target for macrophage-mediated clearance.

Apoptosis is a highly regulated process of programmed cell death. Apoptotic stimuli, which can include bacterial toxins, activate death ligands FADD (Fas-associated death domain) and TRADD (TNF receptor-associated death domain). These activate procaspase 8, which phosphorylate BAD (Bcl-2-associated death domain). These again activate DNase and procaspase 9 (cysteine protease). Procaspase 9 activates caspases 3, 6, 7, which break down nuclear proteins (histone, actin lamins, and poly ADP ribose polymerase (PARP)) leading to apoptosis. C-Myc and p53 proteins induce programmed cell death while the Bcl-2 protein family, normally present in malignant cells, suppresses apoptosis.

Endotoxin

Endotoxin such as LPS, PGN, and LTA are responsible for septicemia. The Lipid A component of LPS is highly toxic. LPS raises the body temperature and activates macrophages to produce IL-1 and TNF-α, which in turn causes

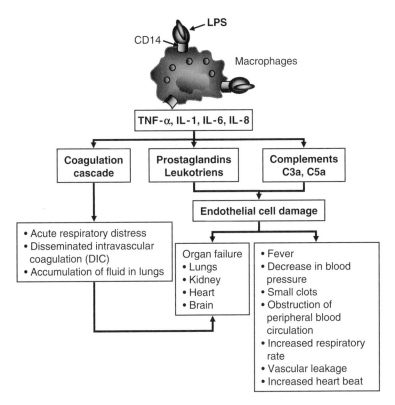

Fig. 4.4 Mode of action of lipopolysaccharide (LPS)

the hypothalamus to release prostaglandins. Increased prostaglandin causes elevated body temperature (fever). In essence, this defense strategy is designed to inhibit bacterial growth. Aspirin blocks prostaglandin release and thus reduces body temperature. Endotoxins can trigger a massive inflammatory response leading to septicemia and septic shock. During septic shock, the circulatory system collapses, blood pressure drops, fever increases, heart rate increases, multiple organs fail, and death follows.

At low concentrations, LPS stimulates macrophages to produce IL-1 and acts as polyclonal activator of B-cells, which in turn produce IL-2 and IL-6. LPS at higher concentrations causes tissue injury and disseminated (widespread) intravascular coagulation (DIC) allowing neutrophils, lymphocytes, and monocytes to stick to the endothelial surfaces (Fig. 4.4). As a result, blood pressure drops, the patient enters shock (referred as septic shock) and death ensues. Prolonged exposure to LPS invokes cachexia, characterized by wasting of muscle and fat cells. LPS is also responsible for appetite suppression resulting in weight loss. LPS and LPS–protein complex binds to the CD14 receptor on monocytes and macrophages and results in release of IL-1, IL-6, IL-8, and TNF-α, which induce prostaglandin and leukotriene production, leading to the damage of endothelial cells. Endotoxins also activate the complement pathway (Alternative pathway) to form C3a and C5a, which also cause severe endothelial cell damage. A membrane attack complex (MAC) forms and cell lysis also occurs (Fig. 4.4)

Genetic Regulation and Secretion Systems for Virulence Factors

Pathogenicity Islands

A genomic segment that carries a cluster of virulence genes is termed as a Pathogenicity Island (PAI), which was first introduced by Prof. Jorg Hacker in late 1980. It was first discovered in an uropathogenic *E. coli* strain 536 and it was demonstrated that the deletion of the PAI resulted in a nonpathogenic phenotype. PAI encodes genes for iron acquisition, adhesions, pore-forming toxins, proteins responsible for apoptosis, secondary messenger pathway toxins, superantigens, lipases, proteases, O antigens, and protein secretion systems designated type I, type II, type III, type IV, and type V secretion systems.

Specific characteristics of a PAI are (1) it carries one or more virulence genes; (2) it is present in pathogenic species but absent in nonpathogenic species; (3) the size range is 10–200 kb; (4) The G + C content of PAI genome (40–60%) is lower than that of the core genome (25–75%); (5) PAI are located close to tRNA genes, indicating that tRNA gene serves as anchor point for insertion of foreign genes (Some bacteriophages also use tRNA as insertion points in the host genome); (6) PAI could be unstable and could be deleted from genome with a frequency higher than the normal rate of mutation; and (7) PAIs are frequently associated with the mobile genetic elements and are flanked by direct repeats (DR), DRs are 16–20 bp long sequence with perfect or nearly perfect sequence repetitions, which serve as integration sites for bacteriophages. PAIs also carry genes for integrases or transposes, probably acquired form the bacteriophages. The insertion sequence (IS) elements are also found in PAI. In addition, PAI can represent integrated plasmids, conjugative transposons, and bacteriophages.

Protein Secretion System

Bacterial secreted proteins perform multiple tasks, including pathogenicity, cell envelop assembly, metabolism, and interaction with host cells. In Gram-negative bacteria, secretion is mediated by a different mechanism than that of Gram-positive bacteria because of the presence of an outer membrane structure in Gram-negative bacteria (see Chap. 2). Therefore, different secretory machineries are required to achieve the same goal. The protein secretion systems in Gram-negative bacteria are designated as type I–V systems.

Type I System

The Type I secretion system (T1SS) has an assembly of an ATP-binding cassette (ABC) transporter protein located within the inner membrane, a periplasmic membrane and an outer membrane. Hemolysins are transported by T1SS.

Type II System

The Type II secretion system (T2SS) consists of 12 subunits that are located in the inner membrane, the periplasm, and the outer membrane. It is responsible for protein secretion for both pathogenic and nonpathogenic bacterial species. The equivalent of T2SS in Gram-positive bacteria is called the Sec (secretory) system. In Gram-positive bacteria, proteins with N-terminal leader sequence are transported across the cytoplasm using the Sec system and the

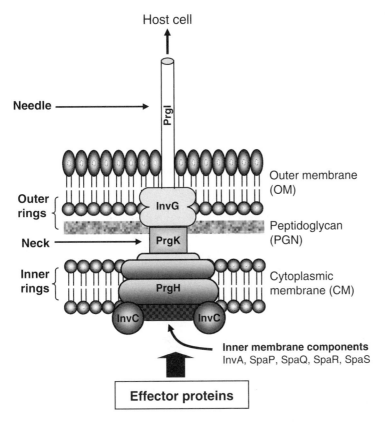

Fig. 4.5 A schematic model of the type III secretion system (TTSS) of *Salmonella enterica* (adapted from Galan, J.E. and Wolf-Watz, H. 2006. Nature 444:567–573; Galan, J.E. and Collmer, A. 1999. Science 284:1322–1328)

leader sequences are cleaved and released outside the cell. In Gram-negative bacteria, T2SS transport the periplasmic derivatives of the substrate proteins across the outer membrane.

Type III System

Although the Type III secretion system (T3SS or TTSS) is termed as a secretion system, its main function to translocate proteins through a third membrane that is the eukaryotic membrane during interaction with host cells. It is called a molecular syringe and its assembly requires the function of more than 20 genes. A typical TTSS apparatus has four major parts: inner rings, neck, outer rings, and a needle (Fig. 4.5). Assembly of the T3SS (TTSS) is similar to that of the flagellum machinery system. The T3SS is involved in virulence; during intimate contact with host cells, bacteria use the T3SS to inject virulence proteins into the host cell to promote invasion, actin-based intra- and intercellular motility, to induce apoptosis, and to interfere with intra-cellular transport process. Genes required for T3SS formation are encoded in plasmids or in PAI. Genes for T3SS in *Salmonella enterica* are located in SPI-1 and SPI-2, in the locus of enterocyte effacement (LEE) for enteropathogenic *E. coli* (EPEC), and in PAI in a large plasmid for *Shigella*.

Type IV System

The Type IV secretion system (T4SS) is similar to T3SS, which is involved in translocation of DNA or DNA–protein complex into eukaryotic host. T4SS is a complex structure composed of ten subunits. The best studied T4SS is in *Agrobacterium tumefaciens*, where it mediates the translocation of DNA–protein into plant cells to induce tumor formation. Among the human pathogens *Bordetella pertussis*, *Legionella pneumophila*, *Brucella* spp., and *Helicobacter pylori*, the T4SS is important.

Type V System

The Type V secretion system (T5SS) is also known as an autotransporter. The secretion system and the substrate are synthesized as preproproteins, and are released into the periplasm via the Sec system. After proteolytic cleavage of the N-terminal leader peptide, the transporter domain of the proprotein is oligomerized to form a β-barrel structure in the outer membrane, which allows the passenger domain to be excreted outside. In pathogenic *E. coli*, T5SS is encoded by EspC PAI and in *Salmonella enterica*, it is encoded by SPI-3.

Regulation of Virulence Genes

Foodborne pathogens encounter a variety of harsh environments in food products and during transit through the gastrointestinal tract. Food-associated environments may include acids, salts, preservatives, peroxides, antimicrobial chemicals, flavoring agents, sugars, and storage temperatures. While in the GI tract, pathogens encounter gastric acid, bile salts, mucus, lysozyme, and natural microbiota. Although these environments are stressful, the pathogens must still be able to express the necessary colonization and invasion factors for their survival and multiplication. A complex regulatory element is thought to play an important role during pathogen interaction with the host system.

Protein synthesis in prokaryotes consists of two major steps: transcription and translation. During transcription the instructions stored in the DNA are transferred to the mRNA. Transcription consists of three steps: initiation, elongation, and termination of RNA synthesis, in which RNA polymerase is involved in the synthesis of mRNA from the DNA. The core RNA polymerase (RNAP) has five subunits; α_1, α_2, β, β' and ω. For the RNAP to bind-specific promoters, another subunit known as sigma factor (σ) is required. The sigma factor greatly increases the specificity of binding of RNAP and also helps in the separation of DNA strands during the initiation of transcription process. The sigma factors detach from the DNA after the initiation process of transcription has begun.

Prokaryotic sigma factors are classified into two structurally unrelated families, σ^N/σ^{54} and σ^{70}. The σ^N is mostly involved in nitrogen metabolism and is also involved in a variety of metabolic processes. The σ^{70} family is larger and more diverse than σ^{54}, and is divided into four groups (Group I–IV) based on their primary amino acid sequences and structures. The Group I sigma proteins/the primary sigma factors (e.g., σ^A of *Bacillus subtilis*) are also known as "housekeeping" sigma factors because they regulate expression of "housekeeping" genes or genes responsible for basic metabolic processes and cell functions. The other groups are known as alternate sigma factors and they regulate specific physiological processes such as survival during stationary phase and during exposure to various stressful environments. Another intriguing aspect is that alternate sigma factors are now shown to regulate virulence genes or virulence-associated genes required for bacterial pathogenesis. Some σ^{70} family members like the σ^S of Gram-negative bacteria (group II), σ^B of Gram-positive bacteria (group III), and extracytoplasmic functioning sigma factors (group IV) are known to contribute to the bacterial stress responses. A few of the alternate sigma factors also play a role in the virulence of bacteria as they respond to the host environment, which is critical in the infection process during the intestinal phase of infection.

Summary

Foodborne pathogens cause three forms of disease: *foodborne infection, foodborne intoxication,* and *foodborne toxicoinfection.* The principle route of infection/intoxication for foodborne pathogens is oral and the primary site of action is the intestine. The infectious dose for foodborne pathogens varies and depends on the type of organism or toxin as well as the type of food (liquid vs. solid). The pathogens that are responsible for infection colonize the gut by producing various adhesion factors including fimbriae, adhesin proteins, and extracellular matrices that allow biofilm formation. Invading pathogens have developed strategies to cross the epithelial barrier. Some use M cells to reach the subcellular location or some pathogens actively penetrate epithelial cells by rearranging host cell cytoskeletal structures. Pathogens localized in subcellular locations multiply, move from cell-to-cell, and induce inflammation and can elicit enough damage to induce diarrhea and gastroenteritis. Some intracellular pathogens induce apoptosis in macrophages, dendritic cells and neutrophils, thus ensuring their survival in host tissues. Pathogens may also translocate to deeper tissues including the liver, lymph nodes, spleen, brain, and placenta. Foodborne intoxication is mediated by exotoxins produced by pathogens in the food, which induces cell damage, fluid and electrolyte losses, and apoptosis following consumption of the contaminated food. The mechanisms of exotoxin action vary, and based on the toxin action, can be classified as A–B type toxins, membrane acting toxins, superantigens, proteases, protein synthesis inhibitors, and signal transduction modulators. The bacterial cell structure-associated endotoxins (LPS, PGN) are generally associated with systemic foodborne infection, and these toxins modulate immune system to induce the release of large quantities of cytokines that promote fever, decrease blood pressure, and induce septic shock. In most pathogens, virulence factors encoded genes are located in various pathogenicity islands or islets, which may be found in plasmids, bacteriophage, or the chromosome. Virulence proteins are exported from cells by various secretory mechanisms and those are designated type I–type V secretory systems. Finally, bacterial virulence gene expression is a complex process that may be controlled by different regulatory elements in the food system as well as in the host. Alternate sigma factors are found to be crucial in virulence gene expressions in foodborne pathogens.

Further Readings

1. Chemielewski, R.A.N. and Frank, J.F. 2003. Biofilm formation and control in food processing facilities. Comp. Rev. Food Sci. Food Safety 2:22–32.
2. Conner, S.D. and Schmid, S.L. 2003. Regulated portals of entry into the cell. Nature 422:37–44.
3. Cossart, P. and Sansonetti, P.J. 2004. Bacterial invasion: the paradigms of enteroinvasive pathogens. Science 304:242–248.

4. Galan, J.E. and Collmer, A. 1999. Type III secretion machines: bacterial devices for protein delivery into host cells. Science 284:1322–1328.
5. Galan, J.E. and Wolf-Watz, H. 2006. Protein delivery into eukaryotic cells by type III secretion machines. Nature 444:567–573.
6. Gouin, E., Welch, M.D., and Cossart, P. 2005. Actin-based motility of intracellular pathogens. Curr. Opin. Microbiol. 8:35–45.
7. Kazmierczak, M.J., Wiedmann, M., and Boor, K.J. 2005. Alternative sigma factors and their roles in bacterial virulence. Microbiol. Mol. Biol. Rev. 69:527–543.
8. Montecucco, C., Papini, E., Schiavo, G. 1994. Bacterial protein toxins penetrate cells via a four-step mechanism. FEBS Lett. 346:92–98.
9. Popoff, M.R. 1998. Interactions between bacterial toxins and intestinal cells. Toxicon 36:665–685.
10. Saier, M.H. Jr. 2006. Protein secretion systems in Gram-negative bacteria. Microbe 1:414–419.
11. Salyers, A.A. and Whitt, D.D. 2002. Bacterial Pathogenesis: A Molecular Approach, 2nd edition, ASM, Washington, DC.
12. Schmidt, H. and Hensel, M. 2004. Pathogenicity islands in bacterial pathogenesis. Clin. Microbiol. Rev. 17:14–56.
13. Schmitt, C., Meysick, K., and O'Brien, A. 1999. Bacterial toxins: friends or foes? Emerg. Infect. Dis. 5:224–234.
14. Sibley, L.D. 2004. Intracellular parasite invasion strategies. Science 304:248–253.
15. Veiga, E. and Cossart, P. 2006. The role of clathrin-dependent endocytosis in bacterial internalization. Trends Cell Biol. 16:499–504.

5

Animal and Cell Culture Models to Study Foodborne Pathogen Interaction

Introduction

Our knowledge of pathogenic mechanism of foodborne pathogens has stemmed from the use of various cultured cell lines and animal models. In early days, animal models were often used to confirm the pathogenic nature of an isolate that was involved in a disease and also to study immune responses. In recent years, however, animals are used as a model to study pathogenic mechanism of a microorganism, immune response to an infection, and to study the efficacy of a vaccine. However, cultured cell lines are now considered indispensable powerful tools in studying the molecular and cellular mechanism of pathogenesis. Both animal and cell culture models are essential for studying pathogenesis, thus one must be familiar with those models and their applications.

To study bacterium and host interactions, it is important to use the strain that cause the disease because of the clonal nature of the pathogens. Pathogens may also loose virulence traits during subculture, therefore one has to avoid multiple subculturing (passage) before pathogenicity testing. During pathogenicity testing, some strains may not exhibit pathognomonic symptoms in either model. Therefore, finding a suitable and sensitive animal or cell culture system is crucial for studying host parasite interactions. Organ cultures such as ligated-ileal loop or embryonated-egg model have been used as a substitute for animal models.

Animal Model

Though humans are ideal model to study human pathogens, it is not possible to use humans for safety, ethical and expense related concerns. However, human volunteers have been used in certain nonfatal diseases. Animal models are used frequently as substitute. Among them, nonhuman primates (monkey, baboon, and chimpanzee) are ideal to mimic many diseases. Again ethical and expense related considerations limit their widespread applications. Rodents are most commonly used animal models. They include mice, rats, rabbits, hamsters, and guinea pigs. Other animals including ferrets, pig or piglets, calves, dogs, and cats are used. Careful considerations should be given while choosing an animal model: Pathogens should infect animals by the same route as humans; and exhibit a similar colonization pattern, similar tissue distribution patterns and same degree of virulence as in humans.

The advantages of using animal models for pathogenicity testing are; generally inbred lines are used, therefore the disease could be reproduced and variability could be reduced. Rodents are small and require less housing space and they breed rapidly thus large numbers could be used for statistical analysis purposes. Rodents are less expensive to house or maintain than the larger animals.

Limitations of use of animal model include, they need greater care than the cell cultures. First of all, animal use requires institutional approval and there are strict guidelines one has to follow to use animals for research purposes. Proper housing facility and veterinary care must be provided for animals. Caregivers or animal handlers must take precautions to avoid accidental exposure to the infective agents. Animals also can attack, scratch, and bite during handling. Furthermore, there are some fundamental differences in physiology between rodents and humans. Rodents are coprophagy, thus reingestion of same bacteria may compromise the experimental data especially for experiment with foodborne pathogens. Resident microflora population in rodents may be different from humans and thus may also influence the data. Most importantly, some animals are not sensitive to pathogens thus may not show the same human-type symptoms or tissue distribution. There are several possible solutions to overcome these problems: use of related bacterial strains which may be infective; use different route of administrations (i.e., ingestion vs. intraperitoneal route or intranasal or intravenous route); use neonatal animals since their immune system is immature; and the use of immunocompromised animals. Immunocompromised conditions can be induced by exposing animals to irradiation, which can destroy immune cells since the immunological stem cells are susceptible to irradiation. Nude mice are genetically engineered mice that do not possess any functional T-cells thus are immunocompromised. Severe combined immunodeficient (SCID) mice lack functional T- and B-cells and are used in some experiments. The immunocompromised animals are expensive to procure and also require greater care in their handling and maintenance (need pathogen free-environment) since they are highly susceptible to pathogens.

Suckling mouse model has been used to study pathogenesis of diarrheagenic microbes including *Vibrio cholerae* and enterotoxigenic *E. coli*. In this model, the test microorganisms or their toxin preparations are administered orally in 2–4 day old suckling mice and mice are sacrificed and examined for fluid accumulation in the gut after 4–24 h. The ratio of gut weight to body weight has been used to determine the effect of toxins. Suckling mice have also been used to study the translocation of protozoan species including *Cryptosporidium parvum* in the gastrointestinal tract and oocyst counts are determined in the intestinal contents.

Organ Culture

Animal organs can be used as an alternative for animal model and should be ideal for investigating certain bacterium host interactions. Examples are, ligated-ileal loop and embryonated eggs. Organs are genetically intact, multiple cell types are represented, and cells retain their original shape and configurations, and the drawback is those are short-lived.

Ligated-Ileal Loop Assay

Generally, diarrheagenic microorganisms (*Salmonella, Vibrio, Shigella*, and *E. coli*) are tested by ligated-ileal loop assay with rabbits or rats. Animals are anaesthetized before initiating the experiment. A small incision is given in the abdomen and portion of the small intestine is pulled out from the abdominal cavity. The knots are placed in every 2–3 cm intervals using strings to create loops. Test materials along with proper controls are injected into the loops and intestine is placed back inside, and the skin incision is sutured. After 18–24 h, the intestine is examined for fluid accumulation (ballooning) due to the diarrheagenic action of a toxin.

Embryonated-Egg Assay

Twelve to fourteen days old embryonated hen's eggs are injected aseptically with test organisms in the chorioallantoic membrane (cavity) and incubated for 3–4 days. Death of embryos or characteristic cytopathic effects is indicative of the infective nature of the test organisms. This model is used for certain bacterial species including *Listeria monocytogenes* and viruses such as influenza virus.

Cultured Cell Lines

Cells derived from animal tissue or insect are attractive models for studying pathogenesis. There are two types of cells; primary and secondary. Primary cells are mortal, consist of mixed cell types and are short-lived. Secondary cells are immortal and consist of one type of cell. The secondary cells are originated either from tumor cells, transformed cells, or hybridoma cells. Transformed cells are generated by introducing viral genes such as those from simian virus 40 (SV40) or Epstein Barr Virus (EBV). Hybridoma cells are made from the fusion of normal and transformed cells. Examples of commonly used secondary cell cultures models are; Caco-2 and HT-29 (colon cells), HepG2 (liver cells), HEp-2 (larynx cell), Henle-407 (small intestine – jejunal), HeLa (cervix), J-774 (macrophage), Vero (monkey kidney), and CHO (Chinese hamster ovary) (Table 5.1).

There are several advantages of cell culture model over animal model for studying microbial pathogenesis. It is a simple and controlled model to study host–bacterium interaction; easy to run experiments with radioactive materials, toxins, or transfection with foreign DNA; cells multiply rapidly thus experiment can be conducted rapidly; the secondary cells are immortal provided nutrients and proper culturing conditions are maintained; and relatively inexpensive compared to animal model.

However, there are several limitations such as – cultured mammalian cells are generally derived from tumor cells; therefore, genetic aberrations have occurred in these cells, and the cells may not be genetically identical to parent cells. Continuous growth of cells will produce mutations and genetic rearrangements in the chromosome therefore these cells may loose traits of original tissue and may loose tissue-specific receptor for bacterial adhesins or

Table 5.1 List of human, animal, and insect cell lines used to study interaction of foodborne pathogens

Bacteria	Cell line	Cell type	Source	Interaction type
Salmonella	CHO	Epithelial	Chinese hamster ovary	Elongation, detachment
	Vero	Fibroblast	Monkey kidney	Lysis, protein synthesis inhibition
	HEp-2	Epithelial	Human laryngeal	invasion
	Henle-407	Epithelial	Human jejunal	Intracellular growth
	J774	Macrophage	Mouse	Intracellular growth
	HeLa	Epithelial	Human cervix	Toxicity, actin polymerization
	HT-29	Epithelial	Human colon	Apoptosis
E. coli	Vero	Fibroblast	Monkey kidney	Lysis, protein synthesis inhibition
	CHO	Epithelial	Chinese hamster ovary	Lysis, toxicity
	Henle-407	Epithelial	Human jejunal	Adhesion
	HEp-2	Epithelial	Human laryngeal	Adhesion, toxicity
	MDBK	Epithelial	Madin-Darby bovine kidney	Invasion
	HeLa	Epithelial	Human cervix	Toxicity, apoptosis
	T84	Epithelial	Human colon	Apoptosis
	Y1	Epithelial	Human adrenal gland?	Rounding, detachment, cAMP
	J774	Macrophage	Mouse	Apoptosis
Shigella	HeLa	Epithelial	Human cervix	Cell death, protein synthesis inhibition
	Vero	Fibroblast	Monkey kidney	Lysis, protein synthesis inhibition
	3T3	Fibroblast	Mouse	Invasion, actin polymerization
	U937	Monocyte	Human	Apoptosis
	Mϕ	Macrophage	Mouse	Invasion, apoptosis
Campylobacter	HeLa	Epithelial	Human cervix	Distended cells (CDT effect)
	Vero	Fibroblast	Monkey kidney	Distended cells (CDT effect)
	AZ-521	Epithelial	Human stomach	Vacuolation
Yersinia	J774	Macrophage	Mouse	Exocytosis, apoptosis
	HEp-2	Epithelial	Human laryngeal	Invasion, lysis
Vibrio	CHO	Epithelial	Chinese hamster ovary	Elongation, cAMP accumulation
Clostridium perfringens	Vero	Fibroblast	Monkey kidney	Membrane permeability
	Caco-2	Epithelial	Human colon	Necrosis, apoptosis, pore formation
	MDDK	Epithelial	Madin-Darby dog kidney	Pore formation

Table 5.1 (continued)

Bacteria	Cell line	Cell type	Source	Interaction type
Listeria monocytogenes	Caco-2	Epithelial	Human colon	Adhesion, invasion, apoptosis
	CHO	Epithelial	Chinese hamster ovary	Detachment, lysis
	Henle-407	Epithelial	Human jejunal	Intracellular growth, death
	Vero	Fibroblast	Monkey kidney	Toxicity, adhesion, invasion.
	HEp-2	Epithelial	Human laryngeal	Invasion
	J774	Macrophage	Mouse	Intracellular growth
	RAW	Macrophage	Mouse	Intracellular growth
	HeLa	Epithelial	Human cervix	Toxicity
	HUVEC	Endothelial	Human umbilical vein endothelial cell	Intracellular growth
	Hep-G	Epithelial	Human liver	Intracellular growth, apoptosis
	3T3	Fibroblast	Mouse	Invasion, plaque formation, lysis
	Ped-2E9	B-cell	Mouse hybridoma	Lysis, apoptosis
	RI-37	B-cell	Human–mouse hybridoma	Lysis, apoptosis
	S2	Epithelial	Drosophila	Invasion and infection model
Avian flu virus (H5N1)	MDCK	Epithelial cells	Madin-Darby canine kidney	Cytopathic effects, virus multiplication
Hepatitis A virus	Vero	Fibroblast	African green monkey kidney	Virus replication and cytopathic effects

Adapted from Bhunia, A.K. and Wampler, J.L. 2005. Foodborne Pathogens: Microbiology and Molecular Biology, Caister Academic

unmasking low affinity receptors. Cells which normally do not bind to bacteria may bind to the cultured cells and can adhere and invade cultured cells. Fundamentally, immortalized cells are different from the normal cells in the body. Sometimes there are loss of shape and distribution of surface antigens. Cultured cells are not polarized while normal cells are polarized in intact animal, i.e., different parts of the cells are exposed to different environments (lumen, adjacent cells, underlying blood vessels, and tissues). Polarized cells contain different sets of proteins important for their functions. Disruption of cells will expose some surface proteins, which are not previously exposed. Hormones can help immortal cells to convert into polarized cells. Host tissue consists of multiple cell types such as epithelial, fibroblasts, muscle cells, and neurons but the cultured cells consist of only one type of cell, therefore bacterial interaction with concerted host cell cannot be studied with cell culture models. In cultured cells, mucus and other secretory components are lacking which normally interact with pathogens. Thus animal models are often used to confirm or verify the findings from the in vitro cell culture models.

Measurement of Virulence

Animal Model

In animal model, infectivity or lethality of a pathogen or toxins could vary depending on the route of administration of the test agent. General rout of administrations are oral, intragastric (i.g.), intravenous (i.v.), intraperitoneal (i.p.), intramuscular (i.m.), subcutaneous (s.c.), and intradermal (i.d.) routes. To assess the pathogenicity of an agent natural route of infection should be used. Generally, the measurements are expressed as infectious dose (ID) or as lethal dose (LD). Infectious dose 50 (ID_{50}) is defined by the number of bacteria required to infect 50% of animals. Lethal dose 50 (LD_{50}) is defined by the number of bacteria required to kill 50% of the animals. Sigmoidal curves are generated with these experiments where LD_{25}, LD_{50}, LD_{75}, or LD_{100} values can be extrapolated (Fig. 5.1). Infectious dose or lethal dose data provides the crude measure of infection that includes cumulative effects of colonization, spread and symptoms and therefore ID_{50} vs. LD_{50} cannot be compared. Moreover, these data provide relative measure of virulence and cannot compare two different diseases such as bacillary dysentery vs. cholera.

Cell Culture Model

Using tissue culture models, pathogen or toxin interaction with different cell lines (Table 5.1) can be studied. Cytotoxicity or cytopathogenicity assays are used to determine the cytotoxic potential of pathogens. Similarly pathogens ability to adhere, to invade and to move from cell-to-cell can be assessed using cell models. Cytotoxicity assay is generally used as a confirmatory test for a pathogen, which has already been identified or detected by other methods. This in vitro assay also allows one to study the mechanism of pathogenesis of various microorganisms or toxins. Interaction of pathogens with cultured cell lines can be measured by different methods described below.

Microscopy
Degree of cell damage can be assessed under a phase-contrast light microscope. The typical cytopathic effect (CPE) consists of cell detachment, floc-

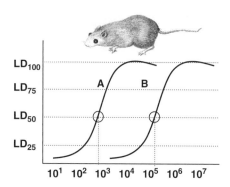

Fig. 5.1 Estimation of LD_{50} values for pathogens extrapolated from a sigmoidal curve generated after infecting a simulated animal model with different concentrations of bacteria (cfu per animal). LD_{50} value for pathogen A is about 10^3 cfu while LD_{50} for pathogen B is 2×10^5 cfu

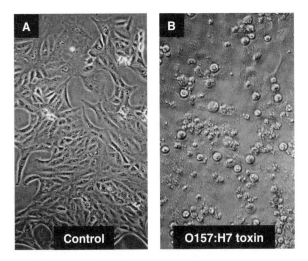

Fig. 5.2 Effect of *Escherichia coli* O157:H7 (EDL933) toxin on Vero cells after 24 h exposure. Toxin exposed cells (panel B) show cytopathic effects characterized by cell rounding, granulation, and detachment (*See Color Plates*)

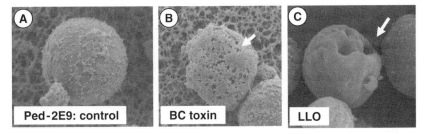

Fig. 5.3 Scanning electron microscopic photographs showing the (**B**) *Bacillus cereus* toxin and (**C**) *Listeria monocytogenes* listeriolysin O (LLO, a hemolysin)-induced membrane pore formation on lymphocyte cell line, Ped-2E9. Panel (**A**) is control

culation, rounding or elongation, cell lysis, cytoplasmic granulation, and cell death (Fig. 5.2). Highest dilutions of the toxin or pathogen showing CPE are considered as the cytotoxic titer for that agent. Furthermore, the nature of cell death (necrosis vs. apoptosis) could be detected by fluorescence microscope after staining cells with acridine orange, ethidium bromide, or propidium iodide. Scanning electron microscope is often used to examine the membrane damage, pore formations, and apoptotic bodies in eukaryotic cells (Fig. 5.3).

One of the most commonly used cytotoxicity assay is Vero cell (monkey kidney cell) assay which is performed to determine the cytotoxins from *Shigella* species, *E. coli*, and *Clostridium difficile*. The cytotoxic or cytopathic effects (CPE) are characterized by cell detachment, flocculation, rounding, and cell death (Fig. 5.2).

Trypan Blue Exclusion Test

Trypan blue is commonly used to asses the viability of mammalian cells. In this assay, cell suspensions are mixed with equal volumes of Trypan blue

(0.4%) and a small volume is placed in a hemacytometer and viable or dead cells are counted under a microscope. Viable cells with intact membrane exclude stain (hence Trypan blue exclusion test) and appear as bright translucent while dead or dying cells with damaged membrane allow stain permeation and appear as blue. Confluent cell monolayers growing on a flask or plates also can be stained with Trypan blue to assess the degree of cell damage induced by toxins or pathogens.

Alkaline Phosphatase Assay

Certain cells, especially those of lymphocyte origin express endogenous membrane anchored alkaline phosphatase (ALP) enzyme (100–165 kDa), which are released from damaged membrane. Membrane active toxin allows the release of ALP from cells, which can be analyzed by addition of a substrate such as *p*-nitrophenyl phosphate (PNPP) or methyl umbelliferyl phosphate (MUP) resulting in colored or fluorescence end products for detection by spectrometers. The drawback of this assay is that not all cell types possess this enzyme. In general epithelial cells carry very low levels while cells of B-lymphocyte origin carry abundant quantities.

Lactate Dehydrogenase Assay

Most cells carry lactate dehydrogenase (LDH) in the cytoplasm. LDH is a low molecular weight enzyme (35 kDa) and may exist in tetrameric form. The enzyme is released from the cells even due to a minor perturbation in membrane integrity thus this assay is very sensitive. LDH release assay has been widely used for studying cell cytotoxicity induced by varieties of microbial or nonmicrobial interactions with mammalian cells. The enzyme activity can be assayed by using an appropriate substrate. LDH converts lactate to pyruvate and NAD receives an electron (H⁺) from NADH and during this process, yellow tetrazolium is converted to red formazan which can be measured spectrophotometrically for quantitative cytotoxic response (Fig. 5.4). Commercial LDH assay kits are available for cytotoxicity assays.

MTT Assay

Eukaryotic cell viability or proliferation could be assayed by metabolic staining method. The yellow tetrazolium MTT (3-[4,5-dimethyl thiazolyl-2]-2,5-diphenyltetrazolium bromide) is a water-soluble compound, which is reduced by metabolically active cells. MTT is converted to water-insoluble purple formazan by the action of dehydrogenase enzyme that generates NADH and NADPH. Purple formazan is then quantified by a spectrophotometer at 570 nm. This assay has been used to measure enterotoxin action of foodborne pathogens including *Bacillus cereus* toxins on eukaryotic cells.

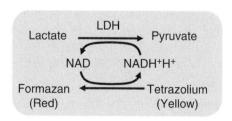

Fig. 5.4 Lactate dehydrogenase (LDH)-mediated color change in the reaction mixture

Cell Death Analysis

Pathogen interaction often results in cellular death, which could be either necrotic (oncosis) or apoptotic. Necrosis, or nonspecific cell death, usually leads to cell swelling with a loss of membrane integrity and initiation of a host inflammatory response. In contrast, apoptosis is a tightly regulated method for removal of damaged or unneeded "self"-cellular debris, especially during growth, development, and cell homeostasis. Apoptosis is characterized by cell shrinkage, chromatin condensation and margination, DNA fragmentation, and membrane blebbing (see Fig. 4.3v; Chap. 4). Many cytokines and cellular proteins regulate cells death via two main pathways (1) by activating tumor necrosis factor receptor (TNFR) and (2) by activating CD95 receptor (also known as Fas). These receptors bind effector molecules and set an intercellular apoptotic cascade in motion by recruiting, binding, and activating Caspase (Cas) proteins or downregulating apoptosis by interacting with inhibitor proteins, such as proteins in the Bcl-2 family.

Apoptosis is usually regarded as a noninflammatory response that targets specific cell without full-scale tissue damage. However, in some cases, pathogens such as *Shigella* and *Salmonella* induce massive macrophage apoptosis and this process is often proinflammatory. Consequently, speculation surrounds the question of whether host cells initiate apoptosis in an attempt to reduce widespread infection or if pathogens themselves may trigger apoptosis through activation of the apoptotic protein cascade as an infection strategy.

Apoptosis or necrosis could be differentiated using a DNA fragmentation assay, by flow cytometry analysis of annexin-V binding to outer leaflet of the apoptotic cells membrane, or through a specific cell staining method such as acridine orange, ethidium bromide, or propidium iodide staining which binds to DNA.

Measurement of Specific Steps in Colonization and Invasion

Animal Model

To study foodborne pathogen colonization, adhesion, invasion, translocation to extraintestinal sites or to study localized tissue damage, animals are administered with pathogens orally or intragastrically. Sometimes animals are fed with antacids to neutralize gastric pH to allow safe bacterial passage to intestinal tract. To determine pathogen colonization and localized cell damage, intestinal sections are collected; histological sections are stained and examined under a microscope. To determine bacterial adhesion, invasion, and translocation to extraintestinal sites, organs and tissues such as liver, spleen, lymph nodes, brain, and placenta (pregnant model) are collected; homogenized and bacterial counts are determined by plating method.

Intravenous or intraperitoneal route of administration of foodborne pathogens have been used to demonstrate the pathogenic potential of the organisms and also to demonstrate the tissue distribution patterns and the nature of infection when intestinal route is bypassed.

In the ileal loop model, toxin preparations or pathogens are injected into the loop to determine fluid accumulation or to study bacterial adhesion and translocation to deeper tissues in a controlled environment. Staining of histological sections of loop can reveal the bacterial adhesion, and invasion and infiltration of inflammatory cells in the site.

Cell Culture Models

Adhesion Assay

Pathogen adhesion is generally performed on confluent cell monolayers formed in the wells of a cell culture plate. Depending on the cell type, some cells has to be grown for a specified times to allow cell polarization before the adhesion assay can be performed. Cell monolayers are inoculated with the test organisms at ratios from 1:0.1 to 1:100 and incubated for 30 min to 1 h at 37 °C and unbound organisms are removed by washing the cell monolayers at least for three times using an appropriate buffer. Bacterial attachment is assessed by performing specific immunostaining, fluorescence staining, or Giemsa staining. Quantitative bacterial counts can be made by first treating the cell monolayers with mild detergent such as Triton-X 100 (0.1–1%) and then by serially diluting and plating the sample on agar plates. Differential fluorescence staining is also used to distinguish intracellular bacteria from surface attached bacteria. Sometimes radioisotope-labeled (^{3}H-thymidine) bacteria are used to determine the binding efficiency.

Invasion Assay

To determine, bacterial invasion, cell monolayers are infected first with the test bacteria for 1–2 h to allow bacterial entry in the cell. Cell monolayers are then washed (three times) with buffer to remove loosely attached or unbound bacteria and treated with antibiotic such as gentamicin (10–100 µg ml^{-1}) for 1–2 h to kill extracellular bacteria. Gentamicin at low concentrations cannot permeate through cell membrane thus intracellular bacteria are protected. Then, cell monolayers are washed and treated with detergent (Triton-X 100) or cold buffer to lyse, and serially diluted and plated to determine the intracellular bacterial counts.

Cell-to-Cell Spread or Plaque Assay

This assay is performed to determine bacterial ability to move from cell-to-cell. Generally fibroblast (L2) or epithelial (Caco-2 or HT-29) cells are used for this purpose. Cell monolayers are first inoculated with the test organisms for 1–2 h, washed and treated with gentamicin (100 µg ml^{-1}) for another 1.5 h. After washing with buffer, the cell monolayers are overlayed with tempered agarose (0.7%) containing gentamicin (10 µg ml^{-1}) and incubated for about 72 h. The monolayers are stained with 0.1% neutral red and examined for plaque (clear zone) formation. Clear plaque reveals the capacity of a pathogen to move from one cell to other and has been used to assess the infectivity of *Shigella* and *Listeria* species.

Summary

Much of our knowledge and understanding of pathogenic mechanism of foodborne pathogens are gained from animal and cell culture models. In modern day, they are indispensable research tools thus one must have apprehension about the availability of different models, their usage, advantages, and limitations. Also, one must choose appropriate model for a pathogen in order to learn the organism's behavior in that environment. Cell culture models, especially the secondary cell lines are most valuable tool and it allows determining virulence potential of pathogens. Cytotoxicity assays are measured by microscopic analysis, and enzyme release assays (lactate dehydrogenase, alkaline phosphatase, or metabolic staining assay). Cell culture models also allow to study bacteria-induced cell death such as apoptosis or necrosis. In addition, these models allow us to study microbial adhesion, invasion, and cell-to-cell movement. Animal models are often used to verify the pathogenic mechanism that has been established in an in vitro cell culture model. Whole animals provide a better picture of dissemination of bacterial pathogens in different organs and tissues and some models show clinical symptoms similar to humans. Sometimes animals are insensitive to human pathogens. In that case immunocompromised animals are employed to study microbial pathogenesis. Pathogen interactions with animals are measured by determining the lethal dose (LD_{50}) or infective dose 50 (ID_{50}). Tissue distribution can be analyzed by microscopic analysis of histological sections or by homogenizing and serially diluting the organs or tissues followed by plating on nutrient agar plates.

Further Readings

1. Bhunia, A.K. and Wampler, J.L. 2005. Animal and cell culture models for foodborne bacterial pathogens. In Foodborne Pathogens: Microbiology and Molecular Biology. Edited by Fratamico, P.M., Bhunia, A.K., and Smith, J.L., Caister Academic, Norfolk, pp 15–32.
2. Salyers, A.A. and Whitt, D.D. 2002. Bacterial Pathogenesis: A Molecular Approach, 2nd edition, ASM, Washington, DC.

6

Staphylococcus aureus

Introduction

The most important species in the genus *Staphylococcus* is *S. aureus*. They are natural inhabitant of human and animal skin but sometimes they can cause infections affecting many organs. Staphylococci form clusters when grown in liquid or solid media hence the name staphylococcus (*Staphyle* means bunch of grapes and *kokkos* means a grain or a berry in Greek) (Fig. 6.1). In 1871, Von Recklinghausen, a German scientist observed cocci in a diseased kidney and called them "micrococci." Later, on the basis of cell arrangements, Billroth (1874) classified them as "monococcus," "diplococcus," "streptococcus," and "gliacoccus." In 1880, Sir Alexander Ogston, a Scottish surgeon and Louis Pasteur, a French scientist confirmed that cocci-forming organisms are capable of causing disease. Later, Ogston coined the name "Staphylococcus." In 1914, Barber discovered that a toxin substance produced by staphylococci was responsible for staphylococcal food poisoning. Staphylococci are mostly associated with community-acquired and nosocomial infections and may be life threatening in immunodeficient conditions. Some staphylococci are methicillin-resistant (MRSA) and vancomycin-resistant, and infection caused by these resistant strains may be fatal because of lack of alternative antibiotics. Most staphylococci are responsible for skin infections such as boil, carbuncle, and furuncle and some cause food poisoning resulting in severe vomiting and diarrhea. Staphylococci also cause mastitis in cow and also cause joint infection leading to edema and arthritis.

Classification

In the Bergy's Manual of Determinative Bacteriology, *Staphylococcus* has been placed in the family of Micrococcaceae. DNA-ribosomal RNA hybridization and comparative oligonucleotide analysis of 16S rRNA has demonstrated that staphylococci form a coherent group at the genus level. Staphylococci are differentiated from other close members of the family with its low G + C content of DNA ranging from 30 to 40 mol%. The genus *Staphylococcus* has been further classified into more than 30 species

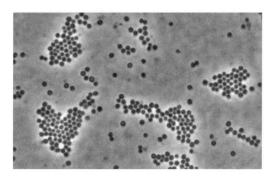

Fig. 6.1 Phase contrast microscopic photograph of *Staphylococcus aureus* cells. Cells appear as clusters (magnification 1,000×)

and subspecies by biochemical analysis and by DNA–DNA hybridization. *Staphylococcus aureus* is the primary species in the genus *Staphylococcus* and is responsible for food poisoning. Other species belong to this genus include *S. intermedius, S. chromogens, S. cohnii, S. caprae, S. caseolyticus, S. delphini, S. epidermidis, S. felis, S. gallinarum, S. haemolyticus, S. hyicus, S. lentus, S. saprophyticus, S. sciuri, S. simulans, S. succinus, S. warneri,* and *S. xylosus*. A majority of them produce enterotoxins.

Morphology

Staphylococcus aureus is a Gram-positive coccus (1 μm in diameter) appearing microscopically as grape-like clusters due to incomplete three planar divisions (Fig. 6.1). They are nonmotile and produce golden yellow colonies. Aureus means golden, i.e., gold coin of Rome. The cell wall of *S. aureus* contains three main components: the peptidoglycan comprising repeating units of *N*-acetyl glucosamine β-1,4 linked to *N*-acetyl muramic acid (Chap. 2, Fig. 2.3); a ribitol teichoic acid bound via *N*-acetyl mannosaminyl-β-1,4-*N*-acetyl glucosamine to a muramyl-6-phosphate; and Protein A, which is covalently linked to the peptidoglycan and particularly is characterized by its ability to bind to Fc component of the immunoglobulin in plasma causing autoagglutination. Most of the other species of staphylococci lack protein A in their cell wall.

Cultural and Biochemical Characteristics

Staphylococcus aureus is catalase-positive, facultatively anaerobic organism, and grows abundantly under aerobic conditions, except for *S. saccharolyticus*, which is a true anaerobe. Under aerobic conditions, it produces acetoin as the end product of glucose metabolism. It ferments mannitol, causes coagulation of rabbit plasma, produces thermonuclease, and is sensitive to lysostaphin. *S. aureus* is salt-tolerant (10–15%) and relatively resistant to drying and heat. *S. aureus* coagulates rabbit plasma relatively quickly while *S. intermedius* and *S. hyicus subsp. hyicus* cause delayed coagulation.

Color Plates

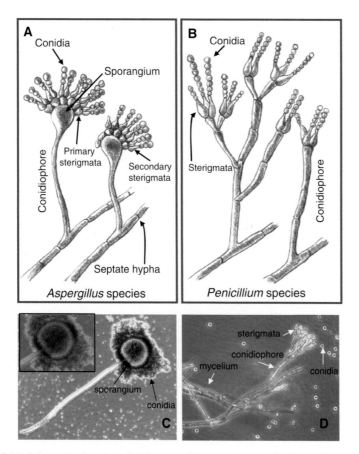

Fig. 2.16 Schematic drawing of (**A**) *Aspergillus* species and (**B**) *Penicillium* species. Panels (**C**) and (**D**) are the light microscopic photograph of *Aspergillus niger* (**C**) and *Penicillium citrinum* (**D**) (magnification 400×)

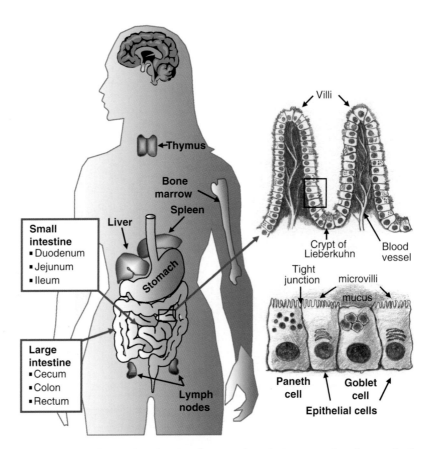

Fig. 3.1 Schematic drawing showing the gastrointestinal tract, various immunologic organs and tissues, and the mucus membrane in human

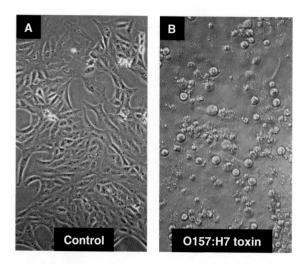

Fig. 5.2 Effect of *Escherichia coli* O157:H7 (EDL933) toxin on Vero cells after 24 h exposure. Toxin exposed cells (panel B) show cytopathic effects characterized by cell rounding, granulation, and detachment

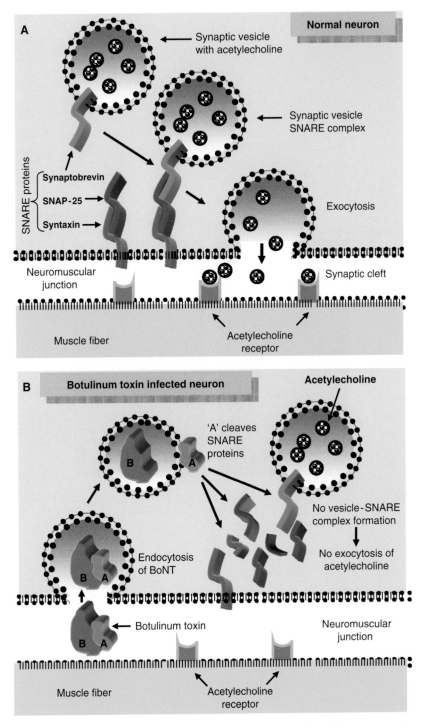

Fig. 8.4 Schematic diagram showing the mechanism of action of botulinum neurotoxin (BoNT) (redrawn from Arnon, S.S. et al. 2001. J. Am. Med. Assoc. 285:1059–1070)

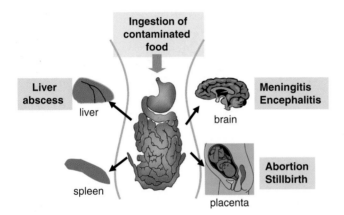

Fig. 9.1 Distribution of *L. monocytogenes* to different tissues and organs

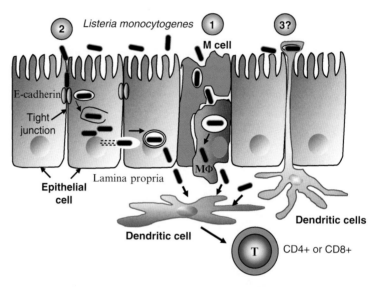

Fig. 9.3 *L. monocytogenes* translocation pathway through intestinal cell lining. Three possible pathways are proposed; (1) translocation through M-cells, (2) active invasion through epithelial cell by internalin/E-cadherin pathway, and (3) translocation by dendritic cells. The latter pathway is yet to be demonstrated experimentally

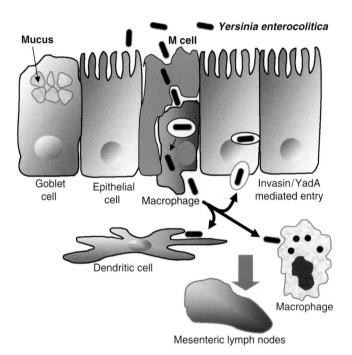

Fig. 13.1 *Yersinia enterocolitica* translocation through intestinal epithelial cells. After entry into the basal layer via M-cells, bacteria invade epithelial calls through interaction with host cell β1-integrin. Macrophage/dendritic cells transport *Yersinia* to mesenteric lymph nodes and also to liver. *Yersinia* prevents phagocytosis, also induces macrophage apoptosis, and prevents cytokine production

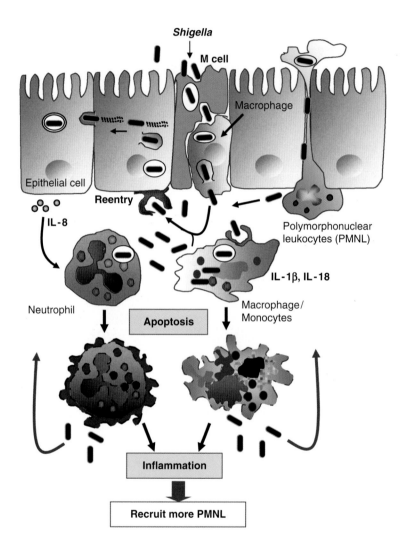

Fig. 15.3 Schematic diagram showing *Shigella* entry through mucosal membrane in the intestine and the induction of inflammation

Virulence Factors

S. aureus produces a family of virulence factors such as adhesion proteins, enterotoxins, superantigens, pore-forming hemolysins, ADP-ribosylating toxins, and proteases (Table 6.1).

Table 6.1 Virulence factors and enzymes produced by *Staphylococcus aureus*

Virulence factors	Receptors
Adhesin proteins	
Spa (protein A)	Fc part of IgG
Bap (biofilm-associated proteins)	
Fbp (fibronectin-binding protein)	Fibronectin, fibrinogen, elastin
ClfA (fibrinogen-binding protein)	Fibrinogen
Cna (collagen adhesin)	Collagen
IsdA, IsdB, IsdC, IsdH (iron-regulated surface proteins)	Hemoglobin, transferrin, hemin
Pls (plasmin-sensitive cell wall protein)	Cellular lipid called ganglioside GM_3
Atl (autolysin amidase)-bacteriolytic action	Fibronectin, fibrinogen, vitronectin
Enolase	Laminin
Teichoic acid	Unknown – binds epithelial cells
Enterotoxins (17)	
Staphylococcal enterotoxin SEA-SER, except SEF	Glycosphingolipid
Pore-forming hemolysins	
Hemolysins α, β, γ, δ	Cholesterol
Superantigens	
TSST (toxic shock syndrome toxin)	MHC class II
Enterotoxins	Glycosphingolipid
Exfoliative toxins (A, B)	
ADP-ribosylating toxins	
Leukocidin	
Pyrogenic exotoxin	
Proteases	
Metalloprotease	
Collagenase	
Hyaluronidase	
Endopeptidase	
Elastase	
Others	
Nuclease	
Lysozyme	
Phospholipases	
Coagulase	

Staphylococci possess multiple adhesion molecules which are collectively known as MSCRAMM (microbial surface components recognizing adhesive matrix molecules). Internalization of the organism is thought to be triggered by MSCRAMM. Adhesion proteins include Bap (biofilm-associated proteins), which is responsible for biofilm formation and colonization in the mammary gland during mastitis. The C-terminus of Bap contains typical cell wall anchoring domain comprising of LPXTG motif, a transmembrane sequence and a positively charged C-terminus. The fibronectin-binding protein (Fbp) binds to fibronectin. Bacterial binding to fibronectin also facilitates internalization into nonprofessional phagocytes, such as kertinocytes, epithelial cells, endothelial cells, and osteoblast. Staphylococci also produce other adhesion factors including ClfA, a fibrinogen-binding protein that activates platelets aggregation and plays a role in staphylococcal arthritis; Pls, a plasmin-sensitive cell wall protein that binds to ganglioside GM_3 of cells and promotes adhesion to nasal epithelial cells and; Cna, a collagen adhesion binds to collagenous tissues, i.e., cartilages.

In addition, *S. aureus* also produces toxic shock syndrome toxins 1 (TSST), responsible for acute illness, fever, erythematous lesion, and hypotension; coagulase; four types of hemolysins (α, β, γ, and δ); membrane active lipase; leukocidin; collagenase; hyaluronidase; metalloprotease; exfoliative toxins; phospholipase (PI-PLC); staphylokinase; and pyrogenic exotoxin.

Food Association and Toxin Production

Staphylococcal food poisoning (intoxication) is one of the most common foodborne illnesses reported worldwide. Nearly one third of all the food poisoning cases in the US were caused by staphylococci during 1970s and 1980s, which in general has decreased over the past two decades. However, it remains the main reported cause of food poisoning in number of countries including Brazil, Egypt, Taiwan, Japan, and most of the other developing countries. Intoxication occurs due to the ingestion of one or more preformed staphylococcal enterotoxins (SEs) in contaminated food. Staphylococcal contamination is associated with creamy food prepared with milk, deli foods, custard (pudding), salad dressing, meats, hams, fish, shellfish, and milk products. Staphylococci can be transmitted through meat grinder's knives and food handlers. *S. aureus* grows when the food is left at room temperature for long periods. Generally the enterotoxins are produced at 10–46 °C. At temperature 60 °C or higher, the organism will not grow; however, below 60 °C the organism will grow and produce toxins.

Enterotoxins

Staphylococcus aureus produces large numbers of extracellular proteins and toxins. The most important toxins are called Staphylococcal enterotoxins (SEs). There are 17 major serologically distinct SEs (SEA through SER with no SEF). In addition, the SEC has three antigenically distinct subtypes: SEC1, SEC2, SEC3, and SEG have a variant form called, SEGv. Many SEs are responsible for food poisoning, acute illness, fever, erythematous lesions, and hypotension. SEs are a heterogeneous group of water soluble single chain globular proteins with molecular weight of about 26–35 kDa. The SE polypeptide chain contains relatively a large number of lysine, aspartic acid, glutamic acid, and tyrosine. The SEs are generally heat resistant and a heat denatured enterotoxin can be renatured by prolonged storage or in presence of urea. Toxins remain active even after boiling for 30 min. In mushroom, they are stable at 121 °C for 28 min. The SEs also are protease resistant and all are capable of causing food poisoning; however, SEA is the most common serotype. It is a 27.1 kDa toxin and its production is not regulated by *agr* (accessory gene regulator). SEB is a 31.4 kDa toxin and is the most heat resistant (stable at 60 °C for 16 h) among all the toxins. SEB also is resistant to gastrointestinal proteolytic enzymes such as chymotrypsin and trypsin. SECs are a group of highly conserved proteins and there are three antigenically distinct subtypes: SEC1, SEC2, and SEC3. Staphylococcal isolates from different animal species produce host specific SECs. SED (24 kDa) is the second most serotypes responsible for food poisoning. SED has the ability to form a homodimer in presence of Zn^{2+} which facilitates its binding to MHC class II molecule on antigen presenting cells and serve as a superantigen (see below).

The genes for enterotoxin production are located either in a bacteriophage, chromosome, or in plasmids. The *sea* and *sep* are located in a bacteriophage; *seb, seh,* and family of *sec* are in the chromosome, and *sed, sej,* and *ser* are located in a plasmid. The *sed* and *sej* genes are colocalized and the same strain always produces these two toxins (SED and SEJ) together. Certain enterotoxin genes are also located in the pathogenicity islands (PAI). There are five staphylococcal PAI: SaPI-1, SaPI-2, SaPI-3, SaPI-4, and SaPI-bov. SaPI-1 contains genes for TSST-1 and SEK and SEQ. Production of enterotoxins is not restricted to *S. aureus* alone, other nonaureus staphylococci are reported to produce enterotoxin.

Enterotoxins are expressed differentially and the toxin production depends on the growth phase of bacteria, bacterial density, pH, and CO_2 levels. SEA and SEJ are synthesized mostly during the exponential phase while SEB, SEC, and SED are produced during transition from exponential to the stationary phase of growth.

Molecular Regulation of Virulence Gene Expression

The pathogenic potential of staphylococci in human can be attributed to the expression of a wide array of virulence factors. Four loci have been implicated in the regulation of expression of virulence factors: accessory gene regulator, *agr*; staphylococcal accessory regulator, *sar; S. aureus* exoprotein expression, *sae*; and exoprotein regulator, *xpr*.

Enterotoxin production is regulated by two-component regulatory system, *agrAC*. In the two-component system, one protein serves as a sensor and transfers signal by phosphorylating the intracellular activator, while another regulates genes to provide response called response regulator. The *agr* locus comprises the global regulatory system, which regulates virulence gene expression in the postexponential phase of growth. It consists of two divergent units driven by promoters P_2 and P_3. The P_2 transcript includes four open reading frames (ORFs) referred to as *agrA, agrB, agrC,* and *agrD*. The P_3 transcript RNA III is the actual effector molecule and activates secretion of enterotoxins and other exoproteins, primarily at the transcriptional level. In the *agr* locus, P2 operon includes the genes for response regulator, *agrA*; histidine kinase, *agrC*; autoinducing peptide, *agrD*; and autoinducing ligand inducer (AIP), *agrB*. The strains that produce high amounts of enterotoxin have a high concentration of RNA III. An intact *agr* locus is necessary for maximum expression of SEC expression. The *agr* locus regulates SEC expression at post-transcriptional level, presumably at the level of translation or secretion. RNA III is produced at lower levels under alkaline pH and SEC production also decreases at this pH levels. In addition, Sar family of proteins also regulates virulence gene expression. SarA is a DNA-binding protein and binds to *agr* promoter stimulating the transcription of RNA II. Activation of RNA II and subsequently RNA III leads to alternating target gene expression. Expression of SEB is positively controlled by *sarA*. Xpr, represents an additional genetic element involved in regulation of some *agr*-regulated proteins. Low levels of SEs production were observed in *xpr* mutants. Reduced levels of RNA II and III were also observed in these mutants.

Mechanism of Pathogenesis

The infectious dose of *S. aureus* is 10^5–10^8 cfu g^{-1} and a toxin concentration of 1 ng g^{-1} of food. The emetic dose (ED$_{50}$) of SEA toxin in Rhesus monkey is 5–20 μg kg^{-1} while 200 ng kg^{-1} may be needed to show intoxication syndrome in human. SEB is highly toxic and a dose of 0.4 μg kg^{-1} is required for humans. Following consumption of *S. aureus* contaminated food, toxins are absorbed and cause typical gastroenteritis while bacteria pass through the intestine without causing any adverse effects on the host. Toxins stimulate vagus nerve endings in stomach linings and stimulate medullary vomiting center to show a violent emetic response (Fig. 6.2). Study with kidney cells indicate that the

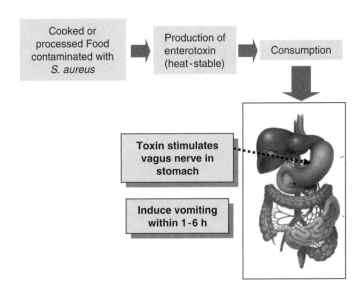

Fig. 6.2 Pathogenic mechanism of intoxication with enterotoxin from *Staphylococcus aureus* (reproduced with permission from Ray, B. and Bhunia, A.K. 2007. Fundamental Food Microbiology, 4th edition, CRC)

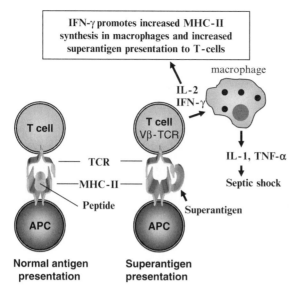

Fig. 6.3 Mechanism of superantigen action for staphylococcal enterotoxins

putative receptor for enterotoxin to be glycosphingolipid. Enterotoxins also elicit damage to the intestinal epithelial cells resulting in the destruction of intestinal villi. The enterotoxins also cause villus distension, crypt elongation, and lymphoid hyperplasia.

The superantigenic property of the staphylococcal enterotoxins distinguishes them from other bacterial toxins. Superantigens are the molecules that have the ability to stimulate an exceptionally high percentage of T-cells. Enterotoxins enter blood circulation and bind to MHC class II molecules on the surface of macrophages. The toxin is presented to the T-cells that carry TCR (T-cell receptor) made with β-chain also called Vβ carrying T-cells. T-cells proliferate and produce large quantities of IL-2 and IFN-γ. Elevated levels of IFN-γ also induce increased MHC class II expression in macrophages which in turn bind more superantigens, and activate more T-cells. Inflammatory cytokines such as IL-1 and TNF-α are produced from activated macrophages and initiate typical toxic shock syndrome with disseminated intravascular coagulation (DIC), low blood pressure, massive shock, and death (Fig. 6.3). Superantigens, unlike others, do not stimulate T-cells indiscrimately, as only specific Vβ sequences are recognized. This mechanism is thought to be responsible for vomiting response, typically following ingestion of these enterotoxins. The release of cytokines might stimulate the neuroreceptors in the intestinal tract and triggers the vomiting center of the brain.

Symptoms

Symptoms of staphylococcal intoxication appear within 1–6h and include nausea, acute vomiting, abdominal pain, diarrhea, headache, cramping, and anaphylactic shock. The disease subsides within 24h. In case of aerosol exposure, sudden onset of fever, chills, headache, and cough occur. Fever may last for several days and the cough can last for 10–14 days.

Prevention and Control

The staphylococcal food intoxication is mostly self-limiting. Fluid therapy and bed rest are recommended for the patients. To prevent *S. aureus* related food poisoning, food should not be left at room temperature for long and should be cooled rapidly to prevent bacterial growth. Hand washing and the use of protective gloves before food handling should reduce the chance of food contamination. Hygienic practices are crucial in preventing staphylococcal food poisoning. Implementation of HACCP and rapid microbiological analysis will aid in controlling pathogens in food manufacturing facility.

Detection

Culture Methods

Conventional culture methods allow isolation of bacteria on Baird-Parker agar plates and the colonies appear as black, shiny, circular, convex, smooth with the entire margin forming a clear zone with an opaque zone (lecithinase halo) around the colonies.

Cytotoxicity-Based Assays

Staphylococcal enterotoxins act as superantigens. Examination of these toxins for the ability to cause mammalian cell damage provides a means to assay these toxins. A bioassay for superantigen on a T-lymphocyte cell line, Raji has been developed. In this assay, SEA-induced cytotoxic action was detected colorimetrically using the CytoTox 96 cell lysis detection kit. This system can detect SEA at picomolar concentrations. Staphylococcus aureus enterotoxins were also detected on Madin-Darby bovine kidney (MDBK), bovine embryo lung (PEB), and dog carcinoma cell line (A-72) for possible cytopathic effects. The PEB cell line was the most susceptible and a cytopathic effect was observed in some cases after 2 h of incubation.

Nucleic Acid-Based Methods

Nucleic acid-based detection systems offer a very good alternative to the conventional culture methods. Nucleic acid probe-based methods have been developed for the detection and enumeration of staphylococci in foods. PCR-based detection of enterotoxin genes including *egc* (enterotoxin gene cluster: SEA to SEE; SEG, SEH, SEI, SEM, SEJ, SEN, and SEO), TSST1, exfoliative toxins A and B (*etaA* and *etaB*), methicillin-resistant (*mecA*) gene, and 16S rRNA from *S. aureus* has been reported. Fluorescence-based real-time PCR (TaqMan-PCR) has been demonstrated for enterotoxins A to D and *mecA* for rapid analysis of large number of samples. A DNA microarray was developed for detection and identification of 17 Staphylococcal enterotoxin (*ent*) genes simultaneously. The assay is based on PCR amplification of the target region of the *ent* genes with degenerate primers, followed by characterization of the PCR products by microchip hybridization with oligonucleotide probes specific for each *ent* gene. The use of degenerate primers allowed the simultaneous amplification and identification of as many as nine different *ent* genes in one *S. aureus* strain.

Quantitative detection of staphylococci or their toxin genes was achieved through the use of quantitative real-time PCR (qRT-PCR). Commercial rapid assay kits that detect *S. aureus* 23S rRNA is available. A commercial array chip called Staphychips also is available for identification of five different staphylococci in an array format.

Immunoassays

Immunoassays based on ELISA are widely used for detection of enterotoxins. Automated commercial detection systems are available. The detection limit is $<0.5-1$ ng g^{-1} of enterotoxin. Enzyme immunoassay is available for the detection of superantigens; SEA, SEB, and SEC, TSST-1, and streptococcal pyrogenic exotoxin A (SPEA) in blood serum. A fluorescence immunoassay has been developed for SEB with detection limit of 100 pg per well and is more sensitive than the conventional ELISA assay. Magnetic bead-based immunoassay has been used to detect SEs in a sandwich format.

Other Rapid Methods

Although above methods are widely applied for staphylococcal detection, some of the other methods which directly detect the whole cell or their metabolites include, direct epifluorescence technique (DEFT), flow cytometry, impedimetry, ATP-bioluminescence, are commonly used for the routine analysis of milk.

Summary

Staphylococcus aureus, a natural inhabitant of human and animal body is mostly associated with community-acquired and nosocomial infections, which can be fatal in immunodeficient patients. Methicillin and vancomycin-resistant *S. aureus* cause serious nosocomial infection. Staphylococci cause mastitis in cow and are also responsible for food poisoning characterized by severe vomiting and diarrhea. *S. aureus* produces a large number of toxins and enzymes, of which the enterotoxins (more than 17 serotypes of toxins are identified) are most important in the production of gastroenteritis. Enterotoxins are heat-stable, and are produced when the temperature of food is at or below 46 °C. Consumption of preformed toxin induce vomiting and diarrhea within 1–6 h. Some enterotoxins are called superantigens, because they form a complex with MHC Class II molecules, activating T-cells to produce excess amounts of cytokines that contribute to diarrhea and fatal toxic shock syndrome. The genes for enterotoxin production are present in pathogenicity islands in the chromosome, in plasmids, and in temperate bacteriophages. Toxin production is regulated by two-component regulatory system called *agrAC* (accessory gene regulator).

Further Readings

1. Balaban, N. and Rasooly, A. 2000. Staphylococcal enterotoxins. Int. J. Food Microbiol. 61:1–10.
2. Bronner, S., Monteil, H., and Prevost, G. 2004. Regulation of virulence determinants in *Staphylococcus aureus*: complexity and applications. FEMS Microbiol. Rev. 28:183–200.
3. Clarke, S.R. and Foster, S.J. 2006. Surface adhesins of *Staphylococcus aureus*. Adv. Microb. Physiol. 51:187–225.
4. Ray, B. and Bhunia, A.K. 2008. Fundamental Food Microbiology, 4th edition, CRC, Boca Raton.
5. Stewart, G.C. 2005. *Staphylococcus aureus*. In Foodborne Pathogens: Microbiology and Molecular Biology. Edited by Fratamico, P.M., Bhunia, A.K., and Smith, J.L., Caister Academic, Norfolk, pp 273–284.

7

Bacillus cereus and *Bacillus anthracis*

Introduction

Bacillus is widely distributed in nature including soils, plants, animals, and humans. A majority of bacilli are nonpathogenic; however, several species produce multiple toxins and can cause various diseases in animals and humans. Of these, three species has significant importance because of their ability to cause disease in humans and other animals. *Bacillus cereus* causes emetic and diarrheal food poisoning. In addition, it can cause systemic fatal infections in immunosuppressed patients and neonates, irreversible eye infection from post-traumatic injury, and metastatic endophthalmitis (i.e., infection spread from other sites to eye). *B. thuringiensis* is pathogenic to insects and has also been associated with foodborne disease. *Bacillus anthracis* causes anthrax which is characterized by septicemia and toxemia in animals and humans with a very high mortality rate. The *B. anthracis* spores are used as bioterrorism agents.

Biology

Bacillus species are Gram-positive, aerobic, spore-forming rods (Fig. 7.1). The size of vegetative cells is in the range of 0.5 × 1.2–2.5 × 10 μm and occurs singly or in chains. Most bacilli are motile. Growth temperature varies between 4 and 50 °C with optimum temperature ranges from 25 to 37 °C. Psychrotolerant bacilli are found in milk and dairy products. Some thermophilic strains can grow at 75 °C. *B. cereus* usually does not grow below pH 4.5. Bacilli produce ellipsoid or cylindrical spores located central to terminal end (Fig. 7.2). The spore contains a signature molecule, dipicolinic acid (2,6-pyridinedicarboxylic acid, DPA), which is essential for sporulation, spore germination, and spore structure. Biophysical and biochemical properties including motility, capsule production, spore structure, hemolysis pattern, urea utilization, nitrate reduction, and sensitivity to penicillin can be used to differentiate *Bacillus* species (Table 7.1). Bacilli are naturally found in soil, dust and water. Milk, meat, spices, meat additives can be contaminated with spores. Bacilli remain viable in nature under harsh conditions because of the ability to form endospores which are difficult to inactivate.

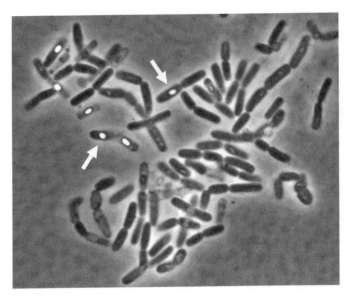

Fig. 7.1 Phase-contrast light microscopic picture of cells of *Bacillus cereus* showing spores (see *arrow*) (magnification 1000X)

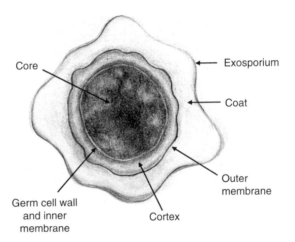

Fig. 7.2 Schematic drawing of transmission electron microscopic picture of a spore from *Bacillus anthracis*

Classification

The genus *Bacillus* contains 51 species and is divided into three groups based on the morphology of spores and sporangia (Groups 1–3). Group 1 is subdivided into 1A and 1B based on the cell size and the presence of poly-β-hydroxybutyrate in the cytoplasm. The organisms belong to Group 1A (*B. cereus* group) are; *B. cereus, B. anthracis, B. thuringiensis, B. mycoides, B. pseudomycoides,* and *B. weihenstephanensis*. Enterotoxin production is linked to *Bacillus*-induced disease and a majority members of Group1A produce enterotoxin. Some strains of other *Bacillus* species such as *B. circulans, B. lentus, B. megaterium, B. oleronius,*

Table 7.1 Summary of biochemical properties of common *Bacillus* species

Bacillus species	Hemolysis	Motility	Capsule	Urea hydrolysis	Nitrate reduction	Sensitivity to penicillin
B. cereus	+	+	v	v	v	–
B. anthracis	–	–	+	–	+	+
B. thuringiensis	+	+	–	+	+	–
B. mycoides	–	–	–	v	v	–
B. megaterium	–	+	–	–	–	–
B. subtilis	–	+	–	v	v	–

[+] positive, – negative, *v* variable

B. licheniformis, B. thuringiensis, B. pumilis, B. sphaericus, B. polymyxa, B. carotarum, and *B. pasteurize* may produce enterotoxin.

B. anthracis is closely related to *B. cereus* but instead of enterotoxin it produces anthrax toxin and is sensitive to penicillin. A *B. cereus* strain has been identified that posses virulence factors, common to *B. anthracis* implicating very close relationship between these two pathogens.

B. thuringiensis produces a wide variety of crystal toxins known as delta endotoxins, which are toxic to insects in an Order-specific manner. Since 1905, these endotoxins have been applied commercially to biologically control pestiferous insects in the Orders Coleoptera, Lepidoptera, and Diptera such as beetles, mosquitoes, butterfly, and moths. The gene for crystal toxin production (*cryA*) is plasmid encoded. A plasmid-cured strain is biologically indistinguishable from *B. cereus*.

Foods Involved

Bacillus cereus-related outbreaks are generally associated with cereal foods, pudding, and soups. Emetic syndrome is generally associated with pasta, rice dishes, beef, poultry, milk pudding, vanilla sauce, and infant formulas. While diarrheal syndrome is associated primarily with meat, fish, soups, vegetables such as corn, cornstarch and mashed potatoes, and dairy products. Milk and dairy products can be contaminated with *Bacillus*, which deteriorates the quality of the milk. *B. cereus* can cause bitty cream by forming a creamy layer on the milk due to the action of lecithinase. *B. cereus* can also form sweet curdle without lowering the pH of the milk. Raw milk is generally contaminated in the farm from soiled udders; thermoduric spores may then survive subsequent pasteurization treatments.

Toxins and Enzymes

Bacillus spp. may produce multiple toxins including emetic toxin and diarrheal enterotoxins. Three diarrheal enterotoxins have been reported – hemolysin BL, nonhemolytic enterotoxin, and cytotoxin K (Table 7.2). In addition, *B. cereus* also produces a cholesterol-binding cytolysin (cereolysin O or hemolysin I), hemolysin II, hemolysin III, and hemolysin IV. The latter

Table 7.2 Toxins produced by *Bacillus cereus*

Toxins	Genes	Molecular weight (kDa)	Activity
Emetic toxin	*ces*	1.2	Emesis (vomiting)
Diarrheal toxin		38–43	Diarrhea
Hemolysin – HBL			Hemolytic, enterotoxic, dermonecrotic
B-component	*hblA*	37.8	
L-component (L1)	*hblD*	38.5	
L-component (L2)	*hblC*	43.5	
Nonhemolytic enterotoxin (Nhe)			Enterotoxic
NheA	*nheA*	41	
NheB	*nheB*	39.8	
NheC	*nheC*	36.5	
Enterotoxin T (BcET)	*bceT*	41	Unknown
Enterotoxin FM (EntFM)	*entFM*	45	Enterotoxic
Cytotoxin K (CytK)	*cytK*	34	β-barrel pore-forming toxin; necrotic dermatitis
Hemolysin I (cereolysin O; CLO)	–	55	Cholesterol-binding pore-forming toxin
Hemolysin II (HlyII)	*hlyII*	382	β-barrel channel-forming toxin
Hemolysin III (HlyIII)	*hlyIII*	24.4	Pore formation, hemolysis

Adapted from Gray, K.M. et al. 2005. J. Clin. Microbiol. 43:5865–5872; Schoeni, J.L. and Wong, A.C.L. 2005. J. Food Prot. 68:636–648

two hemolysins are same as cytotoxin K. The enzymes include phospholipases, lecithinase, sphingomyelinase, collagenase, protease, amylase, and β-lactamase. β-lactamase inactivates penicillin thus make this organism resistant to penicillin.

Emetic Toxin

The molecular mass of emetic toxin is 1.2 kDa and it consists of three repeats of four modified amino acids [D-O-Leu-D-Ala-L-O-Val-L-Val]$_3$ forming a ring structure (dodecadepsipeptide) and is known as cereulide. Cereulide is produced nonribosomally by a large multidomain enzyme complex and it is highly hydrophobic. This toxin is highly heat-stable (121 °C for 90 min), active over a broad pH range (pH 2–11), and is not digested by trypsin. Maximum cereulide production has been reported to occur in cultures incubated between 12 and 22 °C during stationary phase of growth. However, the toxin production is not associated with sporulation. *Bacillus* spores usually survive during cooking or pasteurization and germinate into vegetative cells during when food is temperature abused. Emetic toxin is produced during prolonged food storage. This toxin causes vacuole formation in cultured laryngeal HEp-2 and cervical HeLa cells and causes swelling of spermatozoa head and impedes motility.

Enterotoxins

Hemolysin BL

The best characterized *Bacillus* spp. enterotoxins are 38–43 kDa, heat-labile, and include hemolysin BL (HBL), nonhemolytic enterotoxin (Nhe), and cyto-toxin K (CytK). HBL consists of a single B component (37.8 kDa), and two L components: L1 (38.5 kDa), and L2 (43.5 kDa). All three subunits are highly heterogeneous in *B. cereus* showing variable molecular weights, and all three subunits are required for maximal activity. HBL produces unique discontinu-ous hemolysis pattern characterized by the presence of a clear zone and a zone of incomplete hemolysis on blood agar. The hemolytic activity is highest in guinea pig blood followed by swine, calf, sheep, goat, rabbit, human, and horse. HBL is considered to be an important virulence factor for *B. cereus* because it is hemolytic, cytotoxic, dermonecrotic, and induces vascular per-meability as demonstrated in a rabbit ileal loop assay.

Nonhemolytic Enterotoxin

Another three-component enterotoxin produced by *B. cereus* is the nonhemo-lytic enterotoxin (Nhe), comprised of the NheA (41 kDa), NheB (39.8 kDa), and NheC (36.5 kDa) subunits and these three genes are located in an operon and the transcription of this operon is regulated by a pleiotropic regulator, PlcR that also controls phospholipase C expression. The Nhe carries a high degree of sequence homology with HBL. Much like the HBL complex, all three components are necessary for full activity. NheB was shown to interact with host cell receptor; however, their specific role in gastroenteritis has not been clearly established.

Cytotoxin K

CytK is a 34 kDa toxin and belongs to the family of β-barrel pore-forming toxin similar to a prototype toxin, α-hemolysin from *Staphylococcus aureus*. CytK has been implicated in a *B. cereus* outbreak resulting in bloody diarrhea. CytK has been implicated in *B. cereus*-related deaths due to necrotic enteritis.

Enterotoxin BceT

Originally *B. cereus* enterotoxin T (BceT) was isolated from the B-4ac strain and the molecular weight was established as 41-kDa polypeptide, encoded in *bceT* gene (2.9 kb). Function of BceT in pathogenesis is controversial, some studies suggest it has enterotoxic activity while others dispute its existence and consider it to be an artifact.

Enterotoxin FM

Enterotoxin FM (EntFM) was originally isolated from *B. cereus* FM-1 strain that was responsible for foodborne intoxication. EntFM is a 45 kDa polypep-tide and it is rich in beta structure and contains some unusual sequence arrangements such as repeating Asn residues around residue 280. EntFM has not been implicated as the enterotoxin responsible for a food-poison-ing outbreak; however, *entFM* gene is present in most outbreak-associated strains and is actually the most prevalent enterotoxin gene for all *B. cereus* strains. In laboratory experiment, purified EntFM caused increased vascular permeability and fluid accumulation in rabbit/mouse ileal loop models and was lethal to mice.

Hemolysins

Hemolysin I

Hemolysin I is also known as cereolysin O (CLO), is a 55-kDa cholesterol-binding pore-forming toxin and it has high sequence similarities (57–68%) with other cholesterol-binding pore-forming toxins such as alveolysin, perfringolysin, and streptolysin O from *Paenibacillus alvei, Clostridium perfringens*, and *Streptococcus pyogenes*, respectively. The toxin activity can be inhibited by cholesterol. The contribution of hemolysin I in pathogenesis is not fully understood.

Hemolysin II

Hemolysin II (HlyII) is a large toxin with a putative molecular weight of 382 kDa and is present in *B. cereus* group including *B. thuringiensis*. HlyII belongs to the family of β-barrel channel-forming toxin including α-toxin, leukocidin, and γ-toxin from *S. aureus*, γ-toxin from *C. perfringens* and CytK from *B. cereus* and produces a large pore with an inner diameter of 1.5–2 nm and outer diameter of 6–8 nm. The hemolytic activity is not inhibited by cholesterol.

Hemolysin III

Hemolysin III (HlyIII) is a 24.4 kDa toxin and forms pore on the membrane with a pore diameter of 3–3.5 nm. However, its role in pathogenesis is not known.

Phospholipases

B. cereus produces two phospholipase C enzymes which are specific for phosphatidylcholine (PC-PLC) also known as lecithinase, and phosphatidylinositol (PI-PLC). PC-PLC is a 29.9 kDa enzyme and possesses three zinc atoms in its active site and they contribute to enzyme structural and catalytic properties. PI-PLC is a 34.6 kDa enzyme. These enzymes may be important during respiratory infections resulting in necrosis and tissue hemorrhage in the respiratory tract.

Sphingomyelinase (SMase)

SMase is a 34.2 kDa enzyme and it has two magnesium atoms at its active site. SMase hydrolyzes sphingomyelin to ceramide and phosphorylcholine located in eukaryotic membranes.

Regulation of Toxins

Diarrheagenic toxins are produced during the late exponential to early stationary phase of growth or in the intestine. Glucose acts as a catabolic repressor of HBL while sucrose enhances expression when grown in a modified defined medium (MOD). Expression of *hbl* and *nhe* and 13 other virulence genes is regulated by a pleiotropic 34-kDa regulatory protein, PlcR, which is present in *B. cereus* and *B. thuringiensis*. A truncated PlcR polypeptide is also present in *B. anthracis* due to a nonsense mutation in the *plcR* gene and is probably nonfunctional.

Pathogenesis

Consumption of food contaminated with preformed *B. cereus* enterotoxins results in food-associated intoxication, which is characterized by vomiting, while consumption of bacteria followed by subsequent toxin production de novo results in symptoms and signs of toxicoinfection. The infectious dose of *B. cereus* is highly variable, ranging between 10^5 and 10^8 viable cells or spores per gram to cause intoxication or toxicoinfection.

Emetic Toxin

Emetic syndrome is associated with consumption of contaminated rice dishes, pasta, and noodles. The emetic dose of toxin is estimated to be $30 \mu g \, kg^{-1}$ body weight. The toxin binds to vagus nerve receptors (5-HT_3) in the stomach and subsequently induces vomiting. A 5-HT_3 receptor antagonist, ondansetron hydrochloride, can abolish cereulide-mediated vomiting. Cereulide can also cause liver failure. It acts as K^+ ionophore and inhibits fatty acid oxidation affecting mitochondrial activity in hepatocytes, which results in massive degeneration of heptocytes. The typical symptoms consist of nausea and vomiting and appear within 1–6h following ingestion and the disease resemble signs and symptoms that of staphylococcal food poisoning, also an intoxication.

Diarrheagenic Toxin

Diarrheagenic toxins are produced after colonization of bacteria in the small intestine. Though the mode of action of toxins are not fully understood, it is speculated that the enterotoxins induce diarrhea by stimulating the cAMP system and by forming pores in the membrane. As a result, Na^+, Cl^-, and H_2O are lost from the epithelial cells resulting in electrolyte imbalance. HBL-mediated pore formation is thought to contribute to diarrheal symptoms too. Each member of HBL complex binds to the membrane and it has been hypothesized that HBL complex forms transmembrane pores by oligomerizing all three components of hemolysin (Fig. 7.3). HBL causes fluid accumulation in experimental rabbit ileal loop models resulting in necrosis of villi, edema in submucosal layer, infiltration of lymphocytes in the interstitial spaces, and accumulation of blood.

Similarly, CytK also forms pores with a predicted pore size of approximately 7 Å. The CytK toxin spontaneously oligomerizes and becomes resistant to sodium dodecyl sulfate (SDS). Diarrheagenic symptoms appear within 8–16h after consumption of bacteria and begin with mild diarrhea, nausea, abdominal cramping, followed by watery stools. Vomiting is occasionally induced in patients exposed to diarrheagenic toxin.

Prevention and Control

The major contributing factors for *B. cereus* food-associated illness include improper food holding temperature, contaminated equipment, inadequate cooking, and poor sanitation. Furthermore, appropriate heat treatment used for cooking may not be adequate in destroying *B. cereus* spores. Thus the storage temperature of food should be maintained well below the germination threshold of spores. This can be achieved by uniform quick chilling of the food to near 4°C or holding the food above 60°C. *B. cereus* is capable of producing toxins at refrigerated temperature (≥ 4°C) thus prolonged storage at refrigerator should be avoided. Proper sanitary measures should be adopted to prevent cross-contamination while handling food. Since live cells are necessary to

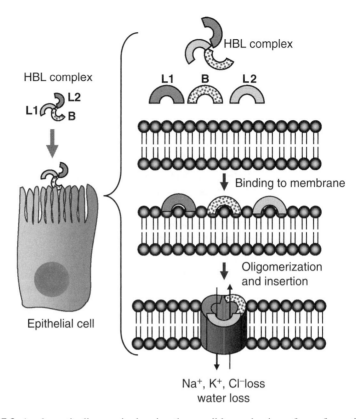

Fig. 7.3 A schematic diagram is showing the possible mechanism of pore formation by *B. cereus* HBL complex on epithelial cells. Each component is believed to participate in the oligomerization process (the model is based on the description of Schoeni, J.L. and Wong, A.C.L. 2005. J. Food Prot. 68:636–648)

cause the symptoms, there should be uniform reheating of a suspected food to above 75 °C before serving.

Detection

Conventional Methods

Traditional plating and biochemical assays are time-consuming and do not indicate *Bacillus* toxin production capabilities. Bacteria can be isolated by plating on selective or differential media containing egg yolk and mannitol, such as polymyxin-egg yolk mannitol-bromothymol agar (PEMBA). Lecithinase activity is determined by observing white precipitate formation surrounding *Bacillus* colonies on egg yolk agar plates and hemolytic activity on blood agar plates. Emetic toxin-producing strains are both hemolytic and nonhemolytic. Spores can be counted indirectly on agar plates. First, food sample is subjected to heat (75–80 °C) for 10–15 min to inactivate the vegetative cells and to promote germination of spores. The sample is then plated to enumerate bacterial colonies and compare the counts with the sample that did not receive heat-shock.

Animal and Cell Culture Method

Rhesus monkey has been used to determine the emetic dose of toxin after oral administration. Rabbit ileal loop (RIL) assays and suckling mice models are also used to determine the action of diarrheagenic toxin (see Chap. 5).

The biological activity of *B. cereus* toxins has been assessed on a number of eukaryotic cell lines. The emetic toxin is assayed colorimetrically by using the HEp-2 cells by using MTT (3-(4,5-dimethylthiazolyl-2)-2,5-diphenyltetrazolium bromide)-based metabolic staining or by microscopic analysis of vacuole formation in HEp-2 cells. Another bioassay based on the loss of motility of boar spermatozoa has been developed. In this assay, *B. cereus* emetic toxin affects the mitochondrial function and oxidative phosphorylation process and inhibits the motility of boar spermatozoa which can be assessed under a microscope. Recently, an emetic toxin assay has been developed based on its mitochondrial respiratory uncoupling activity on rat liver mitochondria.

A number of cell lines such the McCoy, and Vero have been used to detect diarrheal enterotoxin. The toxins cause cytopathic effects causing destruction of cell monolayers. A CHO-based assay detects both the emetic and diarrheal toxins in 24–72 h by MTT-based metabolic-staining assay. *Bacillus* enterotoxin causes severe membrane damage, pore formation and cellular detachment, which can be examined by scanning electron microscopy (Fig. 7.4). A B-cell hybridoma, Ped-2E9-based cytotoxicity assay has been developed to detect *B. cereus* toxin-induced cell damage in 1–2 h either by Trypan blue staining or an alkaline phosphatase release assay.

PCR

Polymerase chain reaction (PCR) assay has been developed to examine the presence of toxin genes in *Bacillus* for strain characterization. Simplex or multiplex PCR-based detection is based on the presence of toxin genes listed in Table 7.2 and offer a high level of sensitivity and specificity, but the PCR assay does not reveal pathogenic potentials of strains since phenotypic expression of genes are not always warranted. PCR-based detection is often hindered by the presence of inhibitors in food matrices, therefore bacterial cells may have to be separated prior to their analysis.

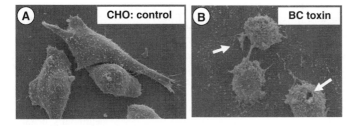

Fig. 7.4 Scanning electron microscopy photograph showing the *Bacillus cereus* toxin (BC toxin)-induced CHO (Chinese hamster ovary) cell damage; pore formation, cell shortening, and detachment from plastic surface (photo courtesy of P.P. Banada)

Antibody-Based Assay

Two commercially available kits, the *Bacillus* Diarrhoeal Enterotoxin Visual Immunoassay (BDE-VIA, Tecra) kit detects only the 41-kDa subunit of NHE and the *Bacillus cereus* Enterotoxin-Reversed Passive Latex Agglutination (BCET-RPLA, Oxoid) detects the L_2 subunit of HBL. Since it is known that different strains of *B. cereus* may produce both, only one or perhaps neither of these subunits, the tests could easily produce false negatives, in a complex medium such as food.

Bacillus anthracis

Biology

Bacillus anthracis is a large Gram-positive bacillus with centrally located ellipsoidal or cylindrical spore. Spores are highly resistant to heat, ultraviolet and ionizing radiations, hydrostatic pressures, and chemical extremes. Vegetative cells typically form long chains. *B. anthracis* is nonmotile, nonhemolytic, and catalase positive (Table 7.1). Virulent strains form capsules while avirulent strains are capsule negative. This zoonotic organism causes anthrax in domestic or wild herbivores, a disease endemic in some parts of Asia, Africa, and South America. Persistence of spores in soil is responsible for occasional outbreaks in animals, although vaccination can prevent infection. Medical intervention (antibiotic treatment) in the very early phase of infection can prevent death. The human form of anthrax is considered an occupational disease and is largely limited to veterinarians and animal handlers because of obvious predisposition. Anthrax also is referred as wool sorter's disease since wool can carry infective spores.

Virulence Factors

Bacillus anthracis pathogenesis depends on the expression of key virulence factors which are encoded in two plasmids, pXO1 (181 kb) and pXO2 (96 kb) (Table 7.3). pXO1 carries genes for bipartite lethal anthrax toxin. Anthrax toxin is an A–B type toxin and the B subunit is known as protective antigen (PA), while subunit A is composed of two alternative catalytic subunits: edema factor (EF) and lethal factor (LF). Protective antigen (PA) is an 83-kDa protein that binds to a host cell receptor, called anthrax toxin

Table 7.3 Virulence gene profiles in *Bacillus anthracis*

Virulence genes located in each plasmid	Genes	Molecular weight (kDa)	Function
pXO1 (181 kb)			
Protective antigen (PA)	*pagA*	83	Facilitate entry of LF and EF into cells
Edema factor (EF)	*cya*	89	Activate cAMP, fluid loss
Lethal factor (LF)	*lef*	90	Cytotoxic to macrophages, neutrophils, dendritic cells
pXO2 (96 kb)			
Capsule	*capABC*		Inhibits phagocytosis

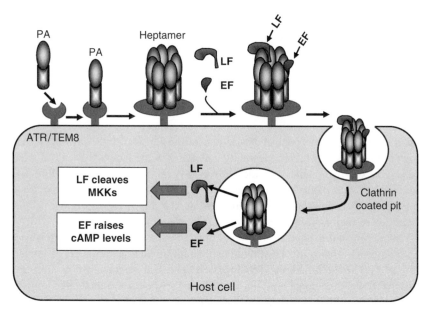

Fig. 7.5 Mechanism of anthrax toxin action on cells. Protective antigen (PA) first binds to anthrax toxin receptor/tumor endothelial marker 8 (ATR/TEM8) and then spontaneously forms heptamer and aids in the translocation of lethal factor (LF) and edema factor (EF) inside the cell. LF cleaves mitogen-activated protein kinases (MKKs) to block signaling pathway while EF increases cAMP level (redrawn from Scobie, H.M. and Young, J.A.T. 2005. Curr. Opin. Microbiol. 8:106–112)

receptor/tumor endothelial marker 8 (ATR/TEM8). PA spontaneously forms heptamer complex and facilitates entry of LF and EF into the cytosol through a clathrin-coated pit (Fig. 7.5). Protective antigen is a good vaccine candidate because of its obvious role in pathogenesis. LF and EF are separated from PA and are released. Lethal factor (LF) is a 90-kDa zinc-dependent metalloprotease that cleaves the N-terminus of several mitogen-activated protein kinases (MAPKs), thus disrupting signal transduction pathways. LF is cytotoxic to endothelial cells, dendritic cells, and macrophages, promoting bacterial survival in the host. Edema factor (EF) is an 89-kDa protein and converts intracellular ATP to cyclic adenylate cyclase (cAMP) and aids in swelling and fluid accumulation in the lungs. Edema factor also inhibits neutrophil-dependent phagocytosis and the oxidative burst of neutrophils.

pXO2 contains genes for capsule (*capABC*) biosynthesis. The capsule, composed of poly-γ-D-glutamic acid, inhibits phagocytosis. Noncapsulated strain has reduced virulence and is used as a vaccine candidate. For example, *B. anthracis* Sterne strain lacks pXO2 and is used as a vaccine strain for veterinary application throughout the world. *B. anthracis* Pasteur strain is pXO1 negative but pXO2 positive. *B. anthracis* Ames strain is negative for both pXO1 and pXO2 and is considered avirulent.

Transcription of virulence genes is regulated by CO_2, carbonate, and temperature. Virulence regulatory genes are located in both plasmids and in a 44.5 kb pathogenicity island. Expression of EF, LF, and PA is regulated by regulatory proteins, AtxA and PagR, while capsule biosynthesis is regulated by *atxA* encoded on pXO1 and *acpA* and *acpB* present in pXO2. A recent report indicated that a *B. cereus* strain contained *B. anthracis* toxin genes and exhibited illness similar to inhalation anthrax.

Pathogenesis

Depending on the route of exposure, human anthrax is classified as cutaneous, inhalation, or gastrointestinal in nature. The cutaneous form, generally associated with occupational contact, has an incubation period of 1–12 days and develops in the arms and hands eventually spreading to the face and neck. Initially, severe blister-like painless papules surrounded by edema develop. Within 5–6 days, ulceration of the papule causes the development of characteristic scabs of cutaneous anthrax. Clinical syndromes include high fever, toxemia, painful swelling of regional lymph nodes, extensive edema, shock, and death.

Inhalation anthrax is the most lethal form of anthrax, leading to death within 2–5 days after exposure. The infective dose for inhalation anthrax is estimated to be 8,000–10,000 spores. The ingested spores are transferred by alveolar macrophages to regional lymph nodes where the spores germinate into vegetative cells. The illness is biphasic and in the early phase, upper respiratory tract infection is evident, showing a flu-like symptom with nonproductive cough. Several days after infection, in the second phase, a sudden onset of high fever, rapidly progressive respiratory failure, and labored breathing, circulatory collapse, and shock are evident. In this phase, vegetative cells multiply and circulate in blood, resulting in septicemia and toxemia. Within 24 h after the initiation of the second phase of infection, elevated levels of IL-1β and TNF-α are produced which are responsible for hypotension, edema, and fatal shock.

In gastrointestinal anthrax, stomach acid possibly facilitates germination of spores. The incubation period is 1–7 days. The vegetative cells can then cause severe damage to epithelial cell linings. Systemic translocation of vegetative cells occurs through M-cells in the Peyer's patches. The symptoms are severe abdominal pain, fever, nausea, vomiting, and bloody diarrhea. Septicemia, shock, and death may follow. The mortality rate of gastrointestinal anthrax is very high, 25–60%.

Treatment and Prevention

Vaccination of animals with *B. anthracis* Sterne strain (devoid of capsule biosynthesis) is very effective in controlling animal anthrax. Early medical intervention with penicillin therapy is most effective. In addition, ciprofloxacin or other quinolones or doxycyclines are also very effective. Supportive therapy should be initiated to restore fluid and electrolyte imbalance and to prevent septic shock.

Detection of *B. anthracis*

Traditional Culture Methods

Culture-based methods are the gold standard for detection and identification of *B. anthracis*. Specimens such as nasal swabs can be directly streaked on blood agar plates (Tryptic soy agar containing 5% sheep red blood cells), incubated overnight at 35 °C, and examined. Samples can be cultured in liquid media before plating onto blood agar plates to isolate colonies. Colonies appear large (4–5 mm in diameter), opaque, white to gray in color, with irregular boundaries. *Bacillus anthracis* colonies are nonhemolytic; therefore, no visible hemolysis is seen around the colonies. Gram-staining shows characteristic large bacilli (4–5 μm in length) often arranged in chains. Nonhemolytic or weakly hemolytic colonies are then transferred to semisolid motility medium to determine motility. *Bacillus anthracis* is nonmotile. Capsule presence is determined by staining cells with India ink followed by microscopic observation. Confirmatory tests of cultures are done by determining the hydrolysis pattern of casein, starch, and gelatin; acid production from salicin, inulin, and mannitol; production of argininedehydrolase and indole, and nitrate reduction (Table 7.1). Lecithinase activity of the strains can be determined on egg yolk agar plates. Blood cultures are positive for anthrax bacilli in cases of inhalation anthrax. Antibiotic susceptibility testing helps determining the antibiotics to be used for treatment of patients.

Summary

Bacillus species are spore-forming aerobic rods and are natural inhabitants of soil, dust, and water and can contaminate milk, meat, rice, and pasta. Endospores are resistant to harsh environmental conditions or processing treatments. The majority of bacilli are nonpathogenic; however, several species produce multiple toxins and can cause various diseases in animals and humans. *Bacillus cereus* causes food-poisoning characterized by a self-limiting gastrointestinal disorder; vomiting and diarrhea. *Bacillus thuringiensis* is pathogenic to insects while *Bacillus anthracis* causes fatal anthrax in animals and humans. *B. cereus* produces two major toxins: emetic and diarrheagenic toxins. Emetic toxin (cereulide) is a low molecular weight circular peptide consisted of three modified amino acids and is resistant to heat, acids, and proteolytic enzymes. Emetic toxin is responsible for intoxication. Hemolysin BL (HBL), nonhemolytic enterotoxin (Nhe), and cytotoxin K are reported to be responsible for diarrheagenic action and these are produced inside the gastrointestinal tract. Diarrheagenic toxin production is regulated by a regulatory protein PlcR. *B. anthracis* causes three forms of disease: cutaneous anthrax, inhalation anthrax, and gastrointestinal anthrax. *B. anthracis* produces a bipartite lethal anthrax toxin and a capsule which are encoded in two virulent plasmids, pXO1 and pXO2. Anthrax toxin is an A–B type toxin with A component, known as protective antigen (PA), which binds to host cell receptor known as anthrax toxin receptor/tumor endothelial marker 8 (ATR/TEM8) and facilitates translocation of catalytic lethal factor (LF) and edema factor (EF) to the host cell cytosol. LF interferes with the signal transduction pathways by inactivating mitogen-activated protein kinases and EF activates increase production of cAMP to allow fluid accumulation and edema. *B. cereus*-related gastrointestinal complications are self-limiting whereas anthrax is highly fatal requiring immediate medial intervention with antibiotic therapy.

Further Readings

1. Ehling-Schulz, M., Fricker, M., and Scherer, S. 2004. *Bacillus cereus*, a causative agent of an emetic type of food-borne illness. Mol. Nutr. Food Res. 48:479–487.
2. From, C., Pukall, R., Schumann, P., Hormazabal, V., and Granum, P.E. 2005. Toxin-producing ability among *Bacillus* spp. outside the *Bacillus cereus* group. Appl. Environ. Microbiol. 71:1178–1183.

3. Granum, P.E. 2005. *Bacillus cereus*. In Foodborne Pathogens: Microbiology and Molecular Biology. Edited by Fratamico, P.M., Bhunia, A.K., and Smith, J.L., Caister Academic, Norfolk, pp 383–419.

4. Gray, K.M., Banada, P.P., O'Neal, E., and Bhunia, A.K. 2005. Rapid Ped-2E9 cell-based cytotoxicity analysis and genotyping of *Bacillus* species. J. Clin. Microbiol. 43:5865–5872.

5. Lattuda, C.P. and McClain, D. 1998. Examination of meat and poultry products for *Bacillus cereus*. In USDA/FSIS Microbiology Laboratory Guidebook. Edited by Day, B.P., USDA/FSIS, Washington, DC.

6. Little, S.F. and Ivins, B.E. 1999. Molecular pathogenesis of *Bacillus anthracis* infection. Microbes Infect. 1:131–139.

7. McKillip, J.L. 2000. Prevalence and expression of enterotoxins in *Bacillus cereus* and other *Bacillus* spp., a literature review. Antonie van Leeuwenhoek 77:393–399.

8. Ngamwongsatit, P., Buasri, W., Pianariyanon, P., Pulsrikarn, C., Ohba, M., Assavanig, A., and Panbangred, W. 2007. Broad distribution of enterotoxin genes (*hblCDA, nheABC, cytK,* and *entFM*) among *Bacillus thuringiensis* and *Bacillus cereus* as shown by novel primers. Int. J. Food Microbiol. (in press).

9. Oncu, S., Oncu, S., and Sakarya, S. 2003. Anthrax – an overview. Med. Sci. Monit. 9:RA276–283.

10. Scobie, H.M. and Young, J.A.T. 2005. Interactions between anthrax toxin receptors and protective antigen. Curr. Opin. Microbiol. 8:106–112.

11. Schoeni, J.L. and Wong, A.C.L. 2005. *Bacillus cereus* food poisoning and its toxins. J. Food Prot. 68:636–648.

12. Rasco, D.A., Altherr, M.R., Han, C.S., and Ravel, J. 2005. Genomics of the *Bacillus cereus* group of organisms. FEMS Microbiol. Rev. 29:303–329.

8

Clostridium botulinum and Clostridium perfringens

Introduction

Clostridium species cause a variety of diseases including food poisoning, neurological disorders, gas gangrene, and food spoilage (Table 8.1). In general, these organisms produce several exotoxins and enzymes, which contribute to the local tissue damage and the pathogenesis. *Clostridium botulinum* is the most important pathogen of this genus and it produces a potent heat-labile botulinum neurotoxin (BoNT). The toxin is responsible for the disease, botulism, which occurs in humans and animals and is considered an agent for bioterrorism. Professor Emile Van Ermengem first isolated *Clostridium botulinum* in 1895 in Belgium from raw ham that was consumed by several people, at a music festival who developed botulism. When injected into laboratory animals, the culture supernatant produced disease confirming the involvement of a deadly neurotoxin. *Clostridium perfringens* is another significant organism of concern, responsible for foodborne toxicoinfection and enteritis in animals.

Classification of *Clostridium* species

Clostridium species are classified based on shape of vegetative cells, cell wall structure, endospore formation, biochemical properties, 16S rRNA sequence homology, mol% G + C content of DNA, and PCR amplification of spacer regions of 16S and 23S rRNA genes. Genome sequence of *C. difficile, C. perfringens, C. tetani, C. botulinum* (type A and nonproteolytic B), and *C. acetobutylicum* are completed which will help understand the molecular nature of these organisms.

Clostridium botulinum

Biology

Clostridium botulinum is an obligate anaerobe. It is a motile, spore-forming Gram-positive rod and spores are located subterminally. It grows in animal intestines and spores are found in soil and plants. The botulinum neurotoxin is

Table 8.1 Classification of *Clostridium* species

Organism	Disease
Clostridium botulinum	Botulism, infant botulism, wound botulism – affects peripheral nerves – flaccid paralysis
Clostridium barati	Infant botulism, hidden botulism
Clostridium butyricum	Infant botulism
Clostridium chauvoei	Black leg – cattle and sheep
Clostridium difficile	Antibiotic-associated membranous colitis (diarrhea) in human
Clostridium histolyticum	Gas gangrene (human)
Clostridium novyi A, B, C, D	Gas gangrene
Clostridium perfringens A, B, C, D, E	Gas gangrene, food poisoning, clostridial myonecrosis, enteritis in animals, enterotoxemias of sheep (struck) in sheep and pigbel in humans
Clostridium septicum	Gas gangrene
Clostridium sordelli	Gas gangrene in humans and animals; liver disease in sheep
Clostridium tetani	Tetanus in humans and animals; affects CNS causing spastic paralysis
Clostridium sporogenes	Nontoxigenic

Adapted from McClane, B. 1995. ASM News 61:465–468

produced at temperatures as low as 3–15 °C. Optimum temperature for growth and toxin production of proteolytic strains is 35–37 °C and for nonproteolytic strains is 26–28 °C. The organism does not grow below pH 4.6. Spores are highly resistant to heat and the decimal reduction time is 0.15–1.8 min at 110–121 °C. The genome size is 3.9–4.1 Mbp and the G + C content is 26–28%. *Clostridium botulinum* spores may have a low significance as a bioweapon because they do not survive well in the healthy adult human gut environment. However, spores present in food products subjected to heat treatment will germinate under favorable conditions to form vegetative cells. Vegetative cells will grow under anaerobic environment in food and produce botulinum toxin that would be cause for a concern. Since the botulinum toxin is sensitive to heat treatment, thorough heating of the food prior to consumption will render food safe.

Sources

Most *Clostridium* spp. survive in soil and grow in animal intestine. *Clostridium* spores can be found in poorly or underprocessed canned foods (such as some home-canned foods), spices, sewage, and plants. The spores are heat-resistant and can survive in the canned food when temperature used below 120 °C. Low acidity (pH above 4.6), low oxygen, and high water content favor spore germination and toxin production. Spices, herbs, and dehydrated mushrooms may serve as potential source of spores. Home-canned vegetables – beans, peppers, carrots, corn, asparagus, potatoes, and fish are implicated in outbreaks. Foil wrapped baked potatoes, cooled at an inadequate rate and extent and served several hours later are responsible for restaurant or at-home outbreaks. Yogurt, cream cheese, canned chilly and jarred peanuts had also caused botulism outbreaks. Condiments – such as sautéed onions, garlic in oil, commercial cheese sauce were implicated in outbreaks. Bees can carry bacterial spores in addition to pollen and honey which may contain clostridial spores and the acceptable limit of clostridial spores in honey is less than 7 per 25 g.

Botulism

C. botulinum produces botulinum toxin which can cause botulism in humans and animals. The toxin is produced at temperatures as low as 3–3.3 °C to 15 °C. The toxin production is influenced by nitrogen and carbon and assembly of toxin is affected by metal ions. Botulinum toxin is heat-sensitive and it can be destroyed at >85 °C for 5 min. The LD_{50} of active botulinum toxin is <0.01 ng in mouse. In adult humans, intravenous or intramuscular administration of 0.9–0.15 µg or oral administration of 70 µg can cause death. There are five types of botulism: foodborne botulism, infant botulism, hidden botulism, inadvertent botulism, and wound botulism. Foodborne, wound and infant botulism are caused primarily by *C. botulinum* toxin A. Toxin B and E types also can cause such botulism but to a lesser frequency or extent.

Foodborne Botulism

Consumption of food contaminated with spores does not generally cause disease in healthy adults. Contaminated food after heating (at least 70 °C) and cooling at a slow rate promotes germination of spores. Spore germination may take place in colon but clostridia are unable to survive because of the resident microflora. Heating removes oxygen thus the anaerobic environment created in the heated food allows germination and growth of organism and subsequent botulinum toxin production. Oral administration of toxin is lethal in adult humans. The incubation period for botulism depends on the rate and the amount of toxin consumed, which can occur as early as 2 h or as long as 8 days, typically it takes 12–72 h. Once consumed, toxin is absorbed, and reaches the blood circulation. Toxin blocks the release of neurotransmitter, acetylcholine, which prevents nerve impulse propagation in peripheral neurons, and induces flaccid paralysis (Fig. 8.1). In flaccid paralysis peripheral nerves are affected while in spastic paralysis, which is caused by the tetanus toxin of *Clostridium tetani*, central nervous system is affected. Once the botulinum toxin is bound

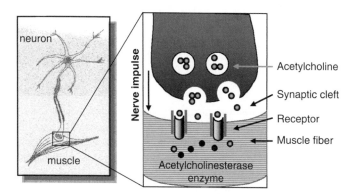

Fig. 8.1 Schematic drawing showing the neuromuscular junction and transmission of nerve impulse

to the neuron, external intervention has little effect. The symptoms appear 4–36 h after ingestion of toxins which include nausea, vomiting, headache, double vision, slurred speech, muscle spasm (dystonia), and muscle weakness. It first affects the upper limbs then the lower limbs and exhibit breathing difficulty because of diaphragm muscle weakness. The heart function also weakens, and death follows. Botulinum toxin affects parasympathetic and sympathetic systems and neuromuscular junctions. Autonomic symptoms include: dry mouth, postural hypotension, urinary retention, and pupillary abnormalities.

Infant Botulism

Clostridium botulinum, C. baratti, and *C. butyricum* are responsible for infant botulism. Honey is the main source for infant botulism. Infants under 1 year of age are mostly susceptible. Since infants have not developed complete colonic microflora populations there are no antagonistic effects from resident flora or from bile salts. In infant, the spores germinate in the intestinal tract and the vegetative cells colonize in the gut. Progression of disease is very slow because of poor absorption of toxin through colonic cell layers. Symptoms are very similar to adult botulism; however, nausea and vomiting are absent. Early signs are weak cry, muscle weakness, difficulty in feeding, i.e., poor suckling ability, hypotonia, and decrease in spontaneous movement. Constipation, tachycardia, and dry mouth, are due to blockade of parasympathetic nerve impulses. Death occurs in severe cases. Overall mortality rate is 5%. Most recover with adequate supportive therapy and interventions. Infant botulism cases are on the rise since 1990. Both bottle-fed and breast-fed infants are susceptible to infant botulism. Honey has been considered as the significant risk factors for infant botulism; therefore children should not be given honey in the first year of life.

Hidden Botulism

This adult form of botulism is hidden from clinicians because of the lack of direct evidence for this type of botulism. Hidden botulism is neither food related, wound associated nor drug use related. This adult form of infant botulism requires bacterial colonization in the gut. Antibiotic treatment due to other illnesses may disturb natural microflora balance in the gut thus allowing *Clostridium* to colonize and produce toxins. Achlorhydria, prior surgery, or Crohn's disease may also aid in the development of this botulism. *C. botulinum* or *C. barati* produces toxin F which is responsible for hidden botulism in adults. Symptoms are very similar to foodborne botulism, i.e., shortness of breath, dizziness, bradycardia, respiratory arrest, decreased voluntary movements, flaccid muscle tone, etc. Hidden botulism is diagnosed by isolation and identification of the *Clostridium* cells from feces.

Wound Botulism

Though the wound botulism is rare, it occurs in patients with traumatic and surgical wounds and intravenous drug users. It is also common in soldiers during war. In recent years, increased numbers of cases has been reported among the intravenous drug users. Wound botulism has also been reported in patients following intranasal cocaine abuse and laboratory workers from inhalation of toxins. Spores lodge in the deep wound or in the injection sites of drug users. Anaerobic environment created by tissue destruction and the growth of aerobic bacteria help germination and growth of *Clostridium*. The incubation period for wound botulism is 4–14 days. Toxin is produced and absorbed through mucus membranes, broken skin or wounds, leading to botulism.

Inadvertent Botulism

In recent years, botulinum toxins are used for treatment of dystonia and other movement-related disorders. The patients treated with intramuscular injection of botulinum toxin have toxin circulating in the blood system and can block neurotransmitter release in adjacent muscle or in the autonomic nervous system. Cosmetic use of BoNT for removal of wrinkles or to improve skin or muscle tones may serve as a possible risk factor for inadvertent botulism.

Mechanism of Pathogenesis

Clostridium botulinum produces three types of toxins: botulinum neurotoxin (BoNT), C2 toxin, and C3 toxin. Botulinum toxin affects neurons while C2 and C3 toxin cause cell damage facilitating the spread of BoNT to tissues.

Botulinum Toxin

The botulinum toxins are serologically classified as A, B, C, D, E, F, and G, which are antigenically distinct. A and B are common in the US; C and D occur in farm animals; E, F, and G are produced by nonbotulinum species. The toxin type G is produced by *C. argentinese* and toxin F is produced by *C. barati*. Some strains can produce a mixture of two BoNTs. The genes for BoNT are chromosomally linked for serotypes A, B, E, and F. Genes for C and D are located in pseudolysogenic bacteriophages and serotype G is located in an 81-MDa plasmid. The BoNT toxin is produced in the culture as a complex consisting of BoNT, hemagglutinin (HA), nontoxin nonhemagglutinin (NTNH), and RNA. The genes for toxin synthesis are present in clusters and are regulated by a regulatory element, BotR (Fig. 8.2). The proteolytic *C. botulinum* strains produce BoNT A, B, and F types while the nonproteolytic strains produce B, C, D, E, and F type neurotoxin. Proteolytic *C. botulinum* strains produce proteolytic enzymes and are able to degrade protein in food (meat) and cause putrefaction and spoilage. The botulinum toxin produced by

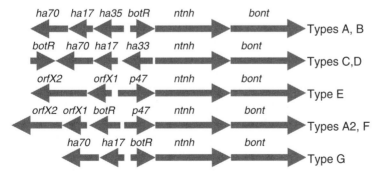

Fig. 8.2 Botulinum neurotoxin synthesis gene clusters of *C. Botulinum*. *ha* hemagglutinin, *ntnh* nontoxic nonhemagglutinin, *bont* BoNT, and *botR* botulinum regulator (adapted from Johnson, E.A. and Bradshaw, M. 2001. Toxicon 39:1703–1722)

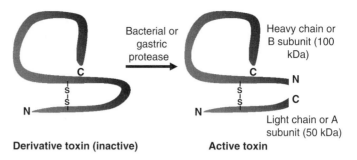

Fig. 8.3 Formation of active botulinum toxin by bacterial or gastric protease

these proteolytic strains is activated by their own protease (Fig. 8.3). While, the nonproteolytic strains do not produce protease hence they do not cause food spoilage, and the activation of BoNT from these strains is mediated by host gastric proteases. Botulinum toxin is produced, when the clostridial cells are lysogenized by the bacteriophage or due to autolysis of cells late in the growth cycle. Toxin production is influenced by nutrient composition, cultural conditions of the growth medium including the presence of metal ions, certain amino acids, peptides, pH, temperature, and the cell density.

Botulinum toxin is an A–B type toxin with a molecular weight of 150 kDa and is derived from a large progenitor toxin called derivative toxin, i.e., inactive form of toxin. Protease from bacteria or stomach cleaves the derivative toxin and converts it into an active form. The active toxin (150 kDa) consists of two subunits: B, also known as heavy chain (100 kDa) and A, known as light chain (50 kDa). A and B are joined by disulphide bond (Fig. 8.3). The BoNT blocks neurotransmitter release results in a flaccid paralysis.

Normally, acetylcholine release from the vesicles at the neuromuscular junction is mediated by SNARE (soluble *N*-ethylmaleimide-sensitive factor attachment protein receptor) proteins, which consists of synaptobrevin, SNAP-25 (synaptosomal-associated protein-25), and syntaxin. SNAP-25 and syntaxin are integral neuronal membrane bound whereas, synaptobrevin is a vesicle-associated membrane protein (VAMP) (Fig. 8.4). During transmission of a nerve impulse, synaptobrevin associated with the synaptic vesicle fuses with the SNAP-25 and syntaxin to form a synaptic fusion complex. Synaptic vesicle with SNARE complex then fuses with the neuron membrane and through exocytosis process, the acetylcholine is released in the synaptic cleft. Acetylcholine binds to acetylcholine receptor located on the muscle fiber, and acetylcholinesterase enzyme depolarizes the acetylcholine and aid in the propagation of nerve impulses (Figs. 8.1 and 8.4).

In the botulism patient, after absorption in the blood, the toxin is transported to the peripheral cholinergic synapses, primarily in the neuromuscular junction. The B subunit of BoNT binds to a specific receptor on neuron, which is a sialic acid containing glycoprotein or glycolipid and it is found only on the neurons. The B subunit forms a channel in the neuron and allows the light chain, A subunit to be internalized. A subunit has a zinc endopeptidase activity, which cleaves SNARE proteins. The A subunit from serotype A, C, and E cleaves SNAP-25, from serotypes B, D, F, and G cleaves synaptobrevin or VAMP, and from serotype C cleaves syntaxin. As a result, the SNARE

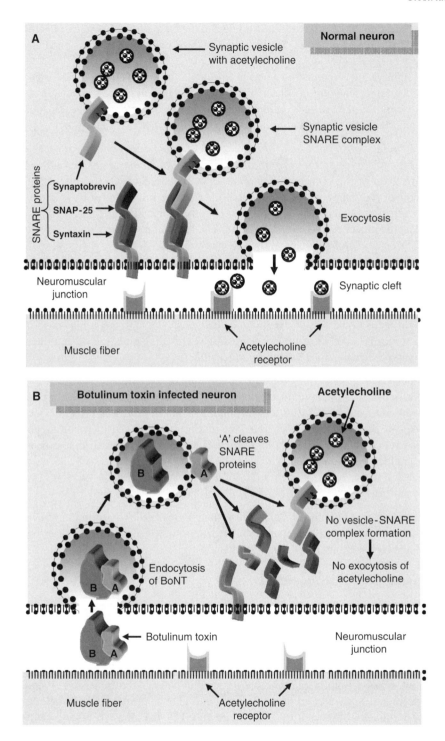

Fig. 8.4 Schematic diagram showing the mechanism of action of botulinum neurotoxin (BoNT) (redrawn from Arnon, S.S. et al. 2001. J. Am. Med. Assoc. 285:1059–1070) (*See Color Plates*)

complex does not fuse with the vesicle and the exocytosis of vesicles contents does not occur. Lack of acetylcholine release prevents transmission of the nerve impulse (Fig. 8.4).

Impaired nerve impulses result in flaccid paralysis, where muscle is tensed and relaxed showing characteristics symptom. In comparison, tetanus toxin prevents the release of γ-aminobutyric acid thus inhibits neuronal transmission. Muscle is tensed and not relaxed and results in spastic paralysis.

C2 Toxin

C2 toxin is an enterotoxin produced by some *C. botulinum* strains. It is an A–B type toxin of 135 kDa consisting of heavy or B-subunit of 90 kDa and the light chain or A-subunit of 45 kDa. About 20 kDa of N-terminus end of B is removed by trypsin to make the protein active. B portion binds to variety of cells, and receptor-mediated endocytosis follows. The A subunit with enzymatic activity translocates across the membrane and catalyzes the ADP-ribosylation of intracellular monomeric G-actin. A-subunit cleaves ADP-ribose from NAD^+ and attaches it to actin, thus interferes with the microfilament formation. It disrupts cytoskeleton structure and causes cell damage. As a result there is an increased vascular permeability, and increased fluid loss from intestinal mucosa. C2 toxin induces necrosis of the epithelial cells and edema in the lamina propria. C2 toxin is not a neurotoxin but possibly helps dissemination of botulinum toxin by increasing the vascular permeability.

C3 Toxin

C3 toxin is a 25-kDa toxin consisting of A-subunit without a B-subunit. C3 toxin ADP-ribosylates Rho, a GTP-binding signal transduction protein, which controls the actin polymerization. This toxin alters the cell morphology and leads to cell death. The pathological significance of C3 in the disease process is unclear.

Symptoms

The clinical symptoms of BoNT are visible 4–36 h after consumption of contaminated foods and are manifested by neurological disorders as described above (see section "Foodborne Botulism").

Prevention and Treatment

Prevention strategies are most important to avoid botulism which include proper canning of food products; boiling of canned foods since homemade canned foods present serious dangers; and restrain from feeding infants with honey; and early diagnosis. Since the botulinum toxin is heat-labile, heating of suspected products (at least at 85 °C for 15 min) before consumption will avert disease. Botulinum toxins are detected in patients serum, wound or stool specimens. Sixty percent of botulism patients show a positive stool sample for *C. botulinum*. Suspected food should be tested by the mouse bioassay, which can detect as low as 0.03 ng of toxin and results are obtained within 1–2 days. Antiserum is used to type the toxin from blood serum or other sources.

Treatment is unsuccessful if toxin has already entered the blood circulation and has bound to the receptors in the neuron. Successful intervention depends on the concentration of toxins being ingested and for how long. If botulism is detected early enough, equine antitoxin can be injected to neutralize toxin that is still circulating in the blood. Toxins which are already bound cannot be neutralized by the antitoxin. Respirators and life support are required to counter neurological damage, and to regenerate the nerve endings. Recovering patients can suffer from permanent neurological disorder (damage). Recovery is prolonged and dry mouth and general fatigue can last for weeks or months. An anticholinesterase drug such as edrophonium chloride may be beneficial to some patients. Guanidine and 4-aminopyridine (4-AP) have been reported to improve the ocular and limb muscle strength but those compounds had little or limited effect on respiratory paralysis. In infant, wound and hidden botulism, antibiotics are needed to clear the bacteria from the system.

Detection

Culture Method

Samples such as blood, stool, or food samples can be tested for the presence of toxin or the *C. botulinum* cells. Food samples are first enriched in the cooked meat medium or trypticase–peptone–glucose–yeast extract (TPGY), which is steamed for 10–15 min to remove dissolved oxygen and incubated at 35 or 28 °C for 5 days under strict anaerobic condition. Gram-staining or observations under a bright field microscopy should reveal the presence of typical cells with "tennis racket" appearance. For selective isolation of *C. botulinum*, enriched culture is either mixed with alcohol or heat treated at 80 °C for 10–15 min and then streaked onto egg yolk agar or liver-veal-egg yolk agar, and incubated at 35 °C for 48 h under anaerobic conditions. The typical *C. botulinum* colonies appear as raised or flat, smooth or rough with some spreading, and irregular edges. On egg yolk agar, colonies show a luster zone (referred to also as a pearly zone), when observed under an oblique light. Often another subculture in TPGY or chopped liver broth under anaerobic conditions for 5 days followed by streaking onto egg yolk agar is needed to isolate pure cultures of *C. botulinum*.

Immunoassays

Immunoassays are widely used for the detection of toxins. In a sandwich assay format, polyclonal antibody to toxin is first bound to the solid surface and then the sample containing toxin is added. A second antibody labeled with enzyme (alkaline phosphatase or horse radish peroxidase) is added. Addition of an appropriate substrate will produce color. This assay is 10–100-fold less sensitive than that of the mouse bioassay and negative results have to be reconfirmed by other assays. Lateral flow immunoassay has been developed which provides results in less than 30 min; however, the assay is less sensitive thus the negative results has to be confirmed by the mouse bioassay.

Mouse Bioassay

Mouse bioassay is considered the gold standard but very expensive. The assay requires a large numbers of mice. Bacterial cell-free culture supernatants are treated with trypsin at 200 μg ml^{-1} and 0.5 ml of each toxin preparation is injected intraperitoneally in mice. Typical symptoms include labored breathing, pinching of the waist and paralysis, which develops in 1–4 days.

BoNT Enzyme Activity Assay

The BoNT has zinc endoprotease activity and cleaves neuronal proteins that regulate the release of the acetylcholine. It cleaves a SNAP-25 while serotype

B cleaves VAMP or synaptobrevin. A synthetic VAMP peptide is used as a substrate to assay for the presence of BoNT enzyme activity colorimetrically. This assay is more sensitive and much faster than the mouse bioassay.

PCR and Oligonucleotide Microarray

A highly specific multiplex PCR assay has been developed for detection of *C. botulinum* serotypes A, B, E, and F from food and fecal materials. The assay is able to detect serotype A, E, and F at 10^2 cells while type B was detected at 10 cells per reaction mixture from naturally contaminated meat, vegetable and fish. An oligonucleotide array has been developed to accurately detect multiple foodborne pathogens including *C. botulinum, E. coli, Listeria monocytogenes, Shigella dysenteriae, Vibrio cholerae, Vibrio parahaemolyticus, Proteus vulgaris*, and *Bacillus cereus*.

Clostridium perfringens

Biology

Clostridium perfringens is a Gram-positive, rod-shaped, spore-forming anaerobic but somewhat aerotolerant bacteria. It is nonmotile and produces a double zone of hemolysis on blood agar plate. The clear zone is produced by perfringolysin O (theta-toxin) and the hazy outer zone is by phospholipase C (alpha-toxin). The optimal growth temperature is 43–45 °C. The generation time is 7.1 min at temperature 33–49 °C. *C. perfringens* does not grow below 15 °C and the pH range for growth is 5–9.0. *C. perfringens* can survive and grow in the presence of curing salts consisted of 300 ppm of sodium nitrite and 4–6% NaCl. *C. perfringens* is a fastidious organism and requires more than 12 different amino acids and vitamins for their growth, thus it grows very well in the meat products. Meat and poultry are generally implicated in most outbreaks. Beef products are responsible for about 40% of *C. perfringens* outbreaks. Roast beef is a major vehicle of outbreak because of improper handling, temperature abuse, inadequate cooling after cooking.

From 1992 to 1997, a total of 248,520 human illnesses due to *C. perfringens* were documented by the CDC. In the year 1993, at least 10,000 cases of *C. perfringens* foodborne disease with 100 deaths were recorded. The impact of this pathogen is likely to be underestimated since it is not part of the CDC's active surveillance program, FoodNet.

Sources

Clostridium perfringens is found in soil, water, sludge, spices, dust, sewage, and contaminated equipment. It is present in low numbers in human/animal intestine and human feces (10^3–10^6 spore per gram). It grows well in meat especially in ground beef. During heating and cooling, if food is allowed to stand at room temperature – it will grow and produce toxin.

Toxins

C. perfringens produces 14 different toxins and can cause food poisoning, and necrotic enteritis in humans, animals, and birds. There are five types of *C. perfringens* (A, B, C, D, and E), based on the production of enterotoxin and

four types of extracellular toxins: alpha (α), beta (β and β2), epsilon (ε), and
iota (ι) (Table 8.2). *C. perfringens* also produces a 60-kDa perfringolysin O
also called theta (θ)-toxin and collagenase also called kappa (κ)-toxin. Type A
strains are predominantly involved in the foodborne toxicoinfection.

Clostridium perfringens Enterotoxin

C. perfringens produces multiple toxins. The primary toxin is called Type A
enterotoxin or *Clostridium perfringens* enterotoxin (CPE). CPE is a 319 amino
acid containing polypeptide of 35 kDa, which is produced during sporulation.
It has no significant similarity with other known toxins. It is heat-labile and
destroyed at temperature 60 °C for 10 min. It is also sensitive to acidic pH, and
pronase but resistant to digestive enzymes, trypsin, and chymotrypsin. CPE
has two important domains; the N-terminal domain (amino acids 26–171) display the cytotoxicity action and the C-terminal domain (amino acids 290–319)
interacts with a cellular receptor protein, claudin (21 kDa), a tight junction
protein located in the junction of intestinal epithelial cells. Intestinal proteases
(chymotrypsin and trypsin) remove a portion of N-terminal sequence up to
amino acid 38 thus making the toxin more potent with increased biologic
activity.

α-Toxin

Alpha-toxin, also known as phospholipase C is produced by *C. perfringens*
Type A. It is a 43 kDa toxin, and is responsible for gas gangrene resulting
in high fever, pain, edema, and myonecrosis. The toxin has phospholipase C
(PLC), sphingomyelinase (SMase), and dermonecrosis activity. The substrates
for PLC and SMase are phosphatidylcholine (PC) and sphingomyelin (SM),
respectively. This toxin has significant homology with PLC from *Listeria
monocytogenes* and *Bacillus cereus*. At high concentration, the α-toxin causes
massive degradation of PC and SM in membranes leading to membrane

Table 8.2 *Clostridium perfringens* classification and their toxin production
profile.

Type	Disease	Enterotoxin	α-toxin	β-toxin	ε-toxin	ι-toxin
A	Gas gangrene, food poisoning, myonecrosis	+	+	−	−	−
B	Dysentery in animals	+	+	+	+	−
C	Necrotic enteritis in humans and animals (pigbel disease)	+	+	+	−	−
D	Pulpy kidney disease	+	+	−	+	−
E	Enteritis in animals	+	+	−	−	+
Gene		*cpe*	*plc*	*cpb, cpb2*	*etx*	*iap, ibp*
Gene location		Plasmid/ chromo- some	chromo- some	plasmid	plasmid	plasmid

Adapted from Brynestad, S. and Granum, P.E. 2002. Int. J. Food Microbiol. 74:195–202 disruption and at low concentration the damage is less severe. The α-toxin induces the production of intracellular mediators, IL-8, TNF-α, platelet activation factors, and the endothelial leukocyte adhesion molecules which possibly contribute to the increased vascular permeability and swelling.

β-Toxin

β-toxin is a 34 kDa toxin, and it oligomerizes to form cation-dependent channels in susceptible membranes. It forms channels of 12 Å in diameter in planar bilayers consisting of phosphatidylcholine and cholesterol. The channel is selective for monovalent cations such as sodium and potassium. The evidence suggests that β-toxin probably acts as a neurotoxin.

Iota-Toxin

The iota-toxin also contributes to the diarrheal disease. It catalyzes ADP-ribosylation of globular actin by inhibiting its synthesis.

Genetic Regulation of Virulence

In *C. perfringens*, a majority of virulence genes are located in plasmids. The enterotoxin, CPE, production is both plasmid and chromosomal linked. About 1–2% of *C. perfringens* isolates that are responsible for food poisoning carry enterotoxin gene (*cpe*). The *cpe* gene in *C. perfringens* that causes food poisoning is usually located in the chromosome. A two-component *virR/virS* locus regulates many virulence genes including phospholipase C (α-toxin), perfringolysin O (θ-toxin) and collagenase (κ-toxin) as well as many nontoxic proteins.

Pathogenic Mechanism

C. perfringens causes infection, not intoxication. Infection dose is about 10^7–10^9 cells. Raw foods or ingredients may be contaminated with spores. Slow cooling after cooking allows spores to germinate since heating removes oxygen and creates anaerobic environment. Also room temperature storage after cooking allows vegetative cells to grow and multiply. Upon consumption, vegetative cells begin to form spores when exposed to the stomach acid and toxin is released following lysis of cells. *C. perfringens* spores with increased heat resistance have been shown to produce increased levels of enterotoxin.

The cytotoxic effect of CPE starts with binding of the toxin to the 21-kDa claudin receptor of tight junction (TJ) proteins and the complex then binds to a 45–50 kDa membrane protein forming a complex of 90 kDa referred as CPE complex. This small complex then interacts with a 70-kDa protein on the membrane and forms a larger complex (155 kDa), which alters the membrane permeability to cause fluid and ion (Na^+, Cl^-) losses. CPE also modulates the architecture of epithelial tight junctions (TJ) by forming a complex with TJ protein occludin and forms a large complex of 200 kDa (Fig. 8.5). Internalization of occludin from TJ to the cytoplasm disrupts the structural integrity and affects the membrane barrier function. As a result, CPE alters the paracellular membrane permeability and contribute to the onset of diarrhea. Histopathological changes include desquamation of epithelial cells, damage to the villi tips, and shortening of villi. These changes affect the fluid and electrolyte transport, decrease the intestinal absorptive capacity, and increase the stool mass.

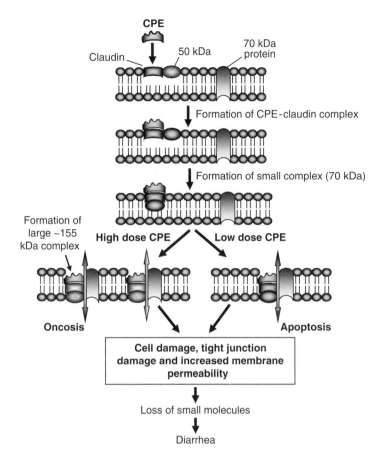

Fig. 8.5 Flow diagram showing the mechanism of action for *Clostridium perfringens* enterotoxin (CPE) (redrawn from McClane, B.A. and Chakrabarti, G. 2004. Anaerobe 10:107–114)

CPE at low concentration (1 µg ml⁻¹) induces apoptosis through activation of caspase 3/7 thus there is little intestinal inflammation. CPE at high concentration (10 µg ml⁻¹) induces necrosis (oncosis), which may contribute to the inflammatory process that is evident during infection.[1]

C. perfringens type A also can cause rare but severe foodborne necrotic enteritis. Necrotic enteritis is primarily caused by certain strains of *C. perfringens* type C and CPE-negative type A in piglet, chicken, calf, lamb, and goats. The necrotizing β-toxin acts on autonomic nervous system causing arterial constriction leading to mucosal necrosis. The toxin also forms multimeric transmembrane pores and facilitates the release of cellular arachidonic acid and inositol. Additionally, the α-toxin with the phospholipase and sphingo-myelinase activity is also responsible for necrotizing effects.

---

[1] Certain aggressive form of endometrial cancer such as uterine serous papillary carcinoma (USPC), which is resistant to chemotherapy express a high levels of claudin-3 and claudin-4. Since CPE interacts with claudin molecules, claudin molecules present a suitable target for CPE-mediated cancer therapy.

Symptoms, Prevention, and Control

Symptoms are manifested by pronounced diarrhea, abdominal cramp, appearing within 8–12 h and most cases resolve within 12–24 h. In healthy individuals, the disease is self-limiting; however, in malnourished person, elderly and very young patients, the organism may colonize, invade, and cause severe ulceration and death. Death may occur due to severe dehydration. Sudden infant death syndrome (SIDS) is associated with this pathogen and the toxins possibly act as superantigens (see Chap. 4) triggering the release of massive amounts of IL-2 by T-lymphocytes.

Treatment and prevention include bed-rest and fluid supplement. Antibiotic therapy may help *Clostridium* to grow and produce more toxins in the colon since antibiotic will inhibit the resident microflora. As a precaution, food should not be allowed to stand at room temperature for an extended period of time. Adequate cooking, holding at hot temperatures ($\geq 60\,^{\circ}C$) or rapid cooling can avoid *C. perfringens*-related food poisoning.

Detection

Culture and PCR Methods

Conventional culture methods are used to isolate *C. perfringens* from stool or food samples. Samples are first suspended in fluid thioglycolate medium and then heat shocked for 70–75 $^{\circ}C$ for 15–20 min and then enriched at 37 $^{\circ}C$ for 18–24 h. Liquid culture is then streaked onto tryptose-sulfite cycloserine agar containing egg yolk (10%) or brain heart infusion agar with 10% sheep blood and incubated in an anaerobic jar at 37 $^{\circ}C$ for 18–24 h to isolate pure colonies.

Multiplex PCR methods have been developed to detect the presence of toxin genes; enterotoxin (*cpe*), alpha (*cpa*), beta (*cpb*), epsilon (*etx*), and iota (*iap*) in *C. perfringens* isolates. This method also allows the typing of isolates based on their toxin production.

Immunoassays

There are two antibody-based assay methods available commercially; reverse passive latex agglutination test from Oxoid and CPE receptor-based enzyme immunoassay from Tech Lab.

Cell Culture Methods

CPE action has been studied using Vero, Caco-2, and MDDK (Madin-Darby canine kidney) cell models. Caco-2 is the most appropriate model since it is an enterocyte-like cell, representing cells of intestinal origin. Loss of membrane integrity due to the CPE action has been assayed by the release of ^{86}Rb. Toxic effects of β-toxin have been studied using human umbilical vein endothelial cells (HUVEC), in which toxin affects membrane permeability resulting in the loss of essential cellular materials.

Summary

Members of the genus *Clostridium* cause varieties of diseases in humans and animals, sometimes with fatal consequences. They are anaerobic spore-forming rod-shaped bacteria and mostly associated with soil and sediments. Two organisms, *Clostridium botulinum* and *C. perfringens* have significant importance in food microbiology because they are responsible for botulism (intoxication) and gastroenteritis (toxicoinfection), respectively. In foodborne botulism, the botulinum toxin is produced in the food during anaerobic growth. Botulinum toxin is an A–B type toxin with a zinc-dependent endopeptidase activity. It cleaves SNARE protein complex, which is responsible for the release of neurotransmitter, acetylcholine from synaptic vesicles into the neuromuscular junction for transmission of nerve impulses. As a result, nerve impulse is blocked resulting in flaccid paralysis. The symptoms appear as early as 2 h after ingestion of toxin and the severity and progression of the disease depends on the amount of toxin ingested. Early medical intervention is needed to save the patient. In contrast, *C. perfringens* produces 14 different toxins and causes toxicoinfection. There are five types of *C. perfringens* (A, B, C, D, and E), classified based on the production of enterotoxin (CPE) and four types of extracellular toxins: alpha (α), beta (β and β2), epsilon (ε), and iota (ι). Type A *C. perfringens* are generally associated with foodborne disease. After consumption of vegetative cells, the bacteria begin to sporulate as they encounter acidic pH of the stomach. The enterotoxin binds to claudin receptor in the tight junction protein and forms a large protein complex with other membrane proteins to form pores in the membrane and alters the membrane permeability to cause fluid and ion (Na^+, Cl^-) losses. CPE also modulates the architecture of epithelial tight junctions (TJ) by forming a complex with TJ protein occludin, forming a large complex of ~200 kDa. As a result, CPE alters the paracellular membrane permeability and promote the onset of diarrhea.

Further Readings

1. Arnon, S.S., Schechter, R., Inglesby, T.V., Henderson, D.A., Bartlett, J.G., Ascher, M.S., Eitzen, E., Fine, A.D., Hauer, J., Layton, M., Lillibridge, S., Osterholm, M.T., O'Toole, T., Parker, G., Perl, T.M., Russell, P.K., Swerdlow, D.L., and Tonat, K. 2001. Botulinum toxin as a biological weapon: medical and public health management. J. Am. Med. Assoc. 285:1059–1070.

2. Bohnel, H. and Gessler, F. 2005. Botulinum toxins – cause of botulism and systemic diseases? Vet. Res. Commun. 29:313–345.

3. Bruggemann, H. 2005. Genomics of clostridial pathogens: implication of extrachromosomal elements in pathogenicity. Curr. Opin. Microbiol. 8:601–605.

4. Brynestad, S. and Granum, P.E. 2002. *Clostridium perfringens* and foodborne infections. Int. J. Food Microbiol. 74:195–202.

5. Cherington, M. 1998. Clinical spectrum of botulism. Muscle Nerve 21:701–710.

6. Johnson, E.A. and Bradshaw, M. 2001. *Clostridium botulinum* and its neurotoxins: a metabolic and cellular perspective. Toxicon 39:1703–1722.

7. Hall, Y.H.J., Chaddock, J.A., Moulsdale, H.J., Kirby, E.R., Alexander, F.C.G., Marks, J.D., and Foster, K.A. 2004. Novel application of an in vitro technique to the detection and quantification of botulinum neurotoxin antibodies. J. Immunol. Methods 288:55–60.

8. Heikinheimo, A. and Korkeala, H. 2005. Multiplex PCR assay for toxinotyping *Clostridium perfringens* isolates obtained from Finnish broiler chickens. Lett. Appl. Microbiol. 40:407–411.

9. Lindstrom, M. and Korkeala, H. 2006. Laboratory diagnostics of botulism. Clin. Microbiol. Rev. 19:298–314.

10. McClane, B. 1995. ASM first international conference on the molecular genetics and pathogenesis of the Clostridia. Progress in understanding clostridial toxins and their genes could reduce the damage they cause. ASM News 61:465–468.

11. McClane, B.A. and Chakrabarti, G. 2004. New insights into the cytotoxic mechanisms of *Clostridium perfringens* enterotoxin. Anaerobe 10:107–114.

12. Novak, J., Peck, M., Juneja, V., and Johnson, E. 2005. *Clostridium botulinum* and *Clostridium perfringens*. In Foodborne Pathogens: Microbiology and Molecular Biology. Edited by Fratamico, P., Bhunia, A.K., and Smith, J., Caister Academic, Norfolk, pp 383–407.

13. Sakurai, J., Nagahama, M., and Oda, M. 2004. *Clostridium perfringens* alpha-toxin: characterization and mode of action. J. Biochem. 136:569–574.

14. Shatursky, O., Bayles, R., Rogers, M., Jost, B.H., Songer, J.G., and Tweten, R.K. 2000. *Clostridium perfringens* beta-toxin forms potential-dependent, cation-selective channels in lipid bilayers. Infect. Immun. 68:5546–5551.

15. Singh, U., Mitic, L.L., Wieckowski, E.U., Anderson, J.M., and McClane, B.A. 2001. Comparative biochemical and immunocytochemical studies reveal differences in the effects of *Clostridium perfringens* enterotoxin on polarized Caco-2 cells versus Vero cells. J. Biol. Chem. 276:33402–33412.

16. Songer, J.G. 1996. Clostridial enteric diseases of domestic animals. Clin. Microbiol. Rev. 9:216–234.

17. Tweten, R.K. 2001. *Clostridium perfringens* beta toxin and *Clostridium septicum* alpha toxin: their mechanisms and possible role in pathogenesis. Vet. Microbiol. 82:1–9.

18. Wise, M.G. and Siragusa, G.R. 2005. Quantitative detection of *Clostridium perfringens* in the broiler fowl gastrointestinal tract by real-time PCR. Appl. Environ. Microbiol. 71:3911–3916.

9

Listeria monocytogenes

Introduction

In 1926, E.G.D. Murray first reported the isolation of *Listeria monocytogenes* from rabbits and since then it was considered as an animal pathogen primarily causing "circling disease" in ruminants (cattle, sheep, and goats), pigs, dogs, and cats. The animals walk in a circle and exhibit uncoordinated posture, and are unable to stand without a support. However, in the late 1970s and early 1980s, this organism emerged as a foodborne pathogen causing numerous outbreaks in humans in North America. It is responsible for rare but fatal systemic disease called listeriosis, affecting primarily immune suppressed populations: pregnant women, infants, AIDS patients, and organ transplant recipients. However, in recent years, gastroenteritis cases resulting from *L. monocytogenes* infection have been reported. *L. monocytogenes* is a facultative intracellular pathogen and has been used as a model organism to study intracellular parasitism. It produces numerous virulence factors that ensure its survival in a host, thus it has been the subject of many recent investigations for its utility to carry foreign genes for vaccination against other diseases including cancers.

About 2,500 people in the US contract invasive listeriosis each year. The mortality rate is highest of all foodborne pathogens, and is usually 20–30%, but has been reported to be as high as 50%. There is a zero tolerance policy for *L. monocytogenes* in ready-to-eat foods in the US. Other countries like Canada and some European countries allow a limit of 100 cfu/25 g food. In most recent years, several outbreaks in the US were linked to ready-to-eat meats. A multistate outbreak involving tainted hotdogs affected 22 states in 1998–1999, causing 101 illnesses, 15 deaths, and 6 miscarriages. In 2000, deli turkey meat was responsible for a multistate outbreak resulting in 30 illnesses, 4 deaths, and 3 miscarriages. In 2000–2001, consumption of Mexican-style soft cheese (Queso Fresco) made with unpasteurized milk resulted in 12 cases of listeriosis and 5 miscarriages in a Hispanic community in North Carolina. In 2002, ready-to-eat turkey deli meat was implicated in a multistate outbreak involving nine states resulting in 54 illnesses, 8 deaths, and 3 stillbirths. In 2003, raw milk cheese was responsible for an outbreak in Texas, and in 2005, a multistate outbreak involving consumption of turkey deli meat affected nine states and caused 12 illnesses. *L. monocytogenes* serotype 4b was implicated in each of these outbreaks. Though *L. monocytogenes* contamination does not always result in outbreaks, product recalls are routinely initiated to prevent outbreaks. The estimated cost associated with *Listeria* contamination is approximately 2 billion dollars per year.

Classification

The genus *Listeria* has six species: *L. monocytogenes, L. innocua, L. ivanovii, L. welshimeri, L. seeligeri*, and *L. grayi. L. ivanovii* has two subspecies: *L. ivanovii subsp. ivanovii* and *L. ivanovii subsp. londoniensis*. Among these, *L. monocytogenes* is pathogenic to humans and animals while *L. ivanovii* is pathogenic to animals and others are considered nonpathogenic. Although *L. seeligeri* is considered nonpathogenic, it possesses a part of the virulence gene cluster, which is present in pathogenic *L. monocytogenes* and *L. ivanovii*. *L. monocytogenes* have 13 distinct O-antigenic patterns, which comprise the serovars; 1/2a, 1/2b, 1/2c, 3a, 3b, 3c, 4a, 4b, 4c, 4d, 4e, 4ab, and 7. Of which, 1/2a, 1/2b, and 4b are responsible for 98% of the outbreaks and serotype 4b is considered the most virulent. A subpopulation of serotype 4b also exists and one or more of those subtypes are considered as epidemic clones. *L. monocytogenes* are grouped into three lineages based on their ribopatterns and their association with outbreaks. Lineage I has the highest pathogenic potential and the serotypes belonging to this group are involved in most epidemic outbreaks. Lineage II has intermediate pathogenic potential and possibly responsible for sporadic outbreaks while the lineage III has low pathogenic risk and rarely cause human infection (Table 9.1). Based on the flagellar (H) antigens, *Listeria* serovars are again classified as A, B, C, D, and E.

Biology

Listeria species are Gram-positive, rod-shaped, and nonspore-forming bacteria. They are $1-2\,\mu m$ long and may exist as single or double cells. Occasionally, *Listeria* may display long chains depending on the growth

Table 9.1 Classification of *Listeria monocytogenes* based on genomic fingerprint patterns and association with epidemic outbreaks

Groups	Outbreaks	Pathogenic potential	Predominant serotypes
Lineage I	Epidemic clones and responsible for most outbreaks	High	1/2b, 3b, 4b, 4d, 4e
Lineage II	Sporadic listeriosis cases	Medium	1/2a, 1/2c, 3c, 3a
Lineage III	Rarely cause human disease	Low	4a, 4c
IIIA (Rham +ve)			4a (avirulent) and 4c (virulent)
IIIB (Rham −ve)			Virulent nonserotype 4a and nonserotype 4c; Serotype 7?
IIIC (Rham −ve)			Virulent 4c

Rham rhamnose fermentation property

conditions and temperatures. *Listeria* is ubiquitous in nature and survives in extreme environments, including broad pH ranges (4.1–9.6), high salts (10%), and in presence of antimicrobial agents. They are psychrophilic and grow at a wide temperature range (1–45 °C). Some species of genus *Listeria* produce hemolysins. *L. monocytogenes, L. ivanovii*, and *L. seeligeri* are hemolytic and produce a zone of β-hemolysis on blood agar plates. The hemolysis produced by *L. ivanovii* is rather pronounced producing two distinct zones; a clear zone nearest to the colony is due to the action of ivanolysin O and the partial zone of lysis in the periphery is caused by the sphingomyelinase (27 kDa) and lecithinase. The zone of hemolysis produced by *L. monocytogenes* or *L. seeligeri* is less intense and may not be readily visible. To differentiate the hemolytic *Listeria*, the CAMP (Christie, Atkinson, Munch-Peterson) test is used. In this assay, *L. monocytogenes* (or *L. seeligeri*) is streaked on blood agar plate, perpendicular to a line of *Staphylococcus aureus*. A clear zone of hemolysis resembling an arrowhead is produced at the junction of two cultures. Similarly, *L. ivanovii* produces a zone of hemolysis when grown in perpendicular to *Rhodococcus equi* on a blood agar plate. Fermentation properties of carbohydrates such as rhamnose, xylose, and mannitol are unique among members of the genus *Listeria*, and are used for identification and characterization of the species (Table 9.2). Pathogenic properties of *Listeria* are also tested by in vitro cell culture assay, invasion assay, and sometimes by in vivo mouse bioassay.

Flagella

Flagella are an essential structural feature of all *Listeria* spp. *Listeria* expresses peritrichous flagella, and displays tumbling motility. The flagellar antigen, also known as H antigen, is classified into A, B, C, D, and E serogroups. The flagellum consists of a 29-kDa subunit flagellin protein, which is encoded by the *flaA* gene. Transcription of *flaA* is regulated by *flaR*. Flagellin is a dimer consisting of two identical proteins with the same molecular mass and assembles on the bacterial surface. The globular subunits are packed around the long axis of the flagellum to provide a stable structure. Flagellar expression is temperature dependent, and maximum expression is observed at 4–30 °C

Table 9.2 Biochemical properties of *Listeria* species

Characteristics	L. monocytogenes	L. innocua	L. ivanovii	L. welshimeri	L. seeligeri	L. grayi
β-hemolysin	+	–	+	–	+	–
CAMP (*S. aureus*)	+	–	–	–	+	–
CAMP (*R. equi*)	–	–	+	–	–	–
Rhamnose	+	v	–	–	–	–
Xylose	–	–	+	+	+	
Mannitol (α-methyl-D-mannoside)	+	+	–	+	–	
Cytotoxicity	+	–	+	–	±	–
Invasion assay	+	–	+	–	–	–
Mouse virulence	+	–	+	–	–	–

CAMP Christie, Atkinson, Munch-Peterson hemolysis assay; + positive; – negative; *v* variable and minimum at 37 °C. It remains unclear whether flagella are important for virulence of *L. monocytogenes* in vivo, especially since expression of flagellin is low at body temperature. Flagellin is immunogenic and the antiflagellar antibodies have been used for serotyping and for bacterial detection.

Sources

Listeria species are ubiquitous in nature and are found in soil, water, sewage, silage, plants, and in the intestinal tract of domestic animals (sheep, cattle, goat, etc.). One to five percent of healthy humans are reported to serve as carriers. Food is the primary vehicle of infection and meat, vegetables, fish, and dairy products are potential source for transmission. Ready-to-eat or minimally processed foods such as hotdogs, luncheon meats, pâté, smoked fish, some soft cheeses, and cheeses made with unpasteurized milk (ethnic products like Mexican-style soft cheese) were responsible for numerous epidemic or sporadic outbreaks. Postprocessing contamination of products is a major concern since those products are eaten without further cooking. *Listeria*-associated gastroenteritis has been linked to rice salad, cold corn and tuna salad, corned beef and ham, cold smoked trout, and cheese.

Disease

Listeria monocytogenes is an intracellular pathogen and affects healthy as well as immunosuppressed populations. In healthy individuals, this organism can cause gastroenteritis and fever. Overall, mortality rate due to infection is about 20–30%, with an annual death rate of about 500 people. The infective dose for this pathogen is not known; however, it is estimated to be about 100 to 10^6 cells depending on the immunological status of the host. In an immunocompromised mouse model, as few as ten bacterial cells have been shown to cause fatal infection. The incubation period for the disease also varies from about 3 days to 3 months in humans depending on the immune status and the number of cells ingested. It causes three forms of disease (1) gastrointestinal form, (2) systemic listeriosis, and (3) abortion and neonatal listeriosis.

Gastrointestinal Form

Since 1993, seven outbreaks of gastroenteritis caused by *L. monocytogenes* were reported worldwide. The serotype 1/2a, 1/2b, and 4b have been implicated in those outbreaks. The foods involved were rice salad, cold tuna salad, cheese, trout, and chocolate milk. The infectious dose associated with those outbreaks was very high at 10^6–10^8 cfu; however, dosage as low as 100 cfu g^{-1} and as high as 10^{11} cfu g^{-1} have been associated with the gastrointestinal form of the disease. Adults are more susceptible than children and people receiving gastric acid suppressive medications are at greatest risk. The incubation period is less than 24 h; however, it may range from 6 h to 10 days. The exact mechanism of gastroenteritis is not known. However, it appears that the organism causes damage to the absorptive villi affecting absorption of nutrients and promoting fluid secretion. The gastroenteritis symptoms are characterized by fever, headache, nausea, vomiting, abdominal pain, and diarrhea. Diarrhea is nonbloody but watery.

Systemic Listeriosis

Listeria monocytogenes causes the disease listeriosis, a rare but fatal disease primarily affecting immunocompromised individuals such as young, old, pregnant, and immunologically challenged hosts. AIDS patients, cancer patients receiving chemotherapy, organ transplant patients, diabetic, alcoholics, and patients with cardiovascular diseases are highly susceptible to this bacterium. The organism may colonize for a short duration in the intestine and passes through the intestinal barrier reaching the blood circulation and the lymphatic system. A majority of the organism is distributed in the liver (90%), spleen (10%), gall bladder and the regional lymph nodes within 24 h. After a brief bacteremic phase, *Listeria* crosses the blood–brain barrier and infects the brain causing meningitis and encephalitis. In pregnant women, *Listeria* crosses the placental barrier and infects the fetus (see below for detail) (Fig. 9.1). Systemic listeriosis is characterized by fever, headache, malaise, septicemia, meningitis, brain stem encephalitis, ataxia, bacteremia, and liver abscess. Sometimes *L. monocytogenes* may be associated with Chrone's disease.

Abortion and Neonatal Listeriosis

In pregnant women, *L. monocytogenes* can cause complications late in the third trimester of pregnancy resulting in stillbirth, or birth of an infected fetus. Pregnancy is associated with suppression of cell-mediated immunity to protect the fetus from rejection, which renders pregnant women extremely susceptible to listeriosis. During pregnancy, systemic listeriosis often results in intrauterine infection. *L. monocytogenes* infects trophoblastic cells in the placenta then rapidly disseminate to trophoblastic structures like syncytiotrophoblast in villous core in the labyrinthine zone and cross the placental barrier. Inflammation at the site is characterized by the infiltration of polymorphonuclear cells. Complications such as preterm labor, spontaneous abortion, stillbirth, or neonatal infection are seen. Although the intracellular lifestyle of *Listeria* has been well studied, relatively little is known about how or why this organism infects the placentofetal unit. It has been shown that *L. monocytogenes* moves from maternal organs to the placenta and then returns to the maternal organs.

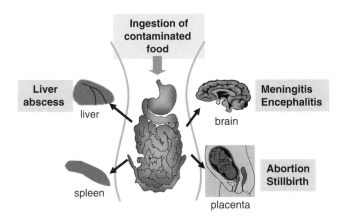

Fig. 9.1 Distribution of *L. monocytogenes* to different tissues and organs (*See Color Plates*)

The placenta is relatively protected from *L. monocytogenes* colonization; however, if a single bacterium enters the placenta and finds a niche, it will colonize in the placenta and act as a source of infection for the mother. It is possible that preterm labor or spontaneous abortion is a protective mechanism for the mother. This allows expulsion of the infected placentofetal unit, removing the source of infection and providing the mother with a greater opportunity for survival. The symptoms of infection begin as flu-like and gradually progress to a severe headache. Decreased fetal movement and early labor will follow.

Two forms of neonatal listeriosis are reported: early and late onset forms. In the early onset form (1.5 days), organism is acquired in the uterus and the infection is widespread exhibiting very high mortality rate. In the late forms (14 days), the infection is possibly acquired during birth from the vaginal tract or from the environment. In this form, meningitis is the predominant clinical feature. The symptoms of neonatal listeriosis include respiratory distress, pneumonia, shortness of breath, hyperexcitability, conjunctivitis, rash, vomiting, cramps, and hypo- or hyperthermia. Mortality rate in neonatal listeriosis is about 36%.

Mechanism of Pathogenesis

The pathogenic mechanism of *L. monocytogenes* is a complex process and it can be subdivided into two phases: intestinal and systemic. The intestinal phase of infection involves bacterial colonization in the intestine and consequent translocation through the mucosal barrier to blood circulation or to the lymphatic system for systemic dissemination. During systemic spread, the organism is transported by dendritic cells or macrophages to the liver, spleen, lymph nodes, brain, and to the placenta (in pregnant women). Many surface associated and secreted proteins have been recognized as important virulence factors (Table 9.3; Fig. 9.2), which are critical for bacterial persistence in the intestinal tract, to enter the host cells for intracellular movement and cell-to-cell spread, and to evade immune system.

The majority of virulence genes are clustered in a 9-kb pathogenicity island comprised of *prfA–plcA–hly–mpl–actA–plcB*. These genes, together with internalin genes (*inlA* and *inlB*), *hpt*, and *bilE* are regulated by the first gene *prfA* (positive regulatory factor A). In addition, expressions of many of these genes that are essential during bacterial transit through gastrointestinal tract such as *gadA, hpt, bilE* are coregulated by an alternate sigma factor, sigma B (σ^B).

Intestinal Phase of Infection and Systemic Spread

Most listeriosis cases are foodborne, thus bacterial translocation through the intestinal tract is prerequisite for infection. After consumption, *Listeria* passes through the stomach and arrives in the small intestine. The glutamate decarboxylase (GAD) system of *L. monocytogenes* protects the bacterium from gastric acid, and furthermore, food particles neutralize pH to ensure safe passage to the small intestine. Bile salt hydrolase (BSH) and the bile exclusion system (BilE) protect bacteria against bile salts and OpuC provides osmotolerance. The organism pierces through the mucosal layer and interacts with and breaches the intestinal epithelial barrier by three possible pathways:

Table 9.3 Major virulence proteins in *Listeria monocytogenes*

Virulence factors	Size (kDa)	Receptor	Function
Protein regulatory factor (PrfA)	27		Regulation of virulence protein expression
Internalin (InlA)	88	E-cadherin (tight junction protein)	Responsible for invasion into intestinal epithelial cells and placenta during pregnancy
Internalin B (InlB)	65	Met (tyrosine kinase), gC1q-R/p32	Entry into hepatocytes and hepatic phase of infection
Virulence protein (Vip)	43	Gp96 (chaperone protein)	Invasion of epithelial cells
Listeria adhesion protein (LAP)	104	Hsp60 (chaperone protein)	Adhesion to intestinal epithelial cells
Autolysin amidase (Ami)	102	Peptidoglycan	Adhesion to host cells
p60 (cell wall hydrolase)	60	Peptidoglycan	Adhesion/invasion
Listeriolysin (LLO) *hlyA*	58–60	Cholesterol	A hemolysin aids in bacterial escape from phagosome inside the cell
Actin polymerization protein (ActA)	90	–	Nucleation of actin tail for bacterial movement inside the cytoplasm
Bile salt hydrolase (BSH)	36	–	Survival in gut
Phospholipase (*plcA* – PI-PLC; *plcB* – PC-PLC)	29–33	–	Lyses of vacuole membrane
Metalloprotease (Mpl)	29	–	Helps synthesis of PLC

through M-cells overlying Peyer's patches, through epithelial cells, and through dendritic cells sampling the lumen of intestine (Fig. 9.3). Dendritic cells and/or macrophages located beneath the epithelial cell lining in lamina propria can process and present antigen to helper T-cells (CD4$^+$) or cytotoxic T-cells (CD8$^+$) cells for immune response or can transport bacteria to extraintestinal sites such as liver, spleen, lymph nodes, the brain and the fetoplacental junction in pregnant women.

Bacterial translocation through naturally phagocytic M-cells is a passive process, while translocation through enterocytes is an active process where bacterial interaction with host cell receptors initiates a cascade of signaling events that modify cellular architecture to promote bacterial passage through cells (Fig. 9.4). The cellular mechanism of *Listeria* pathogenesis involves four major steps (1) attachment and entry, (2) lysis of vacuole (phagosome), (3) intracellular growth, and (4) cell-to-cell spread (Fig. 9.5).

Attachment and Entry

Adhesion is critical in the initial phase of infection. Several adhesion factors have been identified to be involved in this process (Fig. 9.4). Internalin A (InlA) interacts with epithelial cadherin (E-cadherin) to promote bacterial entry into cells and is important during intestinal and uteroplacental infection.

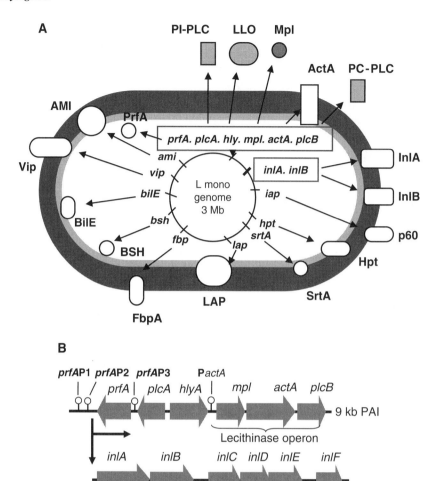

Fig. 9.2 (**A**) Approximate location of *L. monocytogenes* virulence genes in chromosome (3 Mb) and the location of gene products in the extracellular milieu, cell wall (*dark shade*), and cytoplasmic membrane (*light shade*) (adapted and revised from Dussurget, O. et al. 2004. Annu. Rev. Microbiol. 58:587–610). Panel (**B**) is showing virulence gene clusters in 9 kb pathogenicity island (PAI) and the internalin gene family. *Open circles* represent promoters

Internalin B (InlB) interacts with Met, gC1q-R, and proteoglycan receptors, and aids in *Listeria* invasion in hepatocytes and endothelial cells. Synergistic action of both InlA and InlB may be needed to achieve cell invasion to various target sites of bacterial tropism. The mannose-6-phosphate like molecule interacts with insulin-like growth factor II receptor. A virulence-associated invasion protein, Vip, is also involved in the intestinal phase of infection through its interaction with receptor Gp96, a chaperone protein located in the endoplasmic reticulum. Listeria adhesion protein (LAP) interacts with the chaperone protein heat shock protein 60 (Hsp60). Additional adhesion factors including fibronectin-binding protein (Fbp), autolysin amidase (Ami), p60, LpeA (lipoprotein promoting entry), and lipoteichoic acid (LTA) are also involved in the adhesion process.

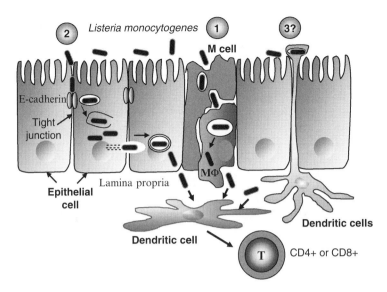

Fig. 9.3 *L. monocytogenes* translocation pathway through intestinal cell lining. Three possible pathways are proposed; (1) translocation through M-cells, (2) active invasion through epithelial cell by internalin/E-cadherin pathway, and (3) translocation by dendritic cells. The latter pathway is yet to be demonstrated experimentally (*See Color Plates*)

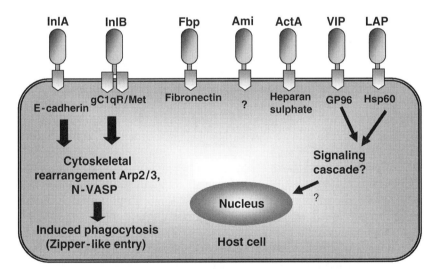

Fig. 9.4 Adhesion and invasion factors and their receptors during *Listeria monocytogenes* interaction

Internalin A

Internalin A is part of the internalin multigene family consisting of Internalin A, B, C, D, E, F, G, H, and J. The *inlA* and *inlB* are located in the same locus while the remainders are on different loci. Internalin A is 800 amino acids long (88 kDa) cell surface protein. It has a signal sequence followed by the

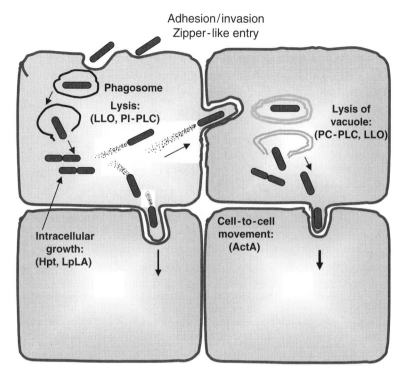

Fig. 9.5 Mechanisms of *L. monocytogenes* entry and cell-to-cell spread (redrawn from Paoli, G.C. et al. 2005. Foodborne Pathogens: Microbiology and Molecular Biology. Edited by Fratamico, P., Bhunia, A.K., and Smith, J.L., Caister Academic, Norfolk, pp 295–325)

presence of leucine-rich repeats (LRRs), which are critical for bacterial entry into eukaryotic cells. The C-terminal end contains a cell wall anchoring motif, LPXTG, which forms crosslinks with peptidoglycan. The LPXTG sequence is the substrate for the enzyme sortase and anchors InlA with the cell wall. InlA synthesis is encoded by *inlA* and is regulated at the transcriptional level by both PrfA-dependent and PrfA-independent mechanisms.

InlA helps *L. monocytogenes* invade epithelial cells following interaction with cellular receptor molecule, E-cadherin, a transmembrane glycoprotein located at the junction of epithelial cells or at the basolateral face of the epithelial cells. Thus, accessibility of InlA to E-cadherin has been an unresolved issue. However, a study suggests that the possible site of entry is at the tip of villi at the site of cell extrusion, which naturally occurs during the renewal of villous epithelial cells exposing cellular E-cadherin to InlA. InlA and E-cadherin interaction is species-specific and the binding is initiated after recognition of the proline residue in position 16 (Pro 16) in the E-cadherin of humans, guinea pigs, and rabbits. In mice, the Pro 16 is replaced with glutamic acid thus, InlA-mediated invasion does not occur in this species. InlA/E-cadherin interaction triggers recruitment of several molecules including β- and α-catenin, which in turn recruit vezatin and myosin VIII that are responsible for actin cytoskeletal rearrangement for bacterial entry. The InlA/E-cadherin pathway is associated with invasion of intestinal and placental epithelial cells and is thought to be involved in the crossing blood–brain barrier. A full-length InlA is required for pathogenicity

Fig. 9.6 Scanning electron microscopic photograph showing the early stage of *L. monocytogenes* invasion to Caco-2 cell (Courtesy of Jennifer Wampler)

and is found in most clinical isolates of *L. monocytogenes*, whereas strains that express truncated InlA due to a nonsense mutation are considered less pathogenic and are predominant among food isolates (Fig. 9.6).

Internalin B

Internalin B is a 630 amino acid polypeptide (~65 kDa) and is located on the surface of cells. It also possess N-terminal signal sequence and LRR repeats, and the C-terminal domain carries three repeats of 80 amino acid long starting with GW (Gly-Trp) repeats (called GW module). The GW module acts as an anchor and remains attached to the membrane lipoteichoic acid of the cell wall. InlB facilitates entry of *Listeria* to varieties of cultured cell lines including Vero, HEp-2, HeLa, hepatocytes (HepG2), and endothelial (HUVEC) cells. InlB binds to cellular receptor, Met, a family of tyrosine kinase. Upon binding with InlB, Met is phosphorylated, and recruits adapter proteins, Cbl, Shc, and Gab 1, which activate phosphatidylinositol (PI)-3 kinase, and a small GTPase, Rac, which in turn activates actin nucleator protein, Arp2/3 for actin polymerization, which alters cytoskeletal configuration to promote bacterial entry by the zipper mechanism (see Chap. 4, Fig. 4.2). The soy protein, genistein, a tyrosine kinase inhibitor, was able to inhibit *Listeria* entry in cells. Cytochalasin D inhibits actin recruitment and prevents bacterial entry. InlB has also been shown to interact with the globular part of 33 kDa complement protein C1q (gC1q-R/p32) and this serves as a coreceptor. GW module interacts with the gC1q-R. InlB interaction is also found to be species-specific. It does not interact with the Met receptor from guinea pig or rabbits.

LAP

Listeria adhesion protein (LAP) is an 866 amino acid polypeptide (~104 kDa) with two major domains. The N-terminal domain contains an acetaldehyde dehydrogenase and the C-terminal end consists of an iron-containing alcohol dehydrogenase domain. LAP is a bifunctional membrane bound enzyme. It does not have an N-terminal leader sequence thus a secretory pathway likely mediates its translocation to the cell surface. LAP interacts with eukaryotic receptor Hsp60, a molecular chaperone. Normally, Hsp60 regulates folding of mitochondrial proteins and proteolytic degradation of misfolded and damaged proteins in cells. LAP displays enhanced affinity for intestinal epithelial cells and plays a major role during the intestinal phase of infection. Nutrient-limiting and anaerobic environment enhance surface expression of LAP, which likely promotes the LAP-mediated translocation during the intestinal phase of infection. The *lap* gene (*lmo1634*) is the leading gene in an operon containing an unknown protein (*lmo1635*), an ATP-binding protein (*lmo1636*), and a membrane protein (*lmo1637*).

Autolysin Amidase

Autolysin amidase (Ami) is an autolytic enzyme and serves as an adhesion protein. The size of Ami varies between serotypes. In serotype 1/2a, Ami is a 917 amino acid long polypeptide, whereas Ami from serotype 4b is 770 amino acid long. It has three domains; a 30 amino acid signal sequence; N-terminal alanine amidase, homologous to Atl autolysin of *Staphylococcus aureus*, and a C-terminal cell wall anchoring (CWA) domain containing GW modules. The N-terminal amidase domain is highly conserved between the two serotypes but the C-terminal sequence is variable. Serotype 1/2a contains eight GW modules while 4b contains six GW modules. Ami uses its CWA domain to anchor to the cell wall where it exerts its autolytic action. It also promotes binding of *L. monocytogenes* to mammalian cells using the CWA domain.

p60

p60 is a major extracellular protein containing 484 amino acid residues (60 kDa) and is also known as cell wall hydrolase (CWH). It was originally reported to be an invasion-associated protein (iap) and thought to be required for invasion into mouse fibroblast cells. The gene coding for p60 is called *iap* and its expression is controlled at the posttranscriptional level. p60 synthesis and expression is not regulated by *prfA*. The *iap* gene was sequenced and showed no homology with other invasin and adhesin proteins but showed homology with a protein from *Streptococcus faecalis*, which is apparently involved in bacteriolytic or autolytic activity. p60 acts as a bacteriolytic protein with murein hydrolase activity. *L. monocytogenes* mutant lacking p60 form long chains and rough colonies. p60 is present in all *Listeria* spp., however, it is heterogeneous among the species. The 120 amino acid residues in the N- and C-terminal ends are conserved in all *Listeria* spp., whereas the middle part, comprising of 240 amino acids, are highly variable. The p60 from *L. monocytogenes* carries two unique sequence regions, which are absent in other *Listeria* spp. p60 production is significantly higher in cells when grown at 37 vs. 26 °C.

Vip

Virulence protein, Vip, has been reported as an invasion protein in *L. monocytogenes*. Vip is a surface protein carrying LPXTG motif responsible for anchoring to the cell wall peptidoglycan, an action that is mediated by the transpeptidase sortase. It contains 399 amino acids (~43 kDa) with an N-terminal signal sequence and a C-terminal sorting signal. It also has a proline-rich region (amino acids 268–318). It is regulated by *prfA* and its expression is poor in BHI at 37 °C. Vip binds to cellular receptor Gp96, a chaperone protein found in the endoplasmic reticulum (ER). Vip-mediated invasion has been seen only in select cell lines, including enterocyte like Caco-2 and mouse fibroblast L2071, but not in Vero cells. It has been suggested that Vip is responsible for invasion during intestinal phase of infection as well as in later stages of infection.

Miscellaneous Adhesion Proteins

Five fibronectin-binding proteins (Fbp) with molecular masses of 55.3, 48.6, 46.7, 42.4, and 26.8 kDa have been identified in *L. monocytogenes*. All Fbp bind to fibronectin, a 450-kDa glycoprotein, present in plasma, in extracellular fluids and on the cell surface. The high molecular weight FbpA is thought to be required for colonization of the mouse liver, and contributes to adherence to human epithelial cells. Lipoteichoic acid (LTA) is a structural component of the Gram-positive cell wall and has been shown to be involved in *L. monocytogenes* adhesion and virulence. Lipoprotein promoting entry (LpeA) is a listerial cell surface-exposed lipoprotein, which shares significant sequence similarity (~60%) to pneumococcal surface adhesin A (PsaA), a known adhesin from *Streptococcus pneumoniae*. LpeA is required for invasion of nonprofessional phagocytic cells but not macrophages.

Lysis of Vacuole (Phagosome)

After entry into the host cell, *L. monocytogenes* is trapped inside a vacuole (phagosome). Before lysosomal fusion with the vacuole can occur, the bacterium escapes by destroying the vacuole with the help of listeriolysin O and a phospholipase enzyme.

Listeriolysin

Listeriolysin O (LLO) is a sulfhydryl (SH)-activated, pore-forming hemolysin with a molecular mass of 58–60 kDa. Cholesterol serves as a receptor and about 30–40 molecules of LLO are required to lyse a single erythrocyte. LLO molecules oligomerize in the membrane forming a pore resulting in the lysis of cell. Maximum cytolytic activity of LLO is at pH 5.5, which coincides with the phagosomal pH and aids in vacuole disruption for the release of the bacterium into cytoplasm. LLO is an essential virulence factor for *L. monocytogenes*. LLO has a PEST (P, Pro; E, Glu; S, Ser; T, Thr)-like sequence, which is the target for degradation by host cell and aids bacterial persistence in the cytoplasm. LLO lacking the PEST sequence induces apoptosis and kills the host cell, thus prevent listerial cell-to-cell movement. All virulent strains are hemolytic; however, avirulent strains may also produce LLO. LLO is also reported to cause apoptosis or cell death in hepatocytes, dendritic cells, and hybrid B-lymphocytes leading to membrane damage, intracellular enzyme release, and DNA degradation.

LLO is encoded by *hly* gene, which is under direct control of *prfA*. Mutation in the *prfA* gene inactivates *hly* and other genes located downstream, such as *plcA* (PI-PLC) and *mpl* (metalloprotease), which are essential for maintenance of the virulence status of *L. monocytogenes*. LLO is a member of the cholesterol-dependent pore-forming toxin, and is related genetically and antigenically to streptolysin O from *Streptococcus pyogenes*, ivanolysin O from *L. ivanovii*, pneumolysin from *Streptococcus pneumoniae*, and seeligerolysin from *L. seeligeri*.

LLO is produced in low concentrations at about 20–200 hemolysin units (HU) ml^{-1} in conventional broth media (one HU equals 1 ng of purified toxin). However, a high level (1,500 HU ml^{-1}) can be achieved after growth in a resin (chelex)-treated broth. LLO production is also regulated by growth temperature, yielding higher amounts at 37 °C in presence of a constant glucose level of 0.2% than at 26 °C.

Phosphatidylinositol-Specific PLC

Two types of phospholipase C are produced by *L. monocytogenes*: phosphatidylinositol-specific phospholipase C (PI-PLC) and phosphatidylcholine-specific phospholipase C (PC-PLC), which are responsible for membrane disruption. PI-PLC works synergistically with LLO to destroy the lipid bilayer membrane of phagosome allowing *L. monocytogenes* escape. PI-PLC is a 33–36 kDa enzyme, encoded by the *plcA* gene which is regulated by *prfA* and is present only in *L. monocytogenes* and *L. ivanovii*. The substrate for PI-PLC is phosphatidylinositol and it has no activity on phosphatidylethanolamine, phosphatidylcholine, or phosphatidylserine. The PC-PLC aids in destruction of the double membrane vacuole during cell-to-cell spread of *Listeria*. The substrate for PC-PLC is phosphatidylcholine.

Intracellular Growth

After escape from vacuole, *L. monocytogenes* multiplies before initiating cell-to-cell movement. Glucose is the preferred carbon source and *Listeria* expresses hexose phosphate translocase (Hpt) to scavenge glucose-1-phosphate, glucose-6-phosphate, fructose-6-phosphate, and mannose-6-phosphate from the host cell cytoplasm. *Listeria* also expresses lipoate protein ligase (LpLA1) to utilize host-derived lipoic acid, a cofactor required for the function of pyruvate dehydrogenase. LpLA1 ligates exogenous lipoic acid to the E2 subunit of the pyruvate dehydrogenase to form E2-lipoamide, which is important during aerobic metabolism.

Cell-to-Cell Spread

ActA

Actin polymerization protein (ActA) initiates actin polymerization to aid bacterial movement inside the cytoplasm. ActA is a 639 amino acid containing cell surface protein of ~90 kDa, and is encoded by *actA* gene, which is under direct control of *prfA*. ActA contains three domains: the N-terminal domain interacts with Arp2/3 complex to initiate actin accumulation. The centrally located proline-rich domain interacts with the members of Enabled (Ena)/vasodilator-stimulated phosphoprotein (VASP) family proteins and helps directional actin assembly and the C-terminal domain anchors to the bacterial cell wall. The actin tail acts as a fixed platform for propulsive movement inside the host cell cytoplasm. Microfilament crosslinking proteins, such as α-actinin, fimbrin, and villin are found around the bacteria as soon as actin filaments are detected on the bacterial surface. Other proteins associated with the actin tail include tropomyosin, plastin (fimbrin), vinculin, talin, and profilin. The mutant strains with defective *actA* expression are unable to accumulate actin, and thus fail to infect adjacent cells.

Microheterogeneity in the molecular mass of the ActA polypeptide among the strains of *L. monocytogenes* serotypes has been reported. *L. ivanovii* also polymerizes actin filaments when present in cytosol and the actin synthesis is encoded by the *iactA* gene, which is homologous to the *L. monocytogenes actA* gene. The IactA is a 1,044 amino acid long polypeptide with a proline-rich repeat sequence, similar to the ActA protein.

The moving bacterium forms a pseudopod-like protruding structure, which extends into the neighboring cell. It is suggested that the protruding structure is engulfed by the neighboring cell facilitating bacterial cell-to-cell spread. Bacteria trapped in the double membrane (the outer membrane is from the newly infected cell and the inner membrane from the previously infected cell), is escaped with the aid of phosphatidylcholine-specific PLC (PC-PLC) and LLO (Fig. 9.5).

Phosphatidylcholine-Specific PLC

PC-PLC is a 29-kDa enzyme that exhibits both lecithinase and sphingomyelinase activity. The enzyme requires zinc as a cofactor and is active at a pH range of 6–7. PC-PLC has weak hemolytic activity but does not lyse sheep red blood cells because these cells are devoid of phosphatidylcholine. The *plcB* gene encodes a 33-kDa proenzyme precursor protein, which is processed to form a mature PC-PLC (29 kDa) with the help of a zinc metalloprotease. Zinc metalloprotease is encoded by *mpl*, a proximal gene in the lecithinase operon. Mutation of *mpl* causes reduced virulence of *L. monocytogenes* in mice. The *plcB* gene is under the direct control of *prfA*.

Regulation of Virulence Genes

As mentioned above, several virulence genes (*prfA–plcA–hly–mpl–actA–plcB*) are clustered in a 9-kb pathogenicity island and these genes are regulated by the first gene, *prfA* (Fig. 9.2). The PrfA protein binds to a palindromic *prfA* recognition sequence (*prfA*-box; -TTAACANNTGTTAA-) located at position – 40 from the transcription start site to initiate the transcription of virulence genes. Three promoters contribute to the *prfA* expression. The two promoters *prfAp1* and *prfAp2* are located immediately upstream of the *prfA* coding region. The third promoter is located upstream of *plcA* gene. PrfA positively regulates its own expression through the activation of *plcA* transcription. All *prfA* regulated virulence genes contain the PrfA-box sequence. PrfA expression is dependent on the growth phase and temperature and it is controlled at the transcription, translation or posttranslation level. During transcription, bicistronic transcripts of *plcA* and *prfA* form loops which increases *prfA* transcription. One *prfA* promoter, *prfA2*, is controlled by the alternative sigma factor B.

Readily metabolizable sugars, like glucose, cellobiose, fructose, mannose, trehalose β-glucosides, and maltose negatively regulate the *prfA*-dependent genes. Such negative regulation is possibly due to the presence of a common catabolite repression (CR) or spontaneous mutation or another molecule responding specifically to β-glucosides. Surprisingly, the expression of PrfA is induced in the presence of cellobiose, suggesting a posttranscriptional modification of PrfA. Addition of charcoal in the media could overcome repression by sugars since charcoal can sequester a diffusible autorepressor substance that is released by *L. monocytogenes.*

Stress response genes such as those necessary for growth during osmotic stress, low pH stress, growth at low temperature, and carbon starvation are regulated by sigma B. Sigma B has been shown to regulate 55 genes including several virulence genes that are necessary during the intestinal phase of infection such as *bsh, gadA, opuCA, inlA,* and *prfA.*

Immunity to *Listeria monocytogenes*

Innate and cell-mediated immunity are most effective against *L. monocytogenes.* During infection, 90% of *L. monocytogenes* cells arrive in the liver and 10% in the spleen within 24–48 h. In the liver, neutrophil influx is high and recognizes infected hepatocytes. Neutrophils, Kupffer cells, and macrophages are responsible for initial control of infection. Cytokines such as TNF-α, IL-1, IFN-β are produced by macrophages, Kupffer cells, to activate macrophages for clearance of bacteria. Complement is activated by the alternative pathway and allows C3b-mediated phagocytosis of *L. monocytogenes.* Cell-mediated immunity involving CD8$^+$ T-cells and activated macrophages provide the best protection against listeriosis. CD8$^+$ T-cell recognizes *Listeria* antigens such as LLO and p60 via MHC class I when presented by infected target cells. In addition, IFN-γ, IL-6, IL-10 produced by CD4$^+$ and CD8$^+$ cells activate macrophages for increased phagocytosis.

Prevention and Control

In 2004, an *L. monocytogenes* risk analysis study conducted jointly by the United States Food and Drug Administration (US-FDA) and the U.S. Department of Agriculture (USDA) grouped various foods into risk categories to provide a warning system and to educate consumers about the risk levels associated with various ready-to-eat foods. The very high-risk food groups include deli meats and frankfurters (unheated). The high-risk food groups contain high fat dairy products, soft unripened cheese, unpasteurized fluid milk, pate and meat spreads, and smoked seafood. Medium-risk products are pasteurized fluid milk, fresh soft cheese, semisoft cheese, soft ripened cheese, cooked RTE crustaceans, deli salads, dry/semidry fermented sausages, and frankfurters (reheated), fruits, and vegetables. The very low-risk products are cultured milk products, hard cheese, ice cream and frozen dairy products, and processed cheese. The persons belonging to high-risk group such as immunocompromised populations or the pregnant women should avoid foods that belong to very high to medium categories food to prevent contraction of listeriosis.

Implementation of Hazard Analysis and Critical Control Point (HACCP) strategies are mandatory for all processing plants that produce ready-to-eat products to reduce or eliminate *L. monocytogenes* from food processing facilities. If a product has been implicated for an outbreak, it is recalled to prevent further incidences. As a precaution, producers are also asked to recall products if any of their products are found to carry *L. monocytogenes* after retail distribution.

Pasteurization or heating of processed foods before consumption has been suggested for certain products. Heating of food at 71 °C for 1 min will kill the bacteria. Improved control of *Listeria* in farm, animal feed, diary cows, silage, and in the processing plant are needed to prevent *Listeria* in products. Routine surveillance of production facilities, implementation of an effective sanitization scheme, and improved product formulation to include antimicrobial inhibitors are some of the best practices to reduce *Listeria* in products.

Antibiotic therapy with gentamicin, ampicillin, amoxicillin, and penicillin G are effective in treating systemic listeriosis. Supplemental therapy consisting of fluid and electrolytes are needed during gastrointestinal forms of listeriosis.

Summary

Listeria monocytogenes is an opportunistic intracellular pathogen. Historically, it has been considered as the causative agent for listeriosis in animals; however, in recent years it is responsible for fatal foodborne diseases in immunocompromised individuals. It is one of the leading foodborne pathogens, and has been implicated in numerous outbreaks in the last three decades. *L. monocytogenes* belonging to lineage I and II are thought to be the most virulent and are generally associated with outbreaks. It causes three forms of disease: gastroenteritis in healthy adults, systemic listeriosis in immunocompromised populations, and abortion and neonatal listeriosis in pregnant mothers and their fetuses. The intestinal phase of infection is a complex process and the mechanism is not fully understood. Similarly, transmission to the brain or to the fetus is less clear. Virulence factors such as GAD, BSH, OpuC, LAP, Vip, and InlA/InlB are important during intestinal phase of infection. *L. monocytogenes* entry into mammalian cells, survival inside the phagosome, escape into the cytoplasm, growth, and cell-to-cell spread are well understood, which are orchestrated by numerous virulence factors such as internalin B (InlB), listeriolysin O (LLO), actin polymerization protein (ActA), phospholipase (PLC), metalloprotease (Mpl), hexose phosphate transport permease (Hpt), and lipoprotein ligase (LpL). Immune response to *L. monocytogenes* is largely dependent on the innate immunity involving neutrophil and macrophages and cell-mediated immunity involving CD8$^+$ T-cell subsets. Humoral immunity possibly has limited or no role in immunity. Risk analysis study has identified several foods to be of high-risk category: hotdogs, sliced deli meats, soft cheeses especially those made with unpasteurized milk and pâté and meat spread, and smoked seafoods. Immunocompromised or high-risk individuals, especially those who are pregnant should avoid these foods.

Further Readings

1. Bakardjiev, A.I., Theriot, J.A., and Portnoy, D.A. 2006. *Listeria monocytogenes* traffics from maternal organs to the placenta and back. PLoS Pathogens 2:623–631.
2. Bhunia, A.K. 1997. Antibodies to *Listeria monocytogenes*. Crit. Rev. Microbiol. 23:77–107.
3. Donnelly, C.W. 2001. *Listeria monocytogenes*: a continuing challenge. Nutr. Rev. 59:183–194.
4. Dussurget, O., Pizarro-Cerda, J., and Cossart, P. 2004. Molecular determinants of *Listeria monocytogenes* virulence. Annu. Rev. Microbiol. 58:587–610.
5. Farber, J.M. and Peterkin, P.I. 1991. *Listeria monocytogenes*, a food-borne pathogen. Microbiol. Rev. 55:476–511.
6. Gahan, C.G.M. and Hill, C. 2005. Gastrointestinal phase of *Listeria monocytogenes* infection. J. Appl. Microbiol. 98:1345–1353.
7. Hamon, M.I., Bierne, H., and Cossart, P. 2006. *Listeria monocytogenes*: a multifaceted model. Nat. Rev. Microbiol. 4:423–434.
8. Kazmierczak, M.J., Wiedmann, M., and Boor, K.J. 2005. Alternative sigma factors and their roles in bacterial virulence. Microbiol. Mol. Biol. Rev. 69:527–543.
9. Lecuit, M., Nelson, D.M., Smith, S.D., Khun, H., Huerre, M., Vacher-Lavenu, M.-C. et al. 2004. Targeting and crossing of the human maternofetal barrier by *Listeria monocytogenes*: role of internalin interaction with trophoblast E-cadherin. Proc. Natl Acad. Sci. USA 101:6152–6157.
10. Monnier, A.L., Join-Lambert, O.F., Jaubert, F., Berche, P., and Kayal, S. 2006. Invasion of the placenta during murine listeriosis. Infect. Immun. 74:663–672.
11. Ooi, S.T. and Lorber, B. 2005. Gastroenteritis due to *Listeria monocytogenes*. Clin. Infect. Dis. 40:1327–1332.
12. Paoli, G.C., Bhunia, A.K., and Bayles, D.O. 2005. *Listeria monocytogenes*. In Foodborne Pathogens: Microbiology and Molecular Biology. Edited by Fratamico, P., Bhunia, A.K., and Smith, J.L., Caister Academic, Norfolk, pp 295–325.
13. Pizarro-Cerda, J. and Cossart, P. 2006. Subversion of cellular functions by *Listeria monocytogenes*. J. Pathol. 208:215–223.
14. Roberts, A.J. and Wiedmann, M. 2003. Pathogen, host and environmental factors contributing to the pathogenesis of listeriosis. Cell. Mol. Life Sci. 60:904–918.
15. Vazquez-Boland, J.A., Kuhn, M., Berche, P., Chakraborty, T., Dominguez-Bernal, G., Goebel, W. et al. 2001. *Listeria* pathogenesis and molecular virulence determinants. Clin. Microbiol. Rev. 14:584–640.

10

Escherichia coli

Introduction

Theodor Escherich, a German bacteriologist reported the isolation of a bacterium called *Bacteria coli* from a fecal sample in 1885. Later in 1888, it was renamed *Escherichia coli*. *E. coli* is a Gram-negative, short-rods (1–2 μm in length), aerobic, and generally motile organism. Some pathogenic strains are also acid resistant. A majority of *E. coli* strains are nonpathogenic and exist harmlessly in the intestinal tract of humans and animals. *E. coli* has been used extensively as a model organism to study bacterial physiology, metabolism, genetic regulation, signal transduction, and cell wall structure and function. Pathogenic *E. coli* strains cause a variety of diseases including gastroenteritis, dysentery, hemolytic uremic syndrome (HUS), urinary tract infection (UTI), septicemia, pneumonia, and meningitis. However, the major concern in recent years has been the increasing numbers of outbreaks of enterohemorrhagic *E. coli*, due to consumption of contaminated meat, fruits, and vegetables primarily in developing countries.

Sources

E. coli belongs to the family of *Enterobacteriaceae* and is common in the intestinal microflora of warm-blooded animals. They are routinely shed into the environment through feces and can contaminate water and soil, and, consequently fruits and vegetables, especially if untreated manures are used as fertilizers. Contaminated fruits and vegetables may even harbor some inside cells of plant tissues. Meats are also a common source of *E. coli* contamination, which may be acquired during slaughter through fecal contact. *E. coli* outbreaks have been associated with meat (especially ground beef), dairy products, mayonnaise, apple cider, sprouts (radish), lettuce, and spinach. *E. coli* outbreak have also been associated with swimming pools and nursing schools.

Classification

Serotypes

"O" antigens are one class of *E. coli* serogroup determinants, consisting of lipopolysaccharide (LPS), and there are 174 O antigens, numbered 1–181, with numbers 31, 47, 67, 72, 93, 94, and 122 omitted. In addition, there are 53 serotypes of "H" or flagellar antigens (H1–H53). Strains that lack flagella are nonmotile (NM). *E. coli* isolates can have a variety of antigen combinations. Thirty serovars are reported to be responsible for diarrheal diseases and the first serogroup identified was O111 and was isolated from a child. "O" antigen identifies serogroup, while "H" identifies serotype. When two strains designated as O111:H4 and O111:H12 means both belong to the same serogroup but different serotypes. There are also 80 capsular antigen or "K" antigen serotypes.

Virotypes

Virotype or pathotype classification is based on the presence of certain virulence factors and their interaction pattern with mammalian cells or tissues such as adhesion to, and/or invasion of mammalian cells, and toxin production (Fig. 10.1). Pathogenic *E. coli* are classified in six virotypes (1) enterotoxigenic *E. coli* (ETEC), (2) enteropathogenic *E. coli* (EPEC), (3) enterohemorrhagic *E. coli* (EHEC), (4) enteroinvasive *E. coli* (EIEC), (5) enteroaggregative *E. coli* (EAEC), and (6) diffusely adhering *E. coli* (DAEC):

1. Enterotoxigenic *E. coli* (ETEC) cells adhere to epithelial cells and produce toxin but do not invade epithelial cells. The predominant serogroups are O6, O8, O11, O15, O20, O25, O27, O78, O128, O148, O149, O159, and O173.
2. Enteropathogenic *E. coli* (EPEC) cells adhere to epithelial cells intimately, produce attachment/effacement lesion and are invasive; however, they do not produce toxin. The notable serogroups are O55, O86, O111, O119, O125, O126, O127, O128, and O142.
3. Enterohemorrhagic *E. coli* (EHEC) also bind strongly to epithelial cells, produce attachment/effacement lesions and produce toxins. The serogroups are O4, O5, O16, O26, O55, O111ab, O113, O117, O157, and O172. Several recently identified serogroups belong to EHEC include O176, O177, O178, O180, and O181.
4. Enteroaggregative *E. coli* (EAEC) adhere to epithelial cells, form aggregates, produce toxin, but do not invade. The serogroups in this virotype are O3, O15, O44, O86, O77, O111, and O127.
5. Enteroinvasive *E. coli* (EIEC) cells also adhere, invade cells, and move from cell-to-cell, but do not produce toxin. The pathogenicity of EIEC resembles infection caused by *Shigella* spp. and the predominant symptom is dysentery. The EIEC serogroups are O28, O29, O112, O124, O136, O143, O144, O152, O159, O164, and O167.
6. The diffusely adhering *E. coli* (DAEC) cells adhere to epithelial cells, but they neither invade nor produce toxin. Details of each virotype are described below.

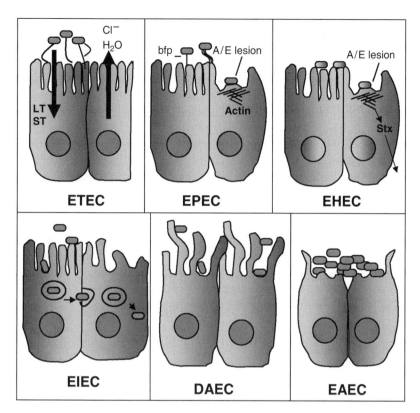

Fig. 10.1 Virotype classifications of *Escherichia coli* based on their interaction with intestinal villous epithelial cells: enterotoxigenic *E. coli* (ETEC), enteropathogenic *E. coli* (EPEC), enterohemorrhagic *E. coli* (EHEC), enteroinvasive *E. coli* (EIEC), diffusely adhering *E. coli* (DAEC), and enteroaggregative *E. coli* (EAEC). *LT* heat-labile toxin, *ST* heat-stable toxin, *bfp* bundle-forming pili, *A/E lesion* attachment/effacement lesion showing actin accumulation, *Stx* Shiga-like toxin

Enterotoxigenic *E. coli*

ETEC was first reported to cause cholera-like illness in adults and children in Calcutta (India) in 1956. ETEC commonly causes infectious diarrhea in people living in the tropical climate and children under the age of two are highly susceptible. Infants in many developing countries suffer from ETEC due to poor sanitary conditions. It also affects travelers since the water and food consumed by them may be contaminated with ETEC. Native people may be resistant because of their frequent exposure to the organism. It is also known as "Traveler's Diarrhea" or "Montezumas Revenge," or "Delhi Belly." ETEC adheres to the mucosal epithelial cells and produces heat-labile toxin (LT) or heat-stable (ST) diarrheal toxins. LT resembles cholera toxin and produces symptoms similar to *Vibrio cholerae*. Infection does not cause any apparent histological changes in the mucosal layers and there is little or no inflammation in the intestine. Symptoms may include watery diarrhea, vomiting, sunken eyes, massive dehydration, and a collapse of circulatory system. The diarrhea lasts for 3–4 days and is self-limited. Diarrhea may be lethal in young children and infants, with a mortality rate of less than 1%.

Virulence Factors and Pathogenesis

Adhesion Factors

ETEC expresses more than 22 colonization factors (CFs), the genes for which are located on the plasmid or chromosome. CFs are proteinaceous fimbrial and afimbrial (fibrillar) structures that allow bacteria to attach to intestinal mucosa. Fimbrial colonization factors are also known as colonization factor antigens (CFA), and three types are reported: CFA/I, CFA/II, and CFA/III. CFA/I is rigid rod shaped; CFA/II is a flexible fimbriae present alone or in association with other rod-shaped fimbriae; and CFA/III is a bundle-forming flexible pilus, also called longus, because of its unusual length (about 40 μm long). One afimbrial colonization factor, CFA/IV is present in ETEC. The tip of CFA is hydrophobic and it is responsible for adhesion to epithelial cells.

Other adhesion factors include TibA, a 104 kDa afimbrial adhesion factor, encoded on the chromosome. TibA aids in bacterial aggregation on epithelial cells and promotes biofilm formation. Tia, a 25 kDa OM (outer membrane) protein encoded in a 46 kb pathogenicity island, is involved in adhesion as well as in invasion. A Tia homolog has been found in other *E. coli* (EPEC and EAEC), suggesting it has a broader role in pathogenesis.

Toxins

ETEC produces two different types of toxins (1) heat-labile toxin (LT; LT-I and LT-II) and (2) heat-stable toxin (ST; STa and STb).

Heat-Labile Toxins (LT-I and LT-II)

LT synthesis is encoded by the *eltAB* operon, and the genetic organization of LT-I is similar to cholera toxin. LT-I is expressed in *E. coli* that cause disease in both humans and animals, while LT-II is found in mainly animal isolates. LT-II has same basic structure and mechanism of action as LT-I. LT-I is an A–B type toxin with molecular mass of 87.5 kDa. The A-subunit is 27 kDa while the B-subunit consists of five identical subunits of 11.7 kDa, and the subunits are arranged in a ring. This $A–B_5$ complex is also called holotoxin, and the amino acid sequence is similar to that of cholera toxin.

After synthesis, the toxin remains associated with the LPS in the OM. During infection of host cell, the B-subunit binds a GTP-binding protein, ganglioside (GM_1) receptor in the epithelial cells (cholera toxin interacts with the same receptor and acts in the same manner; see Chap. 14). The A-subunit then causes ADP-ribosylation of "ganglioside" (Gs) protein (and activates adenylate cyclase resulting in an increase in cyclic adenosine monophosphate (cAMP) level in the cytoplasm. cAMP then activates cAMP-dependent protein kinase A which in turn cause phosphorylation of CFTR (cystic fibrosis transmembrane conductance regulator), a chloride ion transporter protein, increases Cl^- secretion in crypt cells, and decreases absorption of Na^+ and Cl^- by absorptive cells (Fig. 10.2). In addition, the A-subunit is involved in arachidonic acid metabolism leading to the formation of prostaglandin E_2 (PGE_2) and 5-hydroxytryptamine (5-HT), which mediate electrolyte and water release from intestinal cells. LT-II action is similar to LT-I with an exception where it binds to ganglioside GD1 receptor instead of GM_1.

Heat-Stable Toxin (ST)

ST consists of family of small peptide toxin of 2 kDa, which are stable at 100 °C for 30 min. Two major types of ST, (1) STa (STh), methanol soluble

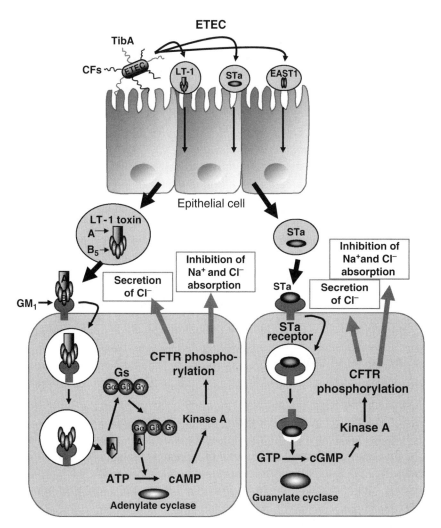

Fig. 10.2 Mechanism of action for enterotoxigenic *E. coli* (ETEC)-mediated diarrhea. After arrival in the intestine, ETEC binds to the epithelial cells using colonization factors (CFs) and/or TibA and produce several toxins including LT-I, LT-II, STa, STb, EAST1, and ClyA. Mechanism of action of LT-I and STa are presented. LT increases cAMP level while ST increases cGMP and both mediate phosphorylation of CFTR (cystic fibrosis transmembrane conductance regulator), a chloride ion transporter protein, which increases Cl$^-$ secretion in crypt cells, and decreases absorption of Na$^+$ and Cl$^-$ by absorptive cells

isolated from human and (2) STb (STp), a methanol insoluble toxin isolated from pig, are found in human ETEC. ST is produced as a 71 amino acid long precursor protein, which is stored in the periplasm. After 18 amino acids are removed, a 54 amino acid long peptide is secreted, and the peptide is further cleaved to a 17–19 amino acid long active peptide with a molecular mass of approximately 2 kDa. STh binds guanylate cyclase C and activates cGMP in the host cell. The cGMP is a signaling molecule and changes in GMP affects ion pumps, inducing increased chloride ion secretion, resulting in increased fluid and ion losses (Fig. 10.2).

STb is found only in porcine ETEC strains; however, some human ETEC may produce STb. STb has no sequence homology with STa and the receptor for STb is sulfatide, a widely distributed glycosphingolipid. After endocyosis, STb activates GTP-binding regulatory protein, which increases the efflux of Ca^{2+}, opens ion channels and activate protein kinase C. Increased Ca^{2+} levels regulate phospholipases (A2 and C) that release arachidonic acid from membrane phospholipids, leading to the formation of prostaglandin E_2 (PGE_2) and 5-HT, which mediate electrolyte and water release from intestinal cells. Unlike STa, STb also stimulates the secretion of bicarbonate from epithelial cells.

Other Toxins

ETEC also secretes EAST1, a novel heat-stable enterotoxin consists of 38 amino acids, which has been isolated from ETEC strains of human and animal origin. It activates cGMP, induces fluid accumulation, and possibly plays a role on the onset of diarrhea. In addition, ETEC also produces EatA, a serine protease autotransporter, which plays a role in pathogenesis by damaging the epithelial cell surface. ClyA (HlyA) is a pore-forming toxin that has been isolated from some ETEC strains. It binds to cholesterol on the membrane and shows lytic activity against erythrocytes, macrophages, and HeLa cells.

Enteropathogenic *E. coli*

Enteropathogenic *E. coli* (EPEC) was the first virotype of *E. coli* to be described and primarily affects children. EPEC is also pathogenic to calves, pigs, rabbits, and dogs. It adheres intimately to epithelial cells and exhibits a "patchy" pattern of adherence. Adherence causes a dramatic change in the ultrastructure of epithelial cells, resulting in the formation of an "attaching and effacing" (EAF) lesion, which is characterized by the formation of a "cup-like" or "pedestal" structures due to the extensive cellular actin rearrangements in the architecture. Microvilli structures gradually disappear and they loose the ability to absorb nutrients. EPEC are highly invasive and cause an inflammatory response and potentially fatal diarrhea in children and infants.

Pathogenesis of EPEC

EPEC interaction with host cells occurs in four stages: expression of adhesion factors, initial localized adherence, signal transduction and intimate contact, and cytoskeletal rearrangement and pedestal formation (Fig. 10.3).

Expression of Adhesion Factors
Initially bacteria bind to epithelial cells but the adhesion is nonintimate. Adhesion experiments conducted with an HEp-2 cell line or in a gnotobiotic pig model demonstrated that the adhesion is mediated by the type IV adhesion fimbriae called bundle-forming pili (BFP) or EPEC adherence factor (EAF), which is similar to Tcp pili of *V. cholerae*. BFP expression is encoded by *bfpA*, which is located on a 92-kb (60 MDa) plasmid. In addition, the *dsbA* gene product, DsbA, is a periplasmic enzyme that mediates the disulfide bond formation between proteins that are involved in localized adherence. EPEC also produce Intimin and a short, surface-associated filament, EspA.

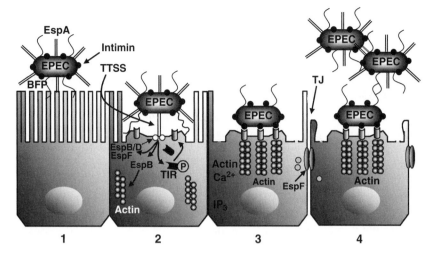

Fig. 10.3 Schematic diagram showing the sequence of events for enteropathogenic *E. coli* (EPEC) on enterocytes during infection. Pathogenic event can be grouped into four stages (1) expression of adhesion factors, (2) initial localized adherence, (3) signal transduction and intimate contact, and (4) cytoskeletal rearrangement and pedestal formation (adapted from Clarke, S.C. et al. 2003. Clin. Microbiol. Rev. 16:365–378; Hayward, R.D. et al. 2006. Nat. Rev. Microbiol. 4:358–370)

Localized Adherence

EPEC adhere to epithelial cells using BFP, Intimin, and EspA, and the type III secretion system (TTSS) injects translocated intimin receptor (TIR) and several effector molecules (EspB, EspD, EspG, EspF, EspH), directly into the host cell. The effector molecules activate cell signaling pathways, allowing actin polymerization and depolymerization to alter cytoskeletal structure. TIR is then phosphorylated by protein kinases and inserted into the host cell membrane to facilitate binding with bacterial Intimin (Figs. 10.3 and 10.5).

Signal Transduction and Intimate Contact

The attachment of EPEC to mammalian cells triggers host cell signal transduction pathways that activate host cell "tyrosine kinase," which causes the release of two signaling molecules: inositol triphosphate (IP_3) and intracellular Ca^{2+}. This activates calcium-dependent "actin depolymerization enzyme" and triggers phosphorylation of host cell proteins–myosin light chains and 90 kDa epithelial Hp90 proteins. Intracellular Ca^{2+} can inhibit Na^+ and Cl^- absorption and stimulate Cl^- secretion. Extensive rearrangement of actin causes abnormalities in cytoskeletal structure, resulting in the formation of the characteristic attaching, and effacing (A/E) lesions, regulated by the genes located in the 35.6-kb pathogenicity island called locus of enterocyte effacement (LEE). Effacement refers to the loss of microvilli. The membrane beneath the bacteria forms a pedestal, a pseudopod-like structure, due to the massive host cell cytoskeletal rearrangements. Cytoskeletal components such as actin filament and its crosslinker, talin and ezrin, as well as α-actinin, and myosin light chain accumulate in the epithelial cell beneath the bacteria. Arp2/3 complex accelerates polymerization of G-actin monomers into filaments and crosslinks with actin. Arp2/3 function is regulated by WASP (Wiskott–Aldrich syndrome protein). In this phase, bacteria lose EspA filaments from the surface.

Intimate contact is mediated by adhesion of Intimin (EaeA), a 94 kDa outer membrane protein, to the TIR receptor (78 kDa) on the host cell membrane. Intimin is encoded by the *eaeA* gene, located in LEE (see below). There are 17 Intimin types based on the heterogeneity in C-terminal sequence of the protein.

Cytoskeletal Rearrangement and Pedestal Formation

Intimate contact results in the formation of a "pedestal-like" structure and attachment and effacement lesion due to massive accumulation of actin filaments. The lesion is further characterized by the deformation and loss of microvilli due to the depolymerization of actin filament in microvilli. The effector protein, EspF translocated through TTSS also affects tight junction (TJ) proteins and mitochondrial function and increases membrane permeability. As a result, there is a malabsorption of nutrients and ions, cell death, and the onset of osmotic diarrhea.

LEE and Regulation of Virulence Genes

Virulence factors for EPEC pathogenesis are located primarily in 34-kb LEE pathogenicity island. LEE is integrated in the chromosome in the vicinity of tRNA gene, *selC*. The LEE has five polycistronic operons (LEE1–LEE5), which contain genes for A/E lesion, Intimin, TIR, the TTSS (molecular syringe), and the exported proteins. The genes for the TTSS are encoded by *espA, espB*, and *espD*, which are part of the translocation apparatus. LEE is also present in EHEC and the organization of genes is similar but the size and the number of genes may vary.

EPEC virulence factors are regulated by the *per* region located on the large plasmid. Expression of A/E activity and production of BFP is optimal at 37 °C during growth at exponential phase.

Enterohemorrhagic *E. coli*

Biology

E. coli belonging to this group cause bloody diarrhea and hemolytic uremic syndrome (HUS), and the disease is prevalent in developed countries. The pathogenesis is attributed to the production of lethal Shiga-like toxins (Stxs), thus EHEC group is also referred as STEC (Shiga toxin-producing *E. coli*). Since Shiga toxin kills Vero cells, this toxin is also referred to as verotoxin and the verotoxin-producing *E. coli* is called VTEC. STEC has been divided into five seropathotypes (A–E), based on the type of outbreaks and the severity of infection (Table 10.1). EHEC is a subset of STEC that carry LEE and display attaching and effacing activity. EHEC does not express BFP, instead the EHEC plasmid carries a homologue of the *lifA* gene encoding lymphostatin, as well as genes encoding TTSS, catalase peroxidase, serine protease, and hemolysin. The principal serotype associated with EHEC group is *E. coli* O157:H7, and the first outbreak caused by this pathogen was reported in 1982–1983.

As opposed to other commensal strains, *E. coli* O157:H7 generally does not ferment sorbitol and does not have β-glucuronidase activity (GUD). It grows rapidly at 30–42 °C, grows poorly at 44–45 °C, and does not grow at 10 °C or below. Strains resistant to pH 4.5 or below (pH 3.6–3.9) have been identified.

Table 10.1 Seropathotypes of STEC

Seropathotypes	Representative serotypes	Frequency of infection	Involvement in outbreaks	HUS and HC
A	O157:H7, O157:NM	High	Common	+
B	O26:H11, O103:H2, O111:NM, O121: H19, etc.	Moderate	Uncommon	+
C	O5:NM, O91:H21, O104:H21, O113: H21, etc.	Low	Rare	+
D	O7:H4, O69:H11, O103:H25, O113: H4, etc.	Low	Rare	−
E	O6:H34, O8:H11, O39:H49, O46:H38, O76:H7, etc.	Not implicated	Not implicated	−

Adapted from Karmali, M.A. et al. 2003. J. Clin. Microbiol. 41:4930–4940

The organism is destroyed by pasteurization temperature and time and killed at 64.3 °C in 9.6 s but the cells survive well in food at −20 °C.

Food Association

The pathogen is generally present in the intestine of animals, particularly in cattle, without causing disease. Stx-producing *E. coli* also have been isolated from the feces of chicken, goats, sheep, pigs, dogs, cats, and sea gulls. Food of animal origin, especially ground beef, has been implicated in many outbreaks in the U.S., Europe, and Canada; however, in late 2006, a major outbreak involving 26 states was associated with spinach and lettuce. The affected people (199 with 3 deaths) were found to have consumed spinach and lettuce in salad or improperly cooked hamburgers. In a 1993 outbreak, affecting over 500 people and causing 4 deaths, consumption of undercooked hamburgers served by a fast-food chain in Washington, Nevada, Oregon, and California was implicated. In addition to ground beef, other foods, such as raw milk, mayonnaise, apple cider, some fruits, uncooked sausages, fermented hard salami, sprouts, and salad have been implicated in outbreaks. EHEC has been routinely isolated from many different types of foods of animal origin, such as ground beef, pork, poultry, lamb, and raw milk (Fig. 10.4). The organism also was isolated in low frequencies from dairy cows, calves, and chickens. However, in some cases, a high percent of feedlot cattle shed *E. coli* O157:H7, some of which are persistent shedders. Interestingly, calves after weaning shed in higher frequency than before weaning.

EHEC Pathogenesis

Colonization

EHEC reach the intestine from contaminated food or water and colonize the intestine. As *E. coli* O157:H7 is acid resistant, it can pass through the stomach unharmed and reach the small intestine, a small infectious dose of 50–100

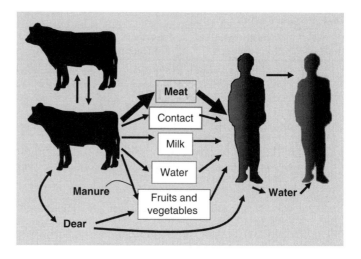

Fig. 10.4 Mode of transmission of *Escherichia coli* O157:H7 to humans. The primary vehicle of transmission is meat but the bacteria can be transported via animal-to-person contact, milk, contaminated water, and fruits and vegetables contaminated with cow manure. Bacteria can move from cow-to-cow and also from wild animals such as deer, caribou, and domestic sheep. Person-to-person transmission occurs directly or via contaminated water such as in the swimming pool

cells is sufficient to cause infection. In addition, pre-exposure of cells to mild acid, as with acidic foods, such as apple cider or fermented hard salami, bacterium becomes more resistant to low pH and ensure better survival during transit through the stomach.

Attachment and Effacement

EHEC causes characteristic attaching and effacing lesion similar to EPEC and occurs in three stages: localized adherence, signaling event, and intimate contact. In the first stage, adhesion of bacteria to the microvilli of the intestinal epithelial cells is mediated by fimbriae (encoded on the 60 MDa plasmid). Unlike EPEC, EHEC does not express BFP. In the second stage, a signal is transmitted to the host cell via TTSS (Fig. 10.5) and phosphorylation of eukaryotic protein occurs, leading to actin polymerization, cytoskeletal rearrangement, and effacement of microvilli. In the third stage, intimate contact is mediated by Intimin (EaeA), a 94 kDa protein encoded by *eaeA* gene located on the LEE pathogenicity island, similar to EPEC. There are 17 Intimin types, based on the heterogeneity in the C-terminal sequence of the protein, and γ1 Intimin is associated with the highly pathogenic STEC strains. Intimin binds to a TIR receptor, and subsequently signaling events amplify the cytoskeletal rearrangement of proteins beneath adherent bacteria. Increased actin filament accumulations are mediated by Arp2/3 complex, which is regulated by N-WASP, and form pedestal with the loss of microvilli structure. A/E pathogenesis causes enterocyte sloughing, inflammation, and possibly diarrhea, which may result from the inhibition of sodium and chloride absorption, activation of chloride channel, loosening of tight junction, increased paracellular permeability, inflammatory response, and cytokine production.

Type III Secretion System and Delivery of Effector Proteins During EHEC Pathogenesis

TTSS plays a crucial role during EHEC and EPEC pathogenesis and it delivers virulence effector proteins directly inside the host cell cytoplasm that are responsible for A/E lesion (Fig. 10.5). The TTSS needle complex is composed of several Esc proteins (EscN, EscR-V, Esc J, EscC, EscF, and EspA) which spans from bacterial cytoplasmic membrane (CM) to outer membrane (OM). The TTSS injects effectors proteins known as Esp (*E. coli* secreted protein) those perform various functions: EspB and EspD form a plasma membrane translocon for effective delivery of effector proteins. EspB also affects cytoskeletal structure by disrupting actin cytoskeleton. Another effector protein, EspH also promotes disruption of actin cytoskeleton. EspG and EspG2 disrupt microtubule and also activate a small GTPase protein Rho. EspF causes membrane disruption in mitochondria, disrupts tight junction proteins (TJ), and causes increased membrane permeability. Bacteria also inject translocated Intimin receptor (TIR; also known as EspE in EHEC) binds to Intimin. After translocation into the host cell, TIR is phosphorylated by host

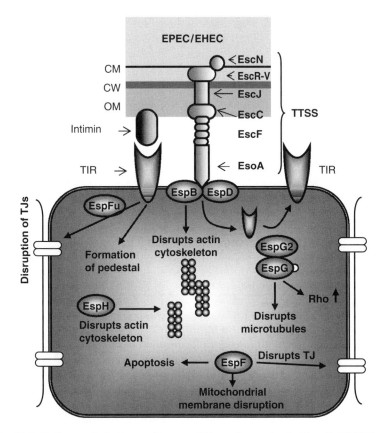

Fig. 10.5 Delivery of virulence effectors of Enterohemorrhagic *E. coli* (EHEC) and enteropathogenic *E. coli* (EPEC) to the host cell by the type III secretion system (TTSS) (figure adapted and redrawn from Hayward, R.D. et al. 2006. Nat. Rev. Microbiol. 4:358–370)

cell enzyme protein kinase A and is integrated into the host cell membrane for interaction with Intimin. As a result, actin accumulation takes place aiding pedestal formation.

Shiga-Like Toxin

STEC produce two types of Shiga-like toxins: Stx1 and Stx2. The Stx1 sequence is highly conserved and exhibits a high sequence similarity to Stx produced by *Shigella dysenteriae* type 1 (see Chap. 15). Antibody developed against the Stx toxin from *S. dysenteriae* type 1 can neutralize Stx1 from STEC but not the Stx2. Recently, a variant of Stx1, called Stx1c is reported. The Stx2 is less related to Stx1 and has several variants; Stx2c, Stx2d, Stx2e, and Stx2f (Table 10.2). Stx2 is highly toxic and has the greatest risk of developing HUS. In fact, an analysis of recent outbreaks of *E. coli* O157:H7 indicate that all the strains involved were positive for Stx2 and this toxin is thought to be essential for STEC (EHEC) pathogenesis. The genes for Stx production are encoded in a temperate bacteriophage related to classic λ phage. The *stx* phages can be induced to enter into lytic cycle, and the resulting free phages can transfer *stx* genes horizontally to *E. coli* or other members of *Enterobacteriace* family.

The Stx molecules are A–B$_5$ heterohexamer toxins of 70 kDa, in which the A-subunit is about 32 kDa and the B-subunit is 7.7 kDa each. Both toxins are called holotoxins, since a single enzymatic A-subunit is present in a non-covalent association with a pentamer of the receptor-binding B-subunits. Both toxins bind to the same receptor, globotriaosylceramide (Gb3), a glycolipid consisting of galactose α (1–4) galactose β (1–4) glucose ceramide and is abundant in the endothelium. Stx1 causes localized damage to the colonic epithelium because of high-binding affinity to Gb3, whereas Stx2 has low affinity for Gb3. Stx1 and Stx2 can reach the circulatory system and the kidneys. Human kidney tubules have high Gb3 and are the major target for toxin-induced damage resulting in characteristic HUS. Other Stx variants, Stx2e and Stx2f, use Gb4 (globotetraosylceramide) as the preferred receptor (Table 10.2).

Following the binding of the B-subunit to receptor, the A-subunit is internalized by receptor-mediated endocytosis and transported to the endoplasmic reticulum (ER). The toxin is activated after cleavage of 4 kDa C-terminal A2 peptide. The resulting active A1 portion have *N*-glycosidase activity and cleaves a purine residue in 28S rRNA, altering the function of ribosome

Table 10.2 Shiga toxin types associated with STEC

Shiga toxin types	Receptor type	Description
Stx1	Gb3	STEC and identical to Stx from *Shigella dysenteriae*
Stx1c	Gb3	Found in *eaeA*-negative STEC, mild diarrhea in humans
Stx2	Gb3	Prototype; HUS in human
Stx2c	Gb3	Diarrhea and HUS in human; common in ovine STEC
Stx2d	Gb3	Found in *eaeA*-negative STEC, mild diarrhea in humans
Stx2d$_{act}$	Gb3	Highly virulent
Stx2e	Gb4	Edema disease in pigs; rare in humans
Stx2f	Gb4	Pigeon isolates; rare in human

Adapted from Gyles, C.L. 2007. J. Anim. Sci. 85(13 Suppl.):E45–E62 such that it no longer can interact with elongation factors EF-1 and EF-2 and inhibiting protein synthesis. A lack of protein synthesis leads to cell death. The severity of infection and damage to the tissues depends on the number of receptors present.

Stx2 is 1,000 times more toxic than Stx1, and Stx2 together with EaeA poses the highest risk of developing HUS. The crystal structure of toxin reveals a greater accessibility for the active site of Stx2 than Stx1, and this possibly contribute to the enhanced cytotoxicity.

Stx has been shown to possess nephrotoxic, cytotoxic, enterotoxic, and neurotoxic activities: Stx causes nephrotoxicity following enteric infection, resulting in massive damage to kidney tubules, bloody urine, and the hemorrhagic uremic syndrome. Since Stx causes chronic kidney damage, there is a need for dialysis and kidney transplant. HUS is also characterized by thrombocytopenia and hemolytic anemia. Though calves are susceptible to EHEC infection, they do not develop HUS because they lack the receptors in the endothelial cells of blood vessels. The Stx also acts as a neurotoxin, causing a neurological disorder called, thrombotic thrombocytopenic purpura (TTP), which is characterized by hemolysis, thrombocytopenia, renal failure, and fluctuating fever. Stx is also reported to have enterotoxin activity resulting in fluid accumulation and diarrhea. Stx also has cytotoxin action mediated by inhibition of cellular protein synthesis and causes programmed cell death. Stx variant toxins are also found in pigs, humans or pigeons and can cause (1) edema disease, (2) bloody diarrhea, (3) hemorrhagic colitis (HC), and (4) HUS.

Inflammation

Inflammation is very prominent in the intestine during infection with *E. coli* O157:H7. Flagellin (H7) is thought to be responsible for inflammatory response, which is characterized by activation of p38 and ERK MAP kinase, nuclear translocation of NF-kappa, and an increased expression of proinflammatory cytokine, IL-8. The flagella bind to toll-like receptor 5 (TLR-5) in epithelial cells, activate NF-κB and induce IL-8 release. The inflammation likely disrupts epithelial barrier function and facilitates Stx passage from the lumen to the submucosal layer. LPS (O157 antigen) activates platelets and together with Stx may cause endothelial cell injury and can contribute to the thrombocytopenia observed in HUS. An LPS-mediated release of cytokines IL-1 and TNF-α from activated macrophages can cause vascular damage during renal failure in HUS patients.

Enterohemolysin

Enterohemolysin (Ehly or Ehx) has been isolated from EHEC and it belongs to the family of RTX (Repeats in ToXin). It is a monomeric pore-forming toxin and its role in pathogenesis is unclear. It is encoded by four genes (*ehxC, ehxA, ehxB,* and *ehxD*) and is located in the 60-MDa plasmid and it is secreted by type I secretion system. RTX may cause localized lesions or affect cells of renal tubules.

Other Virulence Factors

Genome sequence of EHEC O157:H7 revealed the presence of several putative virulence factors: several fimbrial adhesins (Lpf, SfpA), nonfimbrial adhesins

(EfaI, Iha, OmpA, and ToxB), toxins (cytolethal distending toxin; CDT, EHEC hemolysin), proteases (EpeA, EspP/PssA), H7 and H21 flagellin, and urease.

Regulation of Virulence Genes

Regulation of genes located on LEE is complex and involves non-LEE encoded and LEE-encoded genes. The transcriptional regulators, Ler (LEE-encoded regulator) and Grl (global regulator of LEE activator), positively regulate the genes on LEE. EHEC uses a quorum sensing regulatory system to recognize the intestinal environment and activate genes that are required for colonization in the gut. Autoinducers, like epinephrine and norepinephrine, also regulate genes for flagella and motility, which allows bacteria to find a suitable niche in the gut.

The *stx* genes are located in the lysogenic lambdoid phage and are highly expressed when the lytic cascade of the phage is activated. *stx* gene expression is regulated by iron concentration where higher concentration suppress expression.

Symptoms and Complications

Symptoms of EHEC infection occur 3–9 days after ingestion of contaminated food and generally last for 4–10 days. The colitis symptoms include a sudden onset of abdominal cramps, watery diarrhea (which in 35–75% of cases turns bloody), and vomiting. Damage to the blood vessels in the colon is responsible for bloody diarrhea. HUS may develop and is characterized by acute renal failure, microangiopathic hemolytic anemia, and thrombocytopenia. Endothelial cells in the kidney may be damaged by Stx and this leads to the hemolytic uremic syndrome that develops in 5–10% of STEC infected patients. EHEC infection can be fatal, particularly in children under 5 years of age and the elderly. Though the kidney is the primary target, other organs such as the lungs, central nervous system, pancreas, and heart are also affected. A TTP may result from a blood clot in the brain, resulting in seizures, coma, and death.

Enteroaggregative *E. coli*

Characteristics

EAEC causes persistent diarrhea, lasting more than 14 days, in children and adults and is prevalent in developing countries. EAEC causes mostly sporadic cases, but recent data show it also causes outbreaks. EAEC is a highly heterogeneous group, and 40 different O-types have been identified. The persistent diarrhea in children is similar to ETEC and causes mild but significant mucosal damage. EAEC also have pathogenicity islands that carry genes for enterotoxin and mucinase activity.

Adhesion Factors

EAEC express aggregative adherence fimbriae I, II, and III (AAF/I, AAF/II, and AAF/III), Curli, coiled pilus (resembling curled hair), and 18 and 30 kDa outer membrane adhesion proteins. The AAF/1 is a flexible bundle-forming fimbrial structure of 2–3 nm in diameter. The 18-kDa adhesion proteins are thin filamentous (fibrillar) structures and are also called GVVPQ fimbriae (G – glycine, V – valine, P – proline, and Q – glutamine). This sequence is located near the N-terminal end and may be responsible for "clumping" of cells, or adherence to each other, rather than facilitating bacterial attachment to the host. Adherence to cultured HEp-2 cells indicates that EAEC adhesion may be arranged as a characteristic "stacked-brick" pattern. The genes for adhesion are encoded in the 60-MDa plasmid.

Toxins

EAEC produce two types of toxins: ST-like toxin, also called EAST (enteroaggregative ST), the function of which is not clear, and hemolysin-like *E. coli* toxin. This toxin is a 120 kDa pore-forming exotoxin, similar to hemolysin, but it does not lyse RBC. It serves as an important signaling molecule, causing increased host cell Ca^{2+} influx, and actin depolarization, leading to a destabilized host cell cytoskeletal structure.

EAEC Pathogenesis

EAEC adhere to the enterocytes forming aggregates and adherence is characterized by a "stacked-brick" (Fig. 10.1). EAEC pathogen also enhance mucus secretion from goblet cells and trap themselves in mucus-forming biofilms. EAEC does not invade epithelial cells, but produce a 108-kDa cytotoxin that is responsible for histopathological effects. The toxin induces shortening of villi and hemorrhagic necrosis of villous tips, and inflammation that is characterized by the infiltration of mononuclear cells to the submucosa. Infection results in mucoid stool and persistent diarrhea. EAEC infect immunocompromised hosts and bacteria are isolated frequently from AIDS patients stool.

Symptoms and Diagnosis

Symptoms of EAEC infection include watery, mucoid, secretory diarrhea with low grade fever, and no vomiting. Some patients show grossly bloody stools. The HEp-2 adherence assay is the gold-standard for identification of EAEC, although a PCR assay has been developed to detect this group of *E. coli*.

Enteroinvasive *E. coli*

Characteristics

EIEC strains are generally lysine decarboxylase negative, nonmotile, and lactose negative. They are genetically, biochemically, and pathogenetically related to *Shigella* spp. and produce watery diarrhea and dysentery. Sporadic outbreaks are common, however, occasional foodborne outbreaks may occur. An outbreak in the U.S., as early as 1971, was recognized from the consumption of imported camembert cheese contaminated with serotype O124:H17. An outbreak was also reported in a restaurant in Texas involving 370 people.

Disease and Symptoms

Ingestion of as many as 10^6 cells may be necessary for an individual to develop the disease. In colonic mucosa, EIEC first binds and invades epithelial cells, lyses the endocytic vesicle, multiplies in the cytoplasm, moves inside the cytoplasm directionally, and projects toward adjacent cells to spread from cell-to-cell. The genes responsible for invasion are encoded in a 140-MDa plasmid called pInv. A toxin of 63 kDa, encoded by plasmid-borne *sen* gene, has been linked to cause watery diarrhea. Extensive cell damage due to invasion and cell-to-cell spread elicits a strong inflammatory response and bloody mucoid diarrhea, similar to the bacillary dysentery caused by *Shigella*. Human carriers, directly or indirectly, also spread the disease.

The symptoms appear as abdominal cramps, profuse diarrhea, headache, chills, and fever. Some patients may develop dysentery. A large number of pathogens are excreted in the feces. The symptoms can last for 7–12 days, but a person may remain a carrier and shed the pathogens in feces for much longer period of time.

Diffusely Adhering *E. coli*

Diffusely adhering *E. coli* (DAEC) causes infantile diarrhea and produces a diffuse adherence (DA) to cultured HEp-2 cell lines, which is mediated by a fimbrial adhesion, designated F1845. The gene encoding fimbria is located in the chromosome or in a plasmid. DAEC also expresses afimbrial adhesins (Afa) belonging to the Afa/Dr family of adhesins. Some isolates express 100-kDa outer membrane proteins, which are responsible for DA phenotype. LEE has been isolated from DAEC and is thought to carry genes required for attachment/effacement lesions and signaling events in EPEC. DAEC causes watery diarrhea in children without blood or fecal leukocytes. It appears diarrhea is age related and increases with age from 1 year to 4–5 years of age.

Animal and Cell Culture Model Used for Diagnosis of *E. coli*

A ligated rabbit ileal loop (RIL) assay has been used for detection of diarrheal toxins produced by different virotypes/serotypes. For diagnosis of ETEC, calves and piglets are used since no small animal models are available. ETEC causes diarrhea in gnotobiotic pig (e.g., specific pathogen free pig), while EPEC cause attaching and effacing lesions in piglet intestine.

Tissue culture models have been used extensively to study specific traits (See Chap. 5). For example, attachments and effacement phenomenon of EPEC has been studied using HEp-2 (laryngeal cells) and HeLa (cervical cancer cell line) cell lines. Caco-2 cells and HT-29 (colon cancer cells) are used to study ETEC attachment. Interestingly, ETEC does not adhere to HEp-2 cells. HEp-2 cells are also used to study diffuse adherence phenotype of DAEC. Vero cells (African Green Monkey kidney) have been used to study cytotoxicity (See Chap. 5, Fig. 5.1) of EHEC and HEp-2 for attaching/effacing (A/E) assay

Control and Prevention of *E. coli*-Mediated Diarrhea

Proper sanitation, cooking or heating at appropriate temperatures, proper refrigeration, and prevention of cross-contamination should be practiced in order to control the presence of *E. coli* O157:H7 in a ready-to-eat food. EHEC is a heat-sensitive organism and is inactivated at 62.8 °C for 0.3 min in ground beef. The Food Safety Inspection Service (FSIS) in the U.S. has provided the following guidelines to control foodborne illness from this pathogen: use only pasteurized milk; quickly refrigerate or freeze perishable foods; never thaw a food at room temperature or keep a refrigerated food at room temperature over 2 h; wash hands, utensils, and work areas with hot soapy water after contact with raw meat and meat patties; cook meat or patties until the center is gray or brown or internal temperature reaches to 68.3 °C (155 °F); and prevent fecal–oral contamination through proper personal hygiene. Routine surveillance of cattle for the presence of EHEC should be carried out, and cattle should be tested for pathogen presence before slaughter. HACCP should be incorporated into the slaughtering and processing operations. Consumers should be educated for safe handling of raw meats and should avoid cross-contamination of cooked products.

Fatalities from diarrheal diseases are due to the extensive dehydration, and electrolyte imbalance (loss). Oral hydration and the electrolyte replenishment are the most important therapy. Antibiotics can shorten the duration of infection. Antibiotic therapy is less effective, however, in some situations, it is recommended to clear up the infection. For ETEC infection, fluoroquinolones (e.g., ciprofloxacin, norfloxacin, and ofloxacin) are commonly recommended for treatment. As a preventive measure, travelers can use doxycycline and trimethoprim–sulfamethoxazole before a scheduled a trip to the endemic region. Concerns of antibiotic resistance discourage the use of antibiotics as a prophylactic measure; therefore, travelers are advised to avoid potential hazardous food and water. Water should be boiled and food should be properly cooked to prevent infection.

Summary

Most *E. coli* are harmless inhabitant of the intestinal tract, and only a small percentage of strains are considered pathogenic. However, a recent surge in outbreaks and *E. coli* related infections suggest possible increased horizontal or vertical transfer of pathogenic genes among bacterial species. There are six virotypes of *E. coli*, of which, EHEC, EPEC, and ETEC are known to cause severe disease worldwide. Increased insight into their genetic and phenotypic properties of virulence factors and their pathogenic mechanisms should help to formulate appropriate preventive or therapeutic measures. Common themes shared by all *E. coli* virotypes include the following: they adhere to the epithelial cells and cause damage to the cells by initiating signaling events that lead to blockage of protein synthesis, alter cytoskeletal structure, affect ion pumps, increase fluid loss, or cause cell death. In recent years, however, the research focus is geared more toward EHEC group because of their continued association with serious foodborne outbreaks from a wide variety of foods, including fruits, vegetables, meats, and dairy products. Analysis of recent outbreak strains indicates that Stx2 to be the most important toxin of EHEC, causing severe hemolytic uremic syndrome (HUS) and kidney damage. Association of this pathogen with fresh fruits and vegetables present a serious problem, because these products are minimally processed and apparently the processing conditions are inadequate for complete removal or inactivation. Furthermore, these organisms probably have developed strategies to utilize nutrients from plants for prolonged survival inside the plant tissues and they are resistant to washing and disinfections. Diarrheal diseases are preventable by adopting proper sanitary condition during preparation of food, by thorough cooking and by avoiding foods that might be the potential source of the organism. Dehydration and electrolyte loss result from diarrhea, which can be fatal thus hydration is the most important therapy against diarrheal diseases.

Further Readings

1. Beutin, L. 2006. Emerging enterohaemorrhagic *Escherichia coli*, causes and effects of the rise of a human pathogen. J. Vet. Med. B 53:299–305.
2. Clarke, S.C., Haigh, R.D., Freestone, P.P.E., and Williams, P.H. 2003. Virulence of enteropathogenic *Escherichia coli*, a global pathogen. Clin. Microbiol. Rev. 16:365–378.
3. Dean, P., Maresca, M., and Kenny, B. 2005. EPEC's weapons of mass subversion. Curr. Opin. Microbiol. 8:28–34.
4. Donnenberg, M.S., Kaper, B., and Finley, B.B. 1997. Interactions between enteropathogenic *Escherichia coli* and host epithelial cells. Trends Microbiol. 5:109–114.
5. Gyles, C.L. 2007. Shiga toxin-producing *Escherichia coli*: an overview. J. Anim. Sci. 85(13 Suppl.):E45–E62.
6. Hayward, R.D., Leong, J.M., Koronakis, V., and Campellone, K.G. 2006. Exploiting pathogenic *Escherichia coli* to model transmembrane receptor signaling. Nat. Rev. Microbiol. 4:358–370.
7. Jores, J., Rumor, L., and Wieler, L.H. 2004. Impact of the locus of enterocyte effacement pathogenicity island on the evolution of pathogenic *Escherichia coli*. Int. J. Med. Microbiol. 294:103–113.
8. Karmali, M.A., Mascarenhas, M., Shen, S., Ziebell, K., Johnson, S., Reid-Smith, R., Isaac-Renton, J., Clark, C., Rahn, K., and Kaper, J.B. 2003. Association of genomic O island 122 of *Escherichia coli* EDL 933 with verocytotoxin-producing *Escherichia coli* seropathotypes that are linked to epidemic and/or serious disease. J. Clin. Microbiol. 41:4930–4940.
9. Mainil, J.G. and Daube, G. 2005. Verotoxigenic *Escherichia coli* from animals, humans and foods: who's who? J. Appl. Microbiol. 98:1332–1344.
10. Nataro, J.P. and Kaper, J.B. 1998. Diarrheagenic *Escherichia coli*. Clin. Microbiol. Rev. 11:142–201.
11. Qadri, F., Svennerholm, A.-M., Faruque, A.S.G., and Sack, R.B. 2005. Enterotoxigenic *Escherichia coli* in developing countries: epidemiology, microbiology, clinical features, treatment, and prevention. Clin. Microbiol. Rev. 18:465–483.
12. Servin, A.L. 2005. Pathogenesis of Afa/Dr diffusely adhering *Escherichia coli*. Clin. Microbiol. Rev. 18:264–292.
13. Turner, S.M., Scott-Tucker, A., Cooper, L.M., and Henderson, I.R. 2006. Weapons of mass destruction: virulence factors of the global killer enterotoxigenic *Escherichia coli*. FEMS Microbiol. Lett. 263:10–20.

11

Salmonella enterica

Introduction

Daniel E. Salmon first reported the isolation of *Salmonella* from a pig in 1885 and named the organism Bacterium *choleraesuis* (currently known as *Salmonella enterica* serovar Choleraesuis). *Salmonella* causes gastroenteritis and typhoid fever and is one of the major foodborne pathogens of significant public health concern in both developed and developing countries. Worldwide there are 16 million annual cases of typhoid fever, 1.3 billion cases of gastroenteritis, and 3 million deaths due to *Salmonella*. In the US annually there are 2–4 million cases with a death rate of 500–1,000, and an economic loss of about 3 billion dollars. A recent CDC report indicates that the incidence of *Salmonella enterica* serovar Typhimurium decreased significantly (42% decline) from 1996–1998 to 2005; however, the incidence of other serotypes are on the rise such as *S. enterica* serovar Enteritidis and *S. enterica* serovar Heidelberg, each of which increased by 25%, and *S. enterica* serovar Javiana increased by 82%.

Biology

The genus *Salmonella* is a member of the *Enterobacteriaceae* family, and is a Gram-negative, nonspore-forming bacillus. Salmonellae are motile (except *Salmonella* Pullorum and *S.* Gallinarum) and express peritrichous flagella. They are facultative anaerobes that can grow in a temperature range of 5–45 °C with optimum temperature of 35–37 °C. They are able to grow at low pH and are generally sensitive to increased concentrations of salt. *Salmonella* forms long filamentous chains when grown at temperature extremes of 4–8 °C or at 44 °C and also when grown at pH 4.4 or 9.4. All salmonellae are facultative intracellular pathogen and considered pathogenic and can invade macrophages, dendritic and epithelial cells. The virulence genes responsible for invasion, survival, and extraintestinal spread are distributed in *Salmonella* pathogenicity islands (SPI), which are discussed below.

Source and Transmission

Salmonella are present in the intestinal tract of birds, reptiles, turtles, insects, farm animals, and humans. Poultry are a major source for human foodborne salmonellosis, in part due to high-density farming operations which allow colonized birds to quickly spread salmonellae to other birds within a flock. Intestinal colonization by salmonellae increases the risk for contamination during slaughter. Eggs are also reservoirs for *Salmonella*, particularly serovar Enteritidis, as this organism can colonize the ovary of the laying hen. Such transovarian transmission allows bacteria to be present in the egg before the eggshell is formed in the oviduct. As a result, eggs stored at room temperature can contain high concentrations of *Salmonella* (as high as 10^{11} cells per egg).

Human salmonellosis is generally foodborne and is contracted through consumption of contaminated food of animal origin such as meat, milk, poultry, and eggs. Dairy products including cheese and ice cream were also implicated in the outbreak. However, fruits and vegetables such as lettuce, tomatoes, cilantro, alfalfa-sprouts, and almonds have also been implicated in recent outbreaks. A recent (2006–2007) multistate outbreak of salmonellosis was due to the consumption of peanut butter. Animal-to-human or human-to-human transmission can also occur. Organic farming has also increased the risk of foodborne diseases, including salmonellosis.

Classification

Historically, *Salmonella* nomenclature had been based on the places of origin such as *S. miami, S. london, S. richmond, S. dublin, S. indiana, S. kentucky, S. tennessee*, etc. This system of classification is now discontinued and *Salmonella* have been classified based on their susceptibility to different bacteriophages (called phage typing). More than 200 definitive phage types (DT) have been reported. Those include phase type (PT) 1, 4, 8, 13, 13a, 23, DT104, DT108, DT204, etc. Resistance to different antibiotics has also been used as a means of classification. For example, DT104 is resistant to multiple antibiotics including ampicillin, chloramphenicol, streptomycin, spectinomycin, sulfonamides, florfenicol, and tetracycline. It is also reported to be resistant to nalidixic acid and ciprofloxacin. The emerging strain DT204 is resistant to eight to nine antibiotics and is also a major human health concern.

Salmonellae have been grouped based on their somatic (O), flagellar (H), and capsular (Vi) antigenic patterns. There are 2,463 *Salmonella* serotypes, which are now placed under two species: *Salmonella enterica* and *Salmonella bongori*. *S. enterica* contains 2,443 serotypes and *S. bongori* contains 20 serotypes. *S. enterica* now has six subspecies, which are designated by roman numerals: I (enterica), II (salamae), IIIa (arizonae), IIIb (diarizonae), IV (houtenae), and VI (indica). For example, a *Salmonella* isolate is designated as *Salmonella enterica* subspecies I serovar Enteritidis. Under the modern nomenclature system, often, the subspecies information is omitted and the culture is called *S. enterica* serovar Enteritidis and in subsequent appearance, it is written as *S.* Enteritidis. Scientists are thus encouraged to follow this system of *Salmonella* classification and nomenclature to bring uniformity in the reporting and to avoid further confusions.

Table 11.1 The major *Salmonella* groups and their target host

Salmonella	**Pathogen specific to**
Salmonella enterica serovar Typhi	Humans
S. enterica serovar Paratyphi	Humans
S. enterica serovar Typhimurium	Humans
S. enterica serovar Enteritidis	Humans
S. enterica serovar Choleraesuis	Swine
S. enterica serovar Dublin	Cattle
S. enterica serovar Pullorum	Chicken
S. enterica serovar Gallinarum	Chicken

Major Groups

Salmonella (Table 11.1) causes three forms of disease: typhoid fever, gastroenteritis, and bacteremia. *Salmonella enterica* serovar Typhi is the most invasive type and it causes typhoid fever, a systemic disease in humans. *S. enterica* serovar Paratyphi causes typhoid-like infection in humans. *S. enterica* serovar Typhimurium and serovar Enteritidis cause self-limiting gastroenteritis or enterocolitis, which is mostly localized to the gastrointestinal tract and these two are the most common serovars responsible for human infections. *S. enterica* serovar Typhimurium causes typhoid-like infection in mice. *S. enterica* serovar Choleraesuis, a swine-adapted pathogen, causes septicemia (paratyphoid) in pigs. A bovine-adapted *S. enterica* serovar Dublin causes bacteremia, inflammation in the digestive tract and abortion in cows, and serotype Arizonae infects reptiles. Importantly, these serovars (Choleraesuis, Dublin, and Arizonae) occasionally cause infections in humans. *S. enterica* serovars Pullorum and Gallinarum cause infection in poultry.

Pathogenic Mechanism

Gastroenteritis

Salmonella Typhimurium causes serious illness in children, and immunocompromised individuals, often resulting in systemic infection (Fig. 11.1). In healthy individuals, symptoms include fever, diarrhea, abdominal pain, and sometimes vomiting. These symptoms are self-resolving in healthy individuals and usually subside within 3–4 days.

The infectious dose of salmonellae is rather broad and may vary from 1 to 10^9 cfu g^{-1}. Studies conducted with human adult volunteers have indicated that a dose range of 10^5–10^{10} organisms are required to cause disease. The infectious dose decreases if consumed with liquid food that traverses stomach rapidly or food such as milk and cheese that neutralizes the stomach acid. Individuals with underlying conditions (for example, immunocompromised individuals) are susceptible to low dosage of bacteria.

In the intestinal tract, *Salmonella* exhibits tropism for intestinal lymphoid tissue and pass through mucosal microfold (M)-cells present in the follicle-associated epithelium (FAE) overlying Peyer's patches (PP) (Fig. 11.1). M-cells and dendritic cells located in the lamina propria ingest luminal microbiota to maintain gut homeostasis; a process which allows *Salmonella* to cross the

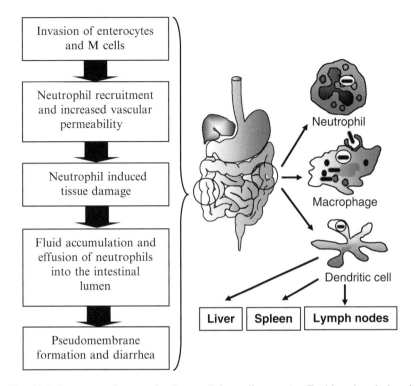

Fig. 11.1 Sequence of events leading to *Salmonella enterica* Typhimurium-induced pathogenesis and diarrhea

epithelial barrier. M-cells located in the solitary intestinal lymphoid tissue (SILT) distributed through out the intestinal lining have also been proposed to be the invasion site for *Salmonella*. SILTs are composed of isolated lymphoid follicles (ILF) that contain B-cells and M-cells. SILTs are ordered structures that can develop into cryptopatches (lymphoid aggregations mostly filled with stem cell-like cells) or larger ILFs that resemble PP and vice versa. In the lamina propria, salmonellae are engulfed by resident dendritic cells or macrophages and replicate inside host cells or induce apoptosis.

Independent of M-cell-mediated entry, salmonellae colonize and invade the apical epithelium in the ileum, cecum and proximal colon, and elicit significant inflammation at the site, which is characterized by neutrophil infiltration, necrosis, edema, and fluid secretion. Massive neutrophil infiltration occurs within 1–3 h of infection. In some cases, salmonellae also disseminate to extraintestinal sites: mesenteric lymph nodes, liver, and spleen (Fig. 11.1). LPS-induced "inflammation" is seen during invasion. LPS also induces abdominal pain, fever, and gastroenteritis. Inflammation and damage to mucosal cells cause diarrhea and fluid loss.

Symptoms appear within 6–24 h as nausea, vomiting, abdominal pain resembling appendicitis, headache, chills, and bloody or nonbloody diarrhea followed by muscular weakness, muscle pain, faintness, and moderate fever. Symptoms can persist for 2–3 days. The mortality rate for salmonellosis is 4.1%. One to five percent of recovering patients may serve as a chronic carrier and shed bacteria from 3 months to 1 year. Systemic forms of the disease may be seen in children or immunocompromised adults, including cancer and AIDS patients.

Pathogenicity Islands

Salmonella virulence gene clusters are located in 12 pathogenicity islands (SPI) and some of them are located near the tRNA genes (Table 11.2). SPIs are thought to be acquired by horizontal gene transfer, while some SPIs are conserved throughout the genus, others are specific for certain serovars. Virulence genes that are involved in the intestinal phase of infection are located in SPI-1 and SPI-2 and the remaining SPIs are required for causing systemic infection, intracellular survival, fimbrial expression, antibiotic resistance, and Mg^{2+} and iron uptake.

SPI-1

SPI-1 is a 43-kb segment that was acquired by horizontal gene transfer from other pathogenic bacteria during evolution. It contains 31 genes responsible for invasion of nonphagocytic cells and components of the Type III secretion system (TTSS) designated as the Inv/Spa-Type III secretion apparatus. The major genes are *invA, invB, invC, invF, invG, hilA, sipA, sipC, sipD, spar, orgA, sopB,* and *sopE*. InvG is an outer membrane protein of the TTSS and it plays a critical role in bacterial uptake and protein secretion. InvA is an inner membrane protein and is involved in the formation of a channel through which the polypeptides are exported. InvH and HilD are accessory proteins involved in the adhesion of *Salmonella*. There are two kinds of effector proteins secreted by the TTSS. One subclass consists of InvJ and SpaO, which are involved in the protein secretion through the TTSS. The other subclass modulates host

Table 11.2 *Salmonella* pathogenicity islands (SPI) at a glance

Islands	Salmonella serovars	Length (kb)	GC (%)	Function
SPI-1	*Salmonella enterica* and *S. bongori*	43	47	TTSS, invasion, iron uptake
SPI-2	*S. enterica*	40	44.6	TTSS, invasion, systemic infection
SPI-3	*S. enterica* and *S. bongori*	17	39.8–49.3	Mg^{2+} uptake, macrophage survival
SPI-4	*S. enterica* and *S. bongori*	27	37–54	Macrophage survival
SPI-5	*S. enterica* and *S. bongori*	7.6	43.6	Enteropathogenicity
SPI-6	*S. enterica subspecies enterica* serovars	59	51.5	Fimbriae
SPI-7	Serovars Typhi, Dublin, Paratyphi	133	49.7	Vi antigen
SPI-8	Serovar Typhi	6.8	38.1	Unknown
SPI-9	*S. enterica* and *S. bongori*	16.3	56.7	Type I secretion system, and RTX-like toxin
SPI-10	Serovars Typhi and Enteritidis	32.8	46.6	Sef fimbriae
SGI-1	Serovars Typhimurium (DT104), Paratyphi, and Agona	43	48.4	Antibiotic resistance genes
HPI	*S. enterica subspecies* IIIa, IIIb, IV	?	?	High affinity iron uptake, septicemia

Adapted from Hensel, M. 2004. Int. J. Med. Microbiol. 294:95–102 cytoskeleton and induces its uptake. SipB and SipC are the major proteins, which interacts with host cytoskeletal proteins to promote *Salmonella* uptake. Inv/Spa is also responsible for macrophage apoptosis. SipA is an actin-binding protein. SopB is inositol phosphate phosphatase and SopE activates GTP-binding proteins. HilA is the central transcriptional regulator of genes encoded located on SPI-1.

SPI-2

SPI-2 is a 40-kb segment that encodes 32 genes. The majority of the genes are expressed during bacterial growth inside the host. It carries genes for Spi/Ssa and TTSS apparatus, i.e., SpiC, which inhibits fusion of the *Salmonella*-containing phagosome and lysosome. The gene products are essential for causing systemic infection and mediate bacterial replication rather than survival within host macrophages.

SPI-3

SPI-3 is a 17-kb locus and has ten genes. It is conserved between *S. enterica* serovar Typhi and Typhimurium and is also found in *S. bongori*. The gene product MgtCB is required for Mg^{2+}-dependent growth, and is essential for survival inside the macrophage.

SPI-4

SPI-4 is a 27-kb locus, which is located next to a putative tRNA gene and contains 18 genes. It is thought to encode genes for Type I secretion system and the gene products are required for survival in the macrophage.

SPI-5

SPI-5 is a 7.6-kb region and encodes six genes. It appears that SPI-5 encodes effector proteins for TTSS. SopB is an inositol phosphatase involved in triggering fluid secretion responsible for diarrhea, is translocated by TTSS. Thus it is thought that SPI-5 is possibly responsible for enteric infection.

SPI-6

SPI-6 is a 59-kb locus and is present in both serovars Typhi and Typhimurium. It contains *saf* gene cluster for fimbriae, *pagN* for invasion, and several genes with unknown function.

SPI-7 or Major Pathogenicity Island (MPI)

This is a 133-kb locus and is specific for serovar Typhi, Dublin, and Paratyphi. It encodes gene for Vi antigen, a capsular polysaccharide, which illicits high fever during typhoid fever. SPI-7 also carries *pil* gene cluster for type IV pili synthesis and encodes gene for SopE effector protein of TTSS.

SPI-8

SPI-8 is a 6.8-kb locus and appears to be specific for serovar Typhi. It carries genes for putative bacteriocin biosynthesis but functional attributes have not been investigated.

SPI-9

SPI-9 is about 16-kb locus and it carries genes for type I secretion system and a large putative RTX (repeat in toxin)-like toxin.

SPI-10

SPI-10 is a 32.8-kb locus and is found in serovars Typhi and Enteritidis and encodes genes for Sef fimbriae.

Salmonella Genomic Island-1 (SGI-1)

SGI-1 is a 43-kDa locus and encodes genes for antibiotic resistance. It was identified in *S*. Typhimurium DT104, Paratyphi and Agona which are resistant to multiple antibiotics. The DT104 strain has been implicated in outbreaks worldwide. The insertion site is flanked by direct repeats (DR) and is not associated with tRNA gene. The genes for five antibiotic resistance phenotypes (ampicillin, chloramphenicol, streptomycin, sulfonamides, and tetracycline) are clustered in a multidrug resistance region and are composed of two integrons.

High Pathogenicity Island (HPI)

HPI encodes genes for siderophore biosynthesis, which are required for iron uptake and found in *S. enterica*. It is also present in *Yersinia enterocolitica* and *Y. pseudotuberculosis*.

Type III Secretion System

The TTSS is also called a molecular syringe and is responsible for contact-dependent secretion or delivery of virulence proteins into host cells. This apparatus has been reported to be present in *Salmonella, Shigella*, and *Escherichia coli*. The genes encoding the formation of this secretion system have a chromosomal location in the centisome 63 in *Salmonella*. The gene products included several proteins that form a needle-like organelle on the bacterial envelope (Fig. 11.2). The TTSS also referred as molecular syringe which has four parts: a needle, outer rings, neck, and inner rings. The needle is made of PrgI and a putative inner rod protein, PrgJ. InvG is part of the outer rings structure, neck consists of PrgK, and the base is made of PrgH that make up the inner rings. The inner membrane components are made of InvC, InvA, SpaP, SpaQ, SpaR, and SpaS proteins (Fig. 11.2).

Adhesion and Colonization

Salmonella expresses different types of fimbriae that promote adhesion to M-cells or colonization of intestinal epithelial cells. Type I fimbriae (Fim) binds to α-D-mannose receptor in the host cell; long polar fimbriae (Lpf) bind to cells in the Peyer's patch; and plasmid-encoded fimbriae (Pef) and curli, a thin aggregative fimbriae aid in adhesion to intestinal epithelial cells. Curli helps bacteria to autoaggregate, which enhances survival in the presence of stomach acid or biocides. In addition, lectin-like adhesion molecules bind to glycoconjugate receptor Gal β (1–3) Gal NAc found on enterocytes.

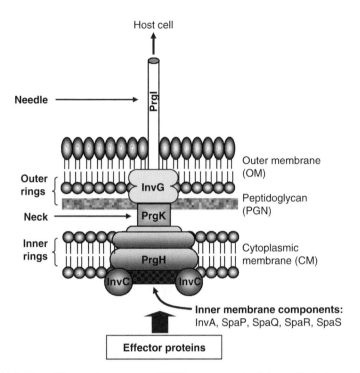

Fig. 11.2 Type III secretion system (TTSS) apparatus in *Salmonella* (adapted from Galan, J.E. and Wolf-Watz, H. 2006. Nature 444:567–573; Galan, J.E. and Collmer, A. 1999. Science 284:1322–1328)

Invasion and Intracellular Growth

Invasion of *Salmonella* is mediated by three mechanisms: phagocytosis by M-cells, phagocytosis by dendritic cells, and induced phagocytosis by epithelial cells (Fig. 11.3).

Phagocytosis by M-Cells

The preferential route of *Salmonella* translocation is through M-cells (Fig. 11.3). *Salmonella* expresses several invasin genes, *invABCD* located in the salmonella pathogenicity island 1 (SPI-1), that promote bacterial attachment and invasion of M-cells to cross the epithelial barrier.

Phagocytosis by Dendritic Cells

Dendritic cells (DCs) located in the lamina propria project their dendrites through the epithelial lining to the lumen to sample the intestinal environment and internalize luminal bacteria to transport them to the basolateral side of the epithelial lining (Fig. 11.3). Phagocytosis of *Salmonella* by dendritic cells thus does not affect the cellular integrity of DC. When enclosed in the DC vacuole, *Salmonella* does not proliferate but still secretes effector proteins, and is transported to extraintestinal sites. DC, rather than macrophages, is thought to be the primary phagocytes in the subepithelial region which are responsible for dissemination of bacteria to extraintestinal sites.

To initiate an adaptive immune response, DC also serves as antigen presenting cells and activate CD4$^+$ or CD8$^+$ T-cells by presenting antigen using MHC

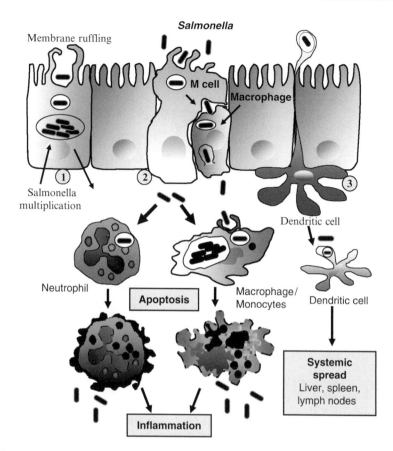

Fig. 11.3 Schematic diagram showing *Salmonella* invasion through host mucosal membrane. *Salmonella* translocates by three possible pathways (1) *Salmonella* induces membrane ruffling by trigger mechanism to allow its own uptake and multiplies inside the vacuole; (2) *Salmonella* also translocates through M-cells to arrive to subcellular location; and (3) *Salmonella* can be taken up by dendritic cells from the lumen and transported to the subcellular location. *Salmonella* also multiplies inside phagocytic cells such as macrophages, neutrophils, and dendritic cells and induce apoptosis and promote inflammation. Dendritic cells are responsible for systemic dissemination of bacteria to lymph nodes, liver, and spleen

class II or Class I molecules, respectively. Recognition of *Salmonella* by DC is mediated by toll-like receptors (TLR) and DC carrying *Salmonella* will activate naïve T-cells to release cytokines and chemokines, which in turn activate NK-cells and other T-cells for IFN-γ production for specific immune response.

Induced Phagocytosis

Binding of *Salmonella* triggers membrane ruffling, a mechanism that allows bacteria to be internalized by the nonprofessional phagocytic cells such as epithelial cells; hence, it is called the "trigger mechanism" (Fig. 11.3 and Fig. 11.4). Membrane ruffling is a well-orchestrated event that allows the formation of a "ruffle-like" lamellipodial appearance on the surface of the host cell membrane. Two major events take place during invasion: GTPase activation and actin polymerization. Contact with host cell initiates a signaling cascade that activates the Rho GTPase regulator that normally maintains the cellular architecture. Rho GTPase converts inactive GDP to active GTP-bound conformations which is

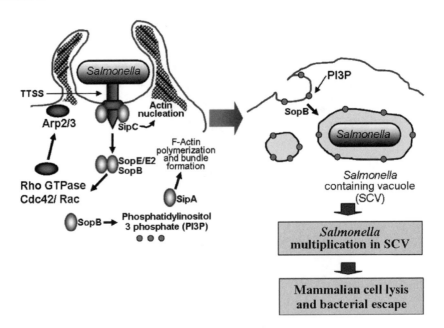

Fig. 11.4 Schematic diagram depicting the mechanism of *Salmonella* invasion by induced phagocytosis. *Salmonella* effectors proteins such as SipC, SipA, SopE/E2, SopB, etc are injected into the host cell by type III secretion system (TTSS), which allow actin polymerization and remodeling to form membrane ruffle and invasion into cell. SopB generates phosphatidyleinositol 3 phosphate (PIP3) decorated phagosme and spacious *Salmonella* containing vacuole (SCV) to allow *Salmonella* residence. [Diagram redrawn from Patel and Galan. 2005. Curr. Opin. Microbiol. 8:10–15]

regulated by yet another regulatory protein called guanine nucleotide exchange factors (GEFs). *Salmonella* Type III effector proteins SopE, SopE2, and SopB activate Rho GTPase, Cdc42, and Rac which activate Arp2/3 complex for initiation of actin polymerization. During this process, globular G-actin is polymerized to highly ordered F-actin necessary for engulfment of bacteria (Fig. 11.4).

Invasion also stimulates tyrosine phosphorylation of epidermal growth factor receptor (EGFR), which triggers the cascade of phosphorylation and dephosphorylation reactions, eventually activating the phospholipase A2 (PLA2). Phosphorylated-PLA2 helps to produce arachidonic acid. The enzyme 5-lipoxygenase converts arachidonic acid into leukotriens, which increases membrane permeability, causing increased fluid accumulation resulting in diarrhea. Leukotriens also controls calcium channels and allows intracellular accumulation of Ca^{2+} influx, which activates actin polymerization and cytoskeletal rearrangement allowing bacterial entry through membrane ruffling. The internalized bacteria are trapped in a membrane bound vesicle called the *Salmonella*-containing vacuole (SCV) (Fig. 11.4).

The gene products required for invasion of epithelial cells are delivered by Type III secretion system (TTSS), the molecular syringe. The genes for the TTSS are encoded in SPI-1 and SPI-2. TTSS type 1 injects SipC and SipA proteins, which are responsible for cytoskeletal rearrangement to promote membrane ruffling, and SopE, SopE2, and SopB which control signaling events that lead to actin rearrangement. TTSS type 2 delivers gene products (SseG, SifA) that promote bacterial survival and replication inside the SCV, the genes for which are located in SPI-2 (Fig. 11.4).

The sequential *Salmonella* uptake and cell lysis events consist of (1) formation of the membrane ruffle which appears as a splash; (2) actin rearrangement to allow cytoskeletal rearrangement; (3) formation of pseudopod to entrap *Salmonella* inside SCV; (4) bacterial multiplication inside SCV; (5) coalescence of multiple SCV-containing bacteria to form a large vesicle; and (6) lysis of the vesicles to allow *Salmonella* release. *Salmonella* then enters the circulation for systemic infection (Fig. 11.4).

Survival in Phagocytes

Salmonella present in the subcellular lamina propria are either engulfed by the macrophages or the dendritic cells for extraintestinal disseminations. *Salmonella* express approximately 40 proteins that aid their survival inside a macrophage. The macrophage attempts to control intracellular *Salmonella* growth via the inducible nitric oxide synthase (iNOS) and NADPH oxidase-dependent respiratory burst. However, salmonellae produce catalase, which inactivates lysosomal H_2O_2 and superoxide dismutase (SOD) that inactivates reactive oxygen.

Salmonella also expresses a two-component signal transduction system, the PhoP/PhoQ system to promote bacterial survival inside the macrophage. PhoQ is a sensor and PhoP is a transcriptional activator that expresses different genes that are required for bacterial survival inside the macrophage, as well as various stresses including carbon and nitrogen starvation, low pH, low O_2 levels, and the action of defensins. In addition, PhoP regulates genes such as *spiC* and *tassC* that prevent lysosome fusion with the *Salmonella*-containing vacuole (SCV). PhoQ regulon activates *pags* genes that are essential for adaptation during the intracellular life cycle.

Regulation of Virulence Genes

In order to cause successful infection, *Salmonella* must survive under diverse environmental conditions that include, acid in stomach, bile salts, oxygen limitations, nutrient starvation, antimicrobial peptides, mucus, and natural microbiota in the intestine. During invasion and growth in macrophages, salmonellae also encounter lysosomal enzymes, hydrogen peroxide, reactive oxygen radicals, iNOS, and defensins. Thus, a large array of gene expression is required to establish an infection in a host. Invasion-associated genes are maximally expressed at 37 °C, at neutral pH, at high osmolarity and during late phase of growth, and are regulated by HilA (hyper-invasive locus), encoded in SPI-1. HilA is the central transcriptional regulator of SPI-1.

RpoS Regulator

At the level of gene transcription, stress responses are controlled by the association of sigma factors such as RpoS/σ^S with the RNA polymerase. The sigma factors interact with the core RNA polymerase to reprogram the promoter recognition specificities to express different sets of genes suitable for survival during environmental changes. Sigma factors are thus considered to be global regulators of stress response that connect many signaling networks with downstream regulatory cascades that ultimately control the expression of genes required for the survival and virulence of the bacteria. RpoS is a well-characterized sigma factor of *Salmonella*, which is produced during changes in growth conditions such as starvation, pH, and temperature changes. RpoS controls the expression of more than 60 proteins such as Nuv, KatE-encoded catalase, acidic phosphatase, etc. It has been reported that production of sigma factors during stress increases the survivability of the cells and promotes crossprotection to additional stresses yet to be encountered. The RpoS may also regulate the virulence genes which play a role during the intestinal phase of infection, as there is an increased likelihood that prior to ingestion, the bacteria might be in stationary phase where RpoS is upregulated.

RpoS also regulates the virulence genes located in plasmid, *spv* (*Salmonella* plasmid virulence), which is essential for systemic infection in mice. RpoS is also believed to regulate unknown chromosomally encoded genes, which play a significant role in bacterial virulence. Recently, it has been demonstrated that the expression of other sigma factors such as σ^E and σ^H respond to extracytoplasmic shock and heat stress, respectively, which may also be dependent on the expression of RpoS.

ATR Response

Acid tolerance response is critical for bacterial transit through the stomach during the gastrointestinal phase of infection or for survival inside the acidic phagosomal environment. Bacteria exposed to milder acidic pH (pH 5) induce a set of genes (as many as 50 genes) that are essential for adaptation of bacteria to more acidic pH (~pH 3). Global regulators such as RpoS, PhoP/Q, OmpR, and Fur play an important role in this process. Two types of ATR responses are identified: log-phase ATR involved during exposure to organic and inorganic acids and stationary-phase ATR induced during late phase of growth.

Treatment and Prevention of Gastroenteritis

S. Typhimurium-induced gastroenteritis is self-limiting. Antibiotic therapy is not recommended for "uncomplicated gastroenteritis." Treatment with chloramphenicol may be needed to clear the *Salmonella* during systemic infection. Prevention of salmonellae includes proper food handling, avoiding cross-contamination, implementing personal hygiene and educating the public about the source and safe handling of foods and proper sanitation. Eggs should be kept refrigerated until eaten to prevent the multiplication of bacteria in the yolk. Proper cooking with a minimum pasteurization temperature of 71.7 °C for 15 s followed by prompt cooling to 3–4 °C or freezing within 2 h would eliminate *Salmonella* from food.

Typhoid Fever

S. enterica serovar Typhi causes systemic febrile illness called typhoid fever, characterized by the ingestion of food/water contaminated with human feces. Individuals recovering from typhoid fever act as a chronic carriers and shed bacteria for months. For this reason, it is important that food handlers abstain

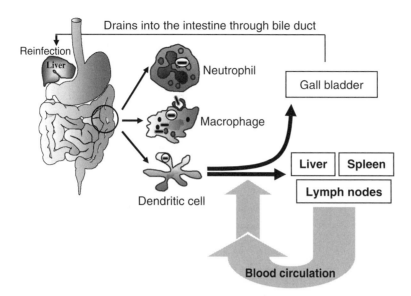

Fig. 11.5 Schematic diagram showing *Salmonella enterica* serovar Typhi spread and interaction with phagocytic cells during systemic infection

from working during and immediately following *S.* Typhi infections. The incubation period for *S.* Typhi infection is 1 week to 1 month. Following consumption of contaminated food, *S.* Typhi translocates via M-cells or dendritic cells and multiply in submucosa layer. Macrophages or DC transfer bacteria to liver and spleen and where bacteria are localized in hepatocytes, splenocytes, reticuloendothelial cell, macrophages, neutrophils, and DCs. Bacteria leave the liver and spleen and enter blood circulation in large numbers and reach gall bladder. From the gall bladder, bacteria could be shed into the intestine for second round of infection through M-cells (Fig. 11.5). In some cases, ulceration of the intestine is seen.

S. Typhi produces Vi antigen, which consists of capsular polysaccharide composed of *N*-acetylglucosamine uronic acid. Vi antigen is antiphagocytic and allows bacterial survival inside phagocytes. Vi antigen also helps to scavenge the reactive oxygen. Vi antigen has been used as a vaccine candidate to provide protection against *Salmonella* infection. Antibodies to Vi antigen are detected in typhoid fever patients.

Symptoms of typhoid fever appear within 1–2 weeks and lasts for 2–3 days. High fever is seen due to high levels of LPS-mediated cytokine release. Other symptoms include malaise, headache, nausea, myalgia, anorexia, constipation, chills, convulsions, and delirium (delusion, restlessness, and temporary disturbance of consciousness). The mortality rate of typhoid fever is as high as 10%. In chronic carriers, the organism is shed from gall bladder for months to years. Antibiotic treatment with fluoroquinolones is the most effective therapy; however, nalidixic acid, ampicillin, and trimethoprim/sulfamethoxazole are also found to be effective. Two types of vaccines are currently employed: injectable form of live attenuated bacteria and Vi antigen, and oral form of attenuated strain of *S.* Typhi Ty21a.

Detection

Culture Methods

The traditional *Salmonella* culture method involves pre-enrichment, selective enrichment, isolation of pure culture, biochemical screening, and serological confirmation, which requires 5–7 days to complete. The USDA and FDA recommended method involves a 6–24h pre-enrichment step in a nonselective broth such as lactose broth, tryptic soy broth, nutrient broth, skim milk, or buffered peptone water. The selective enrichment step requires an additional 24h in Rappaport–Vassiliadis semisolid medium, selenite cystine broth, or Muller Kauffmann tetrathionate broth. Bacterial cells are isolated from selective agar plates such as Hektoen enteric agar (HEA), xylose lysine deoxycholate (XLD), and/or brilliant green agar (BGA). If necessary, biochemical testing is done using triple sugar iron agar and lysine iron agar, which requires an additional 4–24h.

Immunological Methods

Immunological methods including enzyme-linked immunosorbent assay (ELISA), surface adhesion immunofluorescent technique, dot-blot immunoassay, surface plasmon resonance (SPR) biosensor, piezoelectric biosensor, time-resolved immunofluorescence assay (TRF), and fiber optic sensor have been used for detection *Salmonella* and the detection limit for these assays is in the range of 10^5–10^7 cells. These assays require sample enrichment and a concentration step that may include immunomagnetic separation or centrifugation/filtration.

Nucleic Acid-Based Assays

Real-time quantitative polymerase chain reaction using PCR (Q-PCR), reverse transcriptase PCR (RT-PCR), and nucleic acid sequence-based amplification (NASBA) have been used for detection of *Salmonella* from various food matrices. *Salmonella enterica* was detected at 1 cfu ml^{-1} after a culture enrichment of 8–12h in the TaqMan-based Q-PCR using *invA* gene as target. NASBA method has been used for detection of viable *Salmonella* cells and it has been demonstrated to be more sensitive than RT-PCR, and moreover, it requires fewer amplification cycles than the conventional PCR methods.

Summary

Salmonella enterica causes gastroenteritis, typhoid fever, and bacteremia. Worldwide there are 16 million annual cases of typhoid fever, 1.3 billion cases of gastroenteritis, and 3 million deaths. Poultry, egg, meat, dairy products, and fruits and vegetables serve as vehicles of transmission. To induce gastroenteritis, *Salmonella* passes through M-cells overlying Peyer's patches, through dendritic cells or through the epithelial lining in the lower part of small intestine or proximal colon to arrive in the subepithelial location. *Salmonella* is phagocytosed by dendritic cells or macrophages and is transported to extraintestinal sites such as the liver, spleen, and mesenteric lymph nodes. *Salmonella* induces apoptosis of macrophages and DCs and localized infection is characterized by neutrophil infiltration, tissue injury, fluid accumulation, and diarrhea.

Invasion through epithelial cells is a complex process involving multiple virulence factors which orchestrate events that lead to membrane ruffling, actin polymerization, bacterial localization and replication inside a vacuole, and cell lysis. *Salmonella* injects virulence proteins directly inside the epithelial cell cytoplasm using the Type III secretion system (TTSS), a syringe-like apparatus, to induce its own internalization. Genes encoding the TTSS are located in *Salmonella* pathogenicity islands SPI-1 and SPI-2. Genes located in other pathogenicity islands are responsible for survival inside macrophages, enteropathogenicity, iron uptake, and antibiotic resistance. During the intestinal phase of infection, *Salmonella* has to survive under diverse environmental conditions including stomach acid, bile salts, oxygen limitations, nutrient starvation, antimicrobial peptides, mucus, and the presence of natural microbiota. A global regulator, sigma factor like RpoS is thought to regulate the expression of more than 60 proteins, which possibly promote bacterial survival under these conditions. In addition, the transcriptional regulator, HilA, controls genes required for invasion, and PhoP/Q is required for bacterial survival inside macrophages.

Further Readings

1. Andrews, H.L. and Baumler, A.J. 2005. *Salmonella* species. In Foodborne Pathogens: Microbiology and Molecular Biology. Edited by Fratamico, P.M., Bhunia, A.K., and Smith, J.L., Caister Academic, Norfolk, pp 327–339.
2. Biedzka-Sarek, M. and Skurnik, M. 2006. How to outwit the enemy: dendritic cells face *Salmonella*. APMIS 114:589–600.
3. Chiu, C.-H., Su, L.-Hm., and Chu, C. 2004. *Salmonella enterica* serotype Choleraesuis: epidemiology, pathogenesis, clinical disease, and treatment. Clin. Microbiol. Rev. 17:311–322.
4. Coburn, B., Grass, G.A., and Finlay, B.B. 2007. *Salmonella*, the host and disease: a brief review. Immunol. Cell Biol. 85:112–118.
5. Clements, M., Eriksson, S., Tezcan-Merdol, D., and Hinton, J.C.D. 2001. Virulence gene regulation in *Salmonella enterica*. Ann. Med. 33:178–185.
6. Darwin, K.H. and Miller, V.L. 1999. Molecular basis of the interaction of *Salmonella* with the intestinal mucosa. Clin. Microbiol. Rev. 12:405–428.
7. Galan, J.E. and Collmer, A. 1999. Type III secretion machines: bacterial devices for protein delivery into host cells. Science 284:1322–1328.
8. Galan, J.E. and Wolf-Watz, H. 2006. Protein delivery into eukaryotic cells by type III secretion machines. Nature 444:567–573.
9. Groisman, E.A. and Ochman, H. 2000. The path to *Salmonella*. ASM News 66:21–27.
10. Hensel, M. 2004. Evolution of pathogenicity islands of *Salmonella enterica*. Int. J. Med. Microbiol. 294:95–102.
11. Humphrey, T. 2004. *Salmonella*, stress responses and food safety. Nat. Rev. Microbiol. 2:504–509.
12. Maciorowski, K.G., Herrera, P., Jones, F.T., Pillai, S.D., and Ricke, S.C. 2006. Cultural and immunological detection methods for *Salmonella* spp. in animal feeds – a review. Vet. Res. Commun. 30:127–137.
13. Patel, J.C. and Galan, J.E. 2005. Manipulation of the host actin cytoskeleton by *Salmonella* – all in the name of entry. Curr. Opin. Microbiol. 8:10–15.
14. Santos, R.L. and Baumler, A.J. 2004. Cell tropism of *Salmonella enterica*. Int. J. Med. Microbiol. 294:225–233.
15. Tindall, B.J., Grimont, P.A.D., Garrity, G.M., and Euzeby, J.P. 2005. Nomenclature and taxonomy of the genus *Salmonella*. Int. J. Syst. Evol. Microbiol. 55:521–524.

16. Weill, F.-X., Guesnier, F., Guibert, V., Timinouni, M., Demartin, M., Polomack, L., and Grimont, P.A.D. 2006. Multidrug resistance in *Salmonella enterica* serotype Typhimurium from humans in France (1993–2003). J. Clin. Microbiol. 44:700–708.

17. Zhang, S.P., Kingsley, R.A., Santos, R.L., Andrews-Polymenis, H., Raffatellu, M., Figueiredo, J., Nunes, J., Tsolis, RM., Adams, L.G., and Baumler, A.J. 2003. Molecular pathogenesis of *Salmonella enterica* serotype Typhimurium induced diarrhea. Infect. Immun. 71:1–12.

12

Campylobacter and *Arcobacter*

The *Campylobacteraceae* family consists of two genera; *Campylobacter* and *Arcobacter* and members of this family were initially recognized as a veterinary importance causing abortion in animals (Table 12.1). However, in recent years they are implicated in foodborne outbreaks and considered significant human pathogens. *Campylobacter* causes enteritis, bacteremia, endocarditis, periodontal diseases in humans and animals, and the infection often leads to chronic sequelae such as Reiter syndrome, and Guillain-Barré syndrome in humans. *Arcobacter* has been identified relatively recently to cause diarrhea in humans and abortion in animals. *Helicobacter* is closely related to *Campylobacter* but is mostly associated with gastric ulcer in humans (Table 12.1).

Campylobacter

Introduction

Campylobacter (curved rod in Greek) may have been discovered in the late nineteenth century (1886) by Theodor Escherich from an infant who died of cholera and called the disease "cholera infantum." Later in 1913, McFayden and Stockman identified an organism from an aborted sheep and called it, *Vibrio fetus* (now known as *Campylobacter fetus*). Since then *Campylobacter* was considered a significant animal pathogen. In 1972, Dekyser and Butzler isolated a *Campylobacter* strain from blood and feces of a woman who suffered from hemorrhagic enteritis. In the last 30 years, *Campylobacter* has been recognized as a leading pathogen causing diseases in both animals and humans and considered a zoonotic pathogen. A recent report by CDC indicates that between 1998 and 2002 *Campylobacter* was responsible for 61 outbreaks with 1,440 cases in the US. However, it has been estimated that the *Campylobacter*-related infection is the highest among all the foodborne bacterial infections in the US with an estimated 1.9 million cases per year and many of these cases are associated with consumption of chicken.

Table 12.1 Classification of the genus of *Campylobacter* based on biochemical properties

Characteristics	Arcobacter	Campylobacter	Helicobacter
Aerobic growth at 25 °C	+	−	−
Catalase	+	+ (*C. consisus* and *C. upsaliensis* are negative)	+
Oxidase	+	+	+
Urease	−	− (*C. lari* is positive)	− (*H. pylori* is positive)

Adapted from Lehner, A. et al. 2005. Int. J. Food Microbiol. 102:127–135

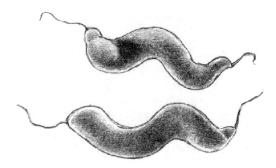

Campylobacter jejuni

Fig. 12.1 Schematic drawing of scanning electron microscopic photographs of *Campylobacter jejuni* showing polar flagella

Biology

Campylobacter species are Gram-negative, nonspore-forming, curved, S-shaped, or spiral helical rods with approximately 0.5–5.0 μm in length. *Campylobacter* displays a single polar flagellum at one or both ends and exhibits corkscrew-like motion (Fig. 12.1). In older cultures, bacteria may actually appear as spherical or coccoid bodies which corresponds to a dormant, viable but nonculturable state (VBNC). These highly successful foodborne pathogens are actually quite fastidious and have a stringent set of growth requirements. *Campylobacter* are microaerophiles that require oxygen concentrations of 3–5% and carbon dioxide of 3–10%. Some *Campylobacter* spp. (*C. coli, C. jejuni, C. upsaliensis,* and *C. lari*) are thermophiles and grow optimally at 42 °C and will not grow below 30 °C. The nonthermophiles include *C. consisus, C. curvas,* and *C. fetus* (give optimum temperature range for these spp.) Growth is further limited by osmotic stress (2% NaCl concentration), desiccation, and pH values less than 4.9. They utilize amino acids instead of carbohydrates for energy. The genome of *C. jejuni* is relatively small, 1.6–1.9 Mbp, indicating the presence of fewer genes compared to other bacterial pathogens, thus reflecting its requirement for complex growth media.

Table 12.2 Classification of *Campylobacter* species based on their biochemical properties

Characteristic	C. jejuni	C. jejuni subsp. doylei	C. coli	C. lari	C. fetus subsp. fetus	C. upsaliensis
Growth at 25 °C	–	±	–	–	+	–
Growth at 35–37 °C	+	+	+	+	+	+
Growth at 42 °C	+	±	+	+	+	+
Nitrate reduction	+	–	+	+	+	+
H$_2$S, lead acetate strip	+	+	+	+	+	+
Catalase	+	+	+	+	+	–
Oxidase	+	+	+	+	+	+
Motility (wet mount)	+	+	+	+	+	+
Hippurate hydrolysis	+	+	–	–	–	–
Nalidixic acid	S	S	S	R	R	S
Cephalothin	R	R	R	R	S	S

+ positive, – negative, *S* sensitive, *R* resistance
Source: FDA/CFSAN – BAM – *Campylobacter* (http://www.cfsan.fda.gov/~ebam/bam-7.html)

Classification

Out of 16 species in the genus *Campylobacter*, 12 are considered pathogenic including *C. fetus, C. coli, C. jejuni, C. upsaliensis, C. consisus, C. curvas,* and *C. lari*. Infection by these pathogens may lead to diarrhea; however, *C. jejuni* has also been implicated in systemic infection. *C. jejuni* is the most recognized *Campylobacter* that is involved in the 95% of the outbreaks and sporadic illnesses. A typing system based on heat-labile antigenic factors has identified 100 serotypes of *C. jejuni, C. coli,* and *C. lari*, whereas the heat-stable typing system based on the LPS O-antigen classifies 60 serotypes. Biochemical properties are used routinely in laboratory for classification of *Campylobacter* spp. (Table 12.2).

Sources

Animals are the main reservoir for *Campylobacter* and can be found in rabbits, birds, sheep, horses, cows, pigs, poultry, and even domestic pets. This organism is also found in vegetables, shellfish, and water. In cases of foodborne disease, poultry products serve as a major source of *Campylobacter* but outbreak investigations have also implicated unpasteurized milk, food handler contamination, and contaminated surface water as infection sources. *Campylobacter* spp. colonize in the caeca of broiler chicken with an average of 10^6–10^7 cfu g^{-1} of cecal content and *C. jejuni* and *C. coli* are reported to be the predominant species.

Antibiotic Resistance

Emergence of antibiotic resistance in thermophilic campylobacters especially in *C. jejuni* and *C. coli* are becoming a major issue worldwide, because of their resistance to fluoroquinolones and macrolide. Antibiotic resistance is also prevalent among poultry isolates of campylobacters, thus FDA has banned the use of fluoroquinolones as growth promoting supplements in poultry production.

Campylobacter species cause self-limiting diarrhea, thus antibiotic treatment is usually not necessary. However, antibiotic is needed for severe cases with prolonged or systemic infections. Campylobacters are zoonotic pathogen thus their transmission to humans will raise a serious concern since the most popular antibiotics will be ineffective against campylobacter treatment.

Disease

In developing countries, *Campylobacter* infection is often limited to children and culminates in watery diarrhea. In industrialized countries *Campylobacter* is the most reported bacteria associated with acute inflammatory enteric infection and accounts for about 2.4 million cases of foodborne illness per year in the United States. Approximately 95% of these human infections are caused by *C. jejuni* or *C. coli*. The infective dose resulting in symptoms is thought to be 500–10,000 organisms and dosage often correlates with the intensity of the attack. Immunocompromised individuals are at higher risk of infection; *Campylobacter* infection is 40–100% more common in AIDS patients than immunocompetent individuals. The incubation period for *Campylobacter* spp. is 1–7 days, but 24–48 h is most common. The infection is marked by inflammatory diarrhea lasting for 3–5 days. This may be accompanied by fever and abdominal cramping mimicking the symptoms of acute appendicitis.

Mechanism of Pathogenesis

Following ingestion, *Campylobacter* spp. reach the lower gastrointestinal tract and invade the epithelial cells in distal ileum and colon, resulting in cell damage and severe inflammation (Fig. 12.2). Virulence factors contributing to

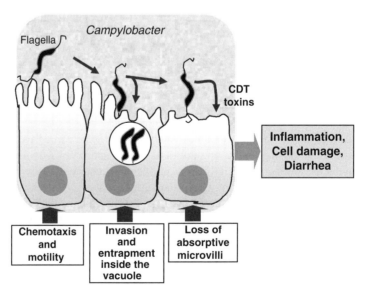

Fig. 12.2 Mechanism of *Campylobacter jejuni*-induced epithelial cell damage in the intestine. Steps in pathogenesis include (1) chemotaxis and motility, (2) adhesion, invasion and growth inside the vacuole, and (3) production of cytolethal distending toxin (CDT) production. As a result cell damage and inflammation lead to fluid loss and diarrhea

chemotaxis and motility, quorum sensing, adhesion, and invasion are required for colonization. Iron acquisition, oxidative stress defense, and resistance to bile salts ensure bacterial survival and growth in the host. Toxin production is another critical aspect in *Campylobacter* pathogenesis.

Intestinal Colonization

Unflagellated *Campylobacter* grow up to three times faster than the wild-type species. However, flagellar expression, encoded by *flaA* and *flaB*, is crucial in crossing the intestinal mucosa. *flaA* mutants show significant decreases in invasion of cultured cells. Other genes, such as *flhA* and *flhB*, also aid flagella expression. Mutation in *flhA* and *flhB* prevents production of FlaA or FlaB, ultimately halting motility and pathogenesis.

   Campylobacter heat-shock proteins (HSP) GroESL, DnaJ, DnaK, and ClpB aid in bacterial thermotolerance during survival in the bird intestine since the gut temperature is about 42 °C. Among the HSPs, only DnaJ was shown to directly contribute to bacterial colonization.

Adhesion and Invasion

Numerous *Campylobacter* adhesion proteins have been identified using both cell culture and in vivo models. *Campylobacter* uses a 37 kDa CadF protein that binds to fibronectin, a 220 kDa glycoprotein commonly found in locations of cell-to-cell contact in the gastrointestinal tract. CadF is required for binding in vivo in the chicken cecum and human INT-407 enterocyte-like cells. *Campylobacter* also uses Peb1 and JlpA for adhesion. Mutation in Peb1 led to significantly decreased invasion of INT-407 cells and decreased colonization in a murine model. Antibody blocking of the surface-expressed lipoprotein, JlpA (42.3 kDa) led to decreased binding in HEp-2 cells. JlpA binds to eukaryotic Hsp90 and induces signal transduction in the host cell. *Campylobacter* also adheres to host cell H-2 antigen for colonization. The outer membrane constituents, lipopolysaccharide (LPS) and lipo-oligosaccharide (LOS) also contribute to bacterial adhesion and serum resistance (Fig. 12.2).

   Binding to host cells, trigger host cell cytoskeletal rearrangements through activation of microfilaments and microtubule that allow bacterial internalization. *Campylobacter* is unable to escape the membrane bound vacuole and replicates at least one cycle inside. Survival inside the vacuole is facilitated by the production of superoxide dismutase to inactivate oxygen radical and the catalase to protect against oxidative stress from host.

Cytolethal Distending Toxin

Cytolethal distending toxin (CDT) is the main toxin produced by *Campylobacter* spp. and is encoded by a three gene (*cdtABC*) operon. Cdt consists of three similar sized molecular weight toxins; CdtA (30 kDa), CdtB (29 kDa), and CdtC (21 kDa). Hence it also is called a tripartite "AB_2" toxin, in which CdtB is the active toxic unit "A" and CdtA and CdtC constitute the "B_2" subunit responsible for binding to cell receptor and internalization of active CdtB. CdtB has nuclease activity that triggers DNA-damage resulting in cell cycle arrest, especially in the G2/M phase of mitosis affecting cell division. CDT also causes distention over 72 h and detachment in target cells such as HeLa, Vero, Caco-2, and CHO cell lines. CDT is thought to disturb the maturation of crypt cells into functional villous epithelial cells thus temporarily cease the absorptive function to induce diarrhea. CDT is heat-labile (70 °C for 30 min) and trypsin-sensitive.

Other toxins produced by *Campylobacter* are; Cholera-like enterotoxin, which activates cAMP; Shiga toxins which inhibit protein synthesis; and hepatotoxin.

Iron Acquisition

The ability of *Campylobacter* to acquire iron from host transferrin (in serum) and lactoferrin (mucus) is important for its survival and pathogenesis. It does not produce any siderophores but uses siderophores (ferrichrome and enterochelin) of other bacteria to acquire iron. However, *Campylobacter* expresses iron uptake systems; hemin and enterochelin. Fur and PerR are two proteins that regulate gene products necessary for iron acquisition.

Guillain-Barré Syndrome

One of the common consequences of *Campylobacter* infection, especially by *C. jejuni* leads to a serious autoimmune disorder that affects peripheral nervous system resulting in flaccid paralysis called Guillain-Barré syndrome (GBS). During infection, sialylation of lipo-oligosaccharide (LOS) occurs. Sialylated LOS mimics the ganglioside structure of myelin sheath of nerve. LOS-reactive antibodies cause demyelination of nerve, block nerve impulses, and culminate into progressive weakness in limbs and the respiratory muscles.

Regulation of Virulence Genes

Iron acquisition is regulated by two proteins; Fur and PerR. Thermotolerance and colonization in the gut is regulated by two-component regulatory system consisting of a histidine kinase (HPK) sensor and a response regulator (RR). The RR is phosphorylated by HPK and regulates the expression of CheY, RacR, and other proteins that are responsible for colonization and thermotolerance (37–42 °C). Furthermore, the FlgS/FlgR two-component signal transduction system regulates the *fla* regulon in *C. jejuni*.

Symptoms

Campylobacter infection is seen in children and young adults and results in acute enterocolitis characterized by severe abdominal cramp, nausea, general malaise, fever, muscle ache, headache, and acute watery to bloody diarrhea lasting for 3–4 days. A majority cases are self-limiting; however *C. jejuni* infection can be severe and even may lead to death in patients with immunocompromised conditions; AIDS, cancer and liver diseases. In developing country, infection is limited to children. Chronic debilitating Guillain-Barré syndrome develops several weeks after the start of the infection characterized by generalized paralysis. Another chronic arthritis condition called Reiter's syndrome also develops in patients affecting the knee joint and the lower back.

Campylobacter infection can cause other inflammatory conditions such as appendicitis, endocarditis, peritonitis (inflammation of peritoneum), meningitis, and cholecystitis (inflammation of gall bladder).

Arcobacter

The genus *Arcobacter* is closely related to the genus *Campylobacter*; however, the members of the genus *Arcobacter* are aerotolerant and are able to grow at temperature below 30 °C. *Arcobacter* (means "arc-shaped" in Latin) is a Gram-negative helical rods (1–3 µm × 0.2–0.4 µm) and sometimes may produce unusually long cells (>20 µm). They express single polar flagellum and display typical corkscrew motility. They are microaerophilic, and grow at temperature ranges of 15–37 °C and a pH range of 6.8–8.0. *Arcobacter* spp. do not grow at 42 °C. They grow well in brain heart infusion agar (BHI) containing 0.6% yeast extract and 10% blood. There are four major species: *A. butzleri*, *A. cryaerophilus*, *A. skirrowii*, and *A. nitrofigilis*, and the former two species are associated with human gastroenteritis. An *Arcobacter*-related outbreak was first reported in 1983 in Italy affecting children in a primary school caused by *A. butzleri* with ten children showing the symptoms of abdominal cramp without diarrhea.

Arcobacter spp. have been isolated from food samples, drinking water, animals, humans, and aborted animal fetuses. *Arcobacter* have been isolated from poultry carcasses but they are not considered as a natural reservoir of this pathogen.

Pathogenic mechanism of *Arcobacter* is not known; however, some species may produce cytolethal distending toxin (CDT) that causes rounding (cytotoxicity) of cultured cell lines. In addition, presence of other cytotoxic factor that causes cell elongation, and vacuole formation is reported. Bacterial adhesion to intestinal cell lines without any invasion has been reported. Hemagglutin, a glycoprotein of 20 kDa has been found in *Arcobacter*, which possibly interacts with a glycan receptor containing D-galactose for bacterial adhesion. *Arcobacter* causes abortion and stillbirth in cows, sheep, and pigs. The organism has been isolated from uterus, oviduct, and placental tissues. The organism is also isolated from stomach of pigs showing gastric ulcer. In humans, especially in children, they cause gastroenteritis and has been routinely isolated from diarrheal patients. *Arcobacter* are also thought to cause chronic enteritis in adults. The symptoms include nausea, abdominal pain, vomiting, fever, chills, and diarrhea.

Prevention and Control

Campylobacter spp. are considered normal inhabitant of livestock, poultry, and wild animals. Therefore, it is rather difficult to control the access of *C. jejuni* to raw foods, particularly foods of animal origin. However, proper sanitation can be used to reduce its load in raw foods during production, processing, and future handling. Preventing consumption of raw foods of animal origin, heat-treatment of a food, when possible, and preventing postheat contamination are important to control *Campylobacter* in foods. Contamination of vegetables can be controlled by applying treated animal manures as fertilizer and washing produce in chlorinated or ozonated water. Good personal hygiene and sanitary practices must be maintained by food handlers to avoid campylobacter-related diarrhea. *Campylobacter* spp. are heat-sensitive with a decimal reduction time at 55 °C is 1 min.

Avoiding fecal contamination during slaughter can reduce pathogen load in food. *Campylobacter* and *Arcobacter* are temperature-sensitive and thus cold storage of meat at or near 4 °C can reduce bacterial counts. Storage in the presence of sodium lactate, sodium citrate, sodium triphosphate can be effective in controlling *Arcobacter*. Heating of food to an internal temperature of 70 °C, and irradiation with 0.27–0.3 kGy for 10 s can inactivate *Arcobacter*.

Experimental approach with fucosylated human milk oligosaccharides has demonstrated the inhibition of *C. jejuni* binding to intestinal H-2 antigen. This and similar strategies could be used to control *Campylobacter* infection in humans.

Since the campylobacteriosis involves self-limiting diarrhea, antibiotic therapy is not required, but maintenance of hydration and electrolyte balance is advised. Antibiotic therapy is however, needed for immunocompromised patients to control bacteremia and sepsis. Erythromycin and newer microlides, azithromycin, and clarithromycin are effective against *C. jejuni* infection. Increased resistance of *Campylobacter* to fluoroquinolone discourages its therapeutic application.

Detection of *Campylobacter* and *Arcobacter*

Several selective isolation media have been formulated to isolate *Campylobacter* spp. from environmental, fecal and food samples. Campylosel medium uses cefoperazone, vancomycin, and amphotericin B as selective agents. The CCDA (charcoal cefoperazone, deoxycholate agar) and CAT (cefoperazone, amphotericin B, teichoplanin) media have been used for isolation of *Campylobacter* at $37\,^\circ$C under microaerophilic conditions, i.e., under oxygen concentrations of 5–10% and a CO_2 concentrations of 1–15%. Sometimes a gas mixture of 15% carbon dioxide, 80% nitrogen, and 5% oxygen is used to create the microaerophilic environment.

PCR-based assays have been developed for detection of *Campylobacter* species and the target gene included flagellin (*flaA*), 16S rRNA, and 16S/23S intergenic spacer region. *Campylobacter* species identification and typing has been done by using Ribotyping, restriction fragment length polymorphism (RFLP), amplified fragment length polymorphism (AFLP), pulsed-field gel electrophoresis (PFGE) and randomly amplified polymorphic DNA (RAPD)-PCR methods.

Arcobacter spp. have been isolated from food samples, poultry carcasses, drinking water, animals, humans, and aborted animal fetuses. An enrichment broth containing cefoperazone, bile salts, thioglycolate, and sodium pyruvate is used. Enrichment at $25\,^\circ$C under aerobic environment is commonly practiced, which requires about 4–5 days. In addition, a commercial medium called modified charcoal cefoperazone deoxycholate (mCCDA) is used for isolation of *Arcobacter* species.

Identification of *Arcobacter* has been done by using various molecular tools that use 16S or 23S rRNA as probes including ribotyping, RFLP, AFLP, PFGE, and RAPD. Multiplex PCR method targeting the 16S and 23S rRNA genes has been developed for the simultaneous detection and identification *Arcobacter* species.

Summary

Historically, *Campylobacter* and *Arcobacter* species are considered as animal pathogens; however, in the last 30 years both were reported to cause several outbreaks in humans causing gastroenteritis. They are fastidious curved rods and have stringent growth requirements. *Campylobacter* is microaerophilic, and several of the species are thermophilic and are unable to grow below $30\,^\circ$C. *Arcobacter* are aerotolerant and can grow below $30\,^\circ$C. Both *Campylobacter* and *Arcobacter* are routinely isolated from livestock, poultry, and water. Outbreak of *Campylobacter* is associated with meat, poultry, and milk. Of 16 species of *Campylobacter, C. jejuni* is responsible for 95% of the outbreaks and is considered the most predominant pathogen. *Campylobacter* pathogenesis depends on the expression of several virulence factors that control their motility, chemotaxis, quorum sensing, bile resistance, adhesion, invasion, toxin production, growth inside cells, and iron acquisition. Bacteria possibly induce their own internalization through signaling events and rearrangement of host cytoskeletal structure, and survival inside the epithelial cells by expressing superoxide dismutase and catalase to deactivate host oxidative stress defense. Cytolethal distending toxin (CDT) arrest cell cycle division and disrupts the absorptive function of villous epithelial cells and promote diarrhea. The *Campylobacter*-induced diarrhea is mostly self-limiting; however; *Campylobacter* may cause fatal infection in immunocompromised patients. Patients suffering from *C. jejuni* infection may also develop Guillain-Barré syndrome characterized by generalized paralysis and muscle pain and the Reiter's syndrome characterized by arthritis in knee joints or lower back. The pathogenic mechanism of *Arcobacter* is unknown, but they cause diarrhea in humans (in children) and abortion and stillbirth in cows, sheep, and pigs.

Further Readings

1. Altekruse, S.F., Stern, N.J., Fields., P.I., and Swerdlow, D.L. 1999. *Campylobacter jejuni* – an emerging foodborne pathogen. Emerg. Infect. Dis. 1:28–36.
2. Baserisalehi, M., Bahador, N., and Kapadnis, B.P. 2006. *Campylobacter:* an emerging pathogen. Res. J. Microbiol. 1:23–37.
3. Bereswill, S. and Kist, M., 2003. Recent developments in Campylobacter pathogenesis. Curr. Opin. Infect. Dis. 16:487–491.
4. Forsythe, S.F. 2006. *Arcobacter.* In Emerging Foodborne Pathogens. Edited by Motarjemi, Y. and Adams, M., CRC, Boca Raton, pp 181–221.
5. Gupta, A., Nelson, J.M., Barrett, T.J., Tauxe, R.V., Rossiter, S.P., Friedman, C.R., Joyce, K.W., Smith, K.E., Jones, T.F., Hawkins, M.A., Shiferaw, B., Beebe, J.L., Vugia, D.J., Rabatsky-Ehr, T., Benson, J.A., Root, T.P., Angulo, F.J., and NARMS Working Group. 2004. Antimicrobial resistance among *Campylobacter* strains, United States, 1997–2001. Emerg. Infect. Dis. 10:1102–1109.
6. Lara-Tejero, M. and Galan, J.E. 2000. A bacterial toxin that controls cell cycle progression as a deoxyribonuclease I-like protein. Science 290:354–357.
7. Lastovica, A.J. and Skirrow, M.B. 2000. Clinical significant of *Campylobacter* and related species other than *C. jejuni* and *C. coli*. In Campylobacter. Edited by Nachamkin I. and Blaser, M.J., American Society for Microbiology, Washington, DC, pp 89–120.
8. Lehner, A., Tasara, T., and Stephan, R. 2005. Relevant aspects of *Arcobacter* spp. as potential foodborne pathogen. Int. J. Food Microbiol. 102:127–135.
9. Mandrell, R.E. and Miller, W.G. 2006. *Campylobacter.* In Emerging Foodborne Pathogens. Edited by Motarjemi, Y. and Adams, M., CRC, Boca Raton, pp 476–521.
10. Moor, J.E., Barton, M.D., Blair, I.S., Corcora, D., Dooley, J.S.G., Fanning, S., Kempf, I. et al. 2006. The epidemiology of antibiotic resistance in *Campylobacter.* Microbes Infect. 8:1955–1966.
11. Murphy, C., Carroll, C., and Jordan, K.N. 2006. Environmental survival mechanisms of the foodborne pathogen *Campylobacter jejuni*. J. Appl. Microbiol. 100:623–632.
12. Park, S.F. 2002. The physiology of *Campylobacter* species and its relevance to their role as foodborne pathogens. Int. J. Food Microbiol. 74:177–188.
13. Smith, J.L. and Bayles, D.O. 2006. The contribution of cytolethal distending toxin to bacterial pathogenesis. Crit. Rev. Microbiol. 32:227–248.

14. Snelling, W.J., Matsuda, M., Moore, J.E., and Dooley, J.S.G. 2005. Under the microscope: *Campylobacter jejuni.* Lett. Appl. Microbiol. 41:297–302.

15. Snelling, W.J., Matsuda, M., Moore, J.E., and Dooley, J.S.G. 2005. Under the microscope: *Arcobacter.* Lett. Appl. Microbiol. 42:7–14.

16. Tee, W. and Mijch, A. 1998. *Campylobacter jejuni* bacteremia in human immunodeficiency virus (HIV)-infected and non-HIV-infected patients: comparison of clinical features and review. Clin. Infect. Dis. 26:91–96.

17. Vandamme, P. 2000. Microbiology of *Campylobacter* infections: taxonomy of the family Campylobacteraceae. In *Campylobacter*. Edited by Nachamkin, I. and Blaser, M.J., ASM, Washington, DC, pp 3–26.

18. van Vliet, A.H.M. and Ketley, J.M. 2001. Pathogenesis of enteric *Campylobacter* infection. J. Appl. Microbiol. 90:45S–56S.

19. Wassenaar, T.M. and Blaser, M.J. 1999. Pathophysiology of *Campylobacter* jejuni infections of humans. Microbes Infect. 1:1023–1033.

20. Wesley, I.V., Wells, S.J., Harmon, K.M., Green, A., Schroeder-Tucker, L., Glover, M., and Siddique, I. 2000. Fecal shedding of *Campylobacter* and Arcobacter spp. in dairy cattle. Appl. Environ. Microbiol. 66:1994–2000.

13

Yersinia enterocolitica and Yersinia pestis

Introduction

The genus *Yersinia* belongs to the family *Enterobacteriaceae* and consists of Gram-negative coccobacilli named after the French scientist Alexandre Yersin. The genus has 11 species (*Yersinia enterocolitica, Y. pseudotuberculosis, Y. pestis, Y. frederiksenii, Y. intermedia, Y. kristensenii, Y. mollaretii, Y. bercovieri, Y. aldovae, Y. rhodei*, and *Y. ruckeri*). The three species are well known to cause infections in humans include: *Yersinia enterocolitica, Y. pseudotuberculosis*, and *Y. pestis*. All three are facultative intracellular pathogens, harbor a 70-kb virulence plasmid (pVY), and exhibit tropism for lymphoid tissues. *Yersinia enterocolitica* is associated with foodborne infections resulting in gastroenteritis, mesenteric lymphadenitis, septicemia, and emerged as a human pathogen during 1930s. The Center for Disease Control and Prevention (CDC) estimates, about 87,000 cases of human diseases occur due to *Y. enterocolitica* infection annually in the US and 90% of those are foodborne. A recent US foodborne outbreak survey report indicates that between 1988 and 2002, there were 8 outbreaks linked to *Y. enterocolitica* with 87 cases. *Yersinia pseudotuberculosis* also causes gastrointestinal disorders, septicemia, and mesenteric adenitis. *Yersinia pestis* causes bubonic or pneumonic plague and the organism can be transmitted through contact with wild rodents and their fleas. Plague is an old-world disease and often referred to as "Black Death" and occurs in the bubonic or pulmonary forms.

Yersinia species can be differentiated based on their biochemical properties (Table 13.1). *Yersinia* grows on MacConkey agar, is catalase positive, oxidase negative, and ferments glucose. Most *Yersinia* species are noncapsulated except *Y. pestis*, which develops an envelope at 37 °C. All three pathogenic species share common antigens (O:3, O:8, O:9). In addition, all three species also share *Yersinia* outer membrane proteins (YOPs), V (immunogenic protein), and W (nonprotective lipoprotein) antigens. The fraction 1 envelope antigen (F1) is produced at 37 °C and has two major components; fraction 1A, polysaccharide, and 1B, a protein. *Yersinia pestis* has been identified as a subspecies of *Y. pseudotuberculosis* based on the 16S rDNA sequence. These two species also share 11 common antigens.

Table 13.1 Classification of *Yersinia* species based on biochemical properties

Characteristics	Y. pestis	Y. pseudotuberculosis	Y. enterocolitica
Motility at 22 °C	−	+	+
Lipase at 22 °C	−	−	v
Ornithine decarboxylase	−	−	v
Urease	−	+	+
Citrate at 25 °C	−	−	−
Voges-Proskauer	−	−	v
Indole	−	−	v
Xylose	+	+	v
Trehalose	+	+	+
Sucrose	−	−	v
Rhamnose	+	+	−
Raffinose	−	−	v

+ positive, − negative, *v* variable
Adapted from Smego, R.A. et al. 1999. Eur. J. Clin. Microbiol. Infect. Dis. 18:1–15

Yersinia enterocolitica

Biology

Yersinia enterocolitica is a Gram-negative short rod, nonspore-forming, and facultative anaerobe. *Y. enterocolitica* grows between 0 and 44 °C, with an optimum growth at 25–29 °C. Growth occurs in milk and raw meat at 1 °C, but at a slower rate. Bacteria can grow in 5% NaCl and at a pH above 4.6 (range pH 4–10). *Yersinia* expresses peritrichous flagella at lower temperature (25 °C) but it is nonflagellated (nonmotile) at 37 °C. Bacterial swimming and swarming motility is thought to be regulated by bacterial quorum sensing ability. *Yersinia* is equipped to maintain biphasic lifestyle; one in the aquatic environment/food system and other in human host. *Yersinia* grows slowly on sheep blood agar, MacConkey agar, Hektoen-Enteric agar producing pinpoint colonies after 24 h of incubation. It ferments sucrose but not xylose or lactose. For selective isolation – cefsulodin–irgasan–novobiocin (CIN) and virulent yersinia enterocolitica (VYE) agar are used.

Classification

Yersinia enterocolitica is classified into five biogroups based on their pathogenicity, and ecologic and geographic distributions: 1A, IB, 2, 3, 4, and 5. *Y. enterocolitica* has about 60 serotypes. Serogroups belong to each biogroup are presented below: IA (O:5; O:6, 30; O:7, 8; O:18; O:46), IB (O:8; O:4; O:13a, 13b; O:18; O:20; O:21), 2 (O:9; O:5, 27), 3 (O:1,2,3; O:5,27), 4 (O:3), and 5 (O:2,3). Worldwide the predominant serogroups that cause most infection are O:3, O:8, O:9, and O:5,27.

Sources

Yersinia enterocolitica is widely distributed in nature, including foods, water, sewage, and animals (cattle, sheep, goats, dogs, cats, rodents); however, pig is the primary reservoir (bacteria present as a commensal) of pathogenic strains that cause infections in humans. Thirty-five to 70% of swine herds and 4.5–100% of individual pigs carry pathogenic *Y. enterocolitica*. Environmental isolates are generally nonpathogenic and belong to the biogroup 1A. The first foodborne outbreak of *Y. enterocolitica* occurred in New York state in 1976 affecting 222 children due to consumption of chocolate milk and was caused by serotype O:8. *Y. enterocolitica* outbreak also is associated with pasteurized and unpasteurized fluid milk. Chitterlings, made of swine intestines also are implicated in several outbreaks in infants in the US. Chitterlings are prepared by boiling the large intestine of swine and are traditional winter-holiday food for many African American families in the US. Though the fats and fecal contents are removed before boiling of the final product, the chitterling preparation requires substantial handling and children in the household may have exposed during its preparation.

Virulence Factors

Yersinia enterocolitica is an invasive enteric pathogen and interestingly not all strains are virulent. Environmental isolates are generally nonpathogenic and pathogenic strains are predominant in pigs. The pathogenic strains vary in serological characteristics. In the US, the most common serovar implicated in yersiniosis is O:8. Pathogenic strains carry several virulence factors encoded in the chromosome and in a 70 kb virulence plasmid (pVY), which are required for adhesion, invasion and colonization of intestinal epithelial cells and lymph nodes, growth and survival inside the macrophages, killing of neutrophils and macrophages, and serum resistance (Table 13.2). The nonpathogenic strains lack the virulence plasmid, pVY.

The chromosomal-linked virulence genes include invasin, attachment invasion locus (Ail), an enterotoxin (Yst), and siderophores such as yersiniabactin and others. The environmental isolates are shown to be negative for *inv* and *ail* genes and are not associated with human disease. The plasmid-linked virulence genes encode Yersinia outer membrane proteins (YOPs) that are responsible for bacterial adhesion, a type III secretion system (TTSS) to deliver virulence proteins to the host cell cytosol to allow bacterial growth inside macrophages, serum resistance, and septicemia. YOPs are present on the cell surface as well as secreted into the medium. YOPs are secreted in the presence of very low levels of Ca^{2+}. YOP expression is temperature dependent and occurs mostly at 37 °C, which is critical during pathogenesis in the host. Description of selected virulence factors are presented below.

Chromosome-Linked Virulence Gene Products

Invasin

Invasin is a 92-kDa outer membrane protein and binds to $\beta1$-integrin receptor and is an important virulence factor for the early phase of intestinal infection. (Invasin is 103-kDa protein in *Y. pseudotuberculosis*.) Invasin is encoded by *inv* gene and its expression is high when Yersinia was grown in media with a neutral pH at 23 °C but low at 37 °C. However, when *Y. enterocolitica* was grown at low pH (equivalent to intestinal pH of 5.6–6) at 37 °C, the invasin expression was enhanced. The *inv* expression is regulated by *rovA* (regulator of virulence A), which is located on chromosome. RovA is a 143 amino acid protein and is present in all three *Yersinia* species. Invasin facilitates bacterial colonization and translocation through M-cells located in the follicle-associated

Table 13.2 *Yersinia enterocolitica* key virulence factors and their genetic origin

Origin	Sizes	Function
Chromosome		
Invasin	92 kDa	Inv binds to β1-integrin and promote adhesion and invasion
Attachment invasion locus (Ail)	17 kDa	Attachment and invasion; serum resistance
Yst (enterotoxin)	3.5–6 kDa	Yst increases cGMP level and fluid secretion
Yersiniabactin (catechol-type)	482 Da	Siderophore, an iron binding protein
Virulence plasmid (pVY)	70 kb	
Ysc (Yop secretion)	28 proteins	A type III secretion system (TTSS)
YopH	51 kDa	Dephosphorylate host proteins, modulate signaling pathway, and prevent phagocytosis
YopM	41.6 kDa	Kinase activity; signaling activity?
YopD	33.3 kDa	Responsible for translocation of YopE and other effector proteins (YopH, YopM, YopO, etc.) across the membrane
YopE		Inactivates Rho family of GTPase as a result disrupts actin cytoskeleton and prevents phagocytosis
YopP	33 kDa	Macrophage apoptosis; alters the expression of cytokines
YopT		Interfere with actin cytoskeleton formation by inactivating Rho GTPase
YadA (adhesion protein)	160–240 kDa	Adhesion to epithelial cells by interacting with β1-integrin; blocks complement-mediated killing; serum resistance
YopB	41.8 kDa	Inhibit cytokine release from macrophages
LcrV	37.2 kDa	Low calcium response

epithelium overlying Peyer's patches. Invasin promotes proinflammatory immune response by activating NF-κB pathway in the epithelial cells. As a result, chemotactic cytokines are made which recruit neutrophils and macrophages at the site of infection. Macrophages ingest *Yersinia* and disseminate them to regional lymph nodes, liver, and spleen.

Attachment Invasion Locus (Ail)
Ail is a 17 kDa membrane protein and is involved in adhesion and in the serum resistance. It possibly acts by binding to complement byproducts thus preventing the formation of membrane attack complex (MAC), which is required for lysis of bacterial cells.

Iron Acquisition
Iron acquisition is achieved by siderophore such as yersiniabactin (catechol-type), which is chromosome-linked. Under iron-starvation, *Yersinia* produces

large amounts of iron receptors, FoxA and FcuA on the outer membrane, which bind the siderophores to sequester iron.

Yersinia Stable Toxin (Yst)

Y. enterocolitica produces an enterotoxin, Yst, which is heat-stable (100 °C for 15 min) and remains active at pH range of 1–11 at 37 °C for 4 h and is methanol soluble. Yst is structurally and functionally homologous to the heat-stable enterotoxin (ST) of enterotoxigenic *Escherichia coli* (see Chap. 10), and is encoded by the chromosomal *yst* gene. Yst is involved in diarrhea. Three subtypes of Yst exist: Yst-a, Yst-b, and Yst-c. The subtypes Yst-a and Yst-b each made of 30 amino acids, while the Yst-c is much larger and consisted of 53 amino acids.

Plasmid (pVY)-Linked Virulence Gene Products

Yersinia Adhesion Protein (YadA)

YadA is an important virulence factor and facilitates the bacterial attachment to host cells and protects *Yersinia* from the nonspecific immune system such as phagocytosis and complement-mediated cell lysis. YadA is an outer membrane protein encoded by the virulence plasmid (pYV). It is present in both *Y. enterocolitica* and *Y. pseudotuberculosis* but is nonfunctional in *Y. pestis*. YadA is a 160–240 kDa protein composed of three monomers, each 44–47 kDa and appears as fibrillar (or lollipop-like) structure covering the entire bacterial cell surface. Each fibrilla is of 50–70 nm in length and 1.5–2.0 Å in diameter. YadA has three parts; N-terminal head, intermediate stalk, and the C-terminal anchor domain. The N-terminal domain contains 25 amino acids long signal sequence. The stalk binds to host cell receptor and resist host immune system, and the C-terminal domain anchors to the bacterial outer membrane.

YadA serves as a major adhesin and binds to several extracellular matrices (ECM) including cell surface-associated collagen, fibronectin, and laminin. YadA also promotes bacterial internalization by interacting with the epithelial β1-integrin proteins by zipper mechanism. Interaction with β1-integrin initiates signaling events that orchestrate actin recruitment to alter the cytoskeletal structure to promote bacterial entry.

YadA is expressed at 37 °C but not at 25 °C. Expression of YadA is regulated by two different gene products: VirF (virulence) and LcrV (low calcium response). VirF senses the optimal temperature, i.e., 37 °C required for protein synthesis, and LcrV regulates the *yadA* expression depending on the availability of extracellular calcium concentration. Furthermore, *yadA* expression is not affected by pH, salt, or sugar concentration.

YadA also disrupts the host cell signaling pathways to block the release of proinflammatory cytokine such as IL-8, which is a chemoattractant for neutrophils. YadA inhibits oxidative burst in polymorphonuclear neutrophils (PMN) and activates YopH, a phosphotyrosine phosphatase that blocks the phagocytic mechanism. YadA inhibits the formation of C3b and membrane attack complex (MAC) by activating photolytic enzyme, factor H, which degrades the C3b.

YopB

YopB suppresses the secretion of macrophage-derived cytokines, IL-1 and TNF-α, which are important in inflammatory response against infection.

Type III Secretion System

The type III secretion system (TTSS) apparatus, called Ysc is essential for secretion of YOPs to extracellular milieu, across the outer membrane and to the host cell cytosol. It is made of 28 proteins and the genes for which are encoded by the pVY plasmid, and their expression is temperature dependent (37 °C). The TTSS delivers six effector proteins into the host cell cytosol: YopH, YopO, YopT, YopP/J, YopE, and YopM. These effector proteins affect signaling events, alter actin cytoskeletal structure to induce bacterial entry by zipper mechanism, phagocytosis, apoptosis, and inflammatory response.

YopH is a 51 kDa protein and interferes with the signal transduction pathway to block phagocyte-mediated killing. Phagocytes (macrophages and neutrophils) are important components of the innate immunity and provide the first line of defense against *Yersinia* infection. The process of phagocytosis involves actin rearrangement to form pseudopods for bacterial internalization. YopH dephosphorylates host phosphotyrosine containing proteins that are involved in actin polymerization to form pseudopods. YopH also interferes with the calcium signaling in neutrophils, and downregulates respiratory bursts in the macrophages and neutrophils. In adaptive immune response, YopH also suppresses B-cell activation and production of cytokines by the T-cells.

YopO/YpkA is an 82 kDa secreted protein and possesses kinase activity, and displays cytotoxic action. YopT possibly has Rho GTPase activity. Rho families of GTPases (Rac1, RhoA, and Cdc42) regulate actin cytoskeleton formation and interference of GTPase activity prevents phagocytosis of bacteria. YopP/J is a 33-kDa protein that blocks inflammation and induce apoptosis in macrophages by inhibiting MAPK (mitogen-activated protein kinase) signaling pathways and NF-κB pathway. As a result cytokine TNF-α and IL-8 production is downregulated. YopE has cytotoxic action causing rounding and detachment from extracellular matrix and also regulates inflammatory response. YopM is a 41 kDa protein with kinase activity with possible role in cellular signaling events.

Pathogenic Mechanism

Foods such as chocolate milk or water that are incriminated for yersiniosis are generally cycled through refrigeration or below 25 °C. Generally, a high dose (10^7–10^9 cells) is required for the disease. Once ingested, bacteria travel to the small intestine and the primary site of infection is terminal ileum and proximal colon. It is speculated that initially bacteria use chromosomally encoded virulence gene products to colonize the intestine until the temperature shift to 37 °C and then initiate the expression of pYV-encoded gene products. Bacteria bind to mucus membrane using invasin, Ail, and YadA and enter through M-cells overlying the Peyer's patches (Fig. 13.1). Invasin interacts with the β1-integrin receptor located abundantly on the M-cells on the luminal side. YadA also aids in the invasion by interacting with the β1-integrin as well as the collagen, fibronectin, and laminin. Engulfed bacteria are then released from the M-cells in the basal layer in the lamina propria, multiply within the lymphoid follicle, and cause necrosis and

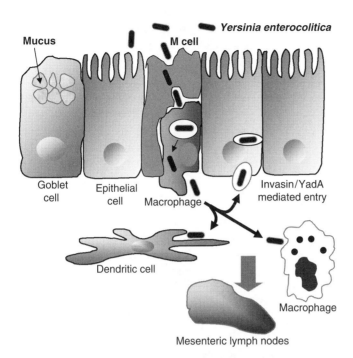

Fig. 13.1 *Yersinia enterocolitica* translocation through intestinal epithelial cells. After entry into the basal layer via M-cells, bacteria invade epithelial calls through interaction with host cell β1-integrin. Macrophage/dendritic cells transport *Yersinia* to mesenteric lymph nodes and also to liver. *Yersinia* prevents phagocytosis, also induces macrophage apoptosis, and prevents cytokine production (*See Color Plates*)

abscess in Peyer's patches. Bacteria are able to reinvade epithelial cells using the β1-integrin receptor located at the basolateral side. From Peyer's patches, bacteria spread to regional mesenteric lymph nodes causing characteristic lymphadenitis. Bacteria also disseminate to liver, spleen, and lungs and survive by resisting phagocytosis by macrophages and polymorphonuclear leukocytes such as neutrophils. Survival inside macrophages is facilitated by the delivery of YOPs to the macrophages by TTSS that interfere with the cellular signaling events by blocking the phagocytosis process, phagocytic oxidative burst and by inducing apoptosis (Fig. 13.2). *Yersinia* also blocks the release of proinflammatory cytokines (TNF-α, IFN-γ, IL-8) from macrophages and other immune cells thus inhibiting inflammation. IL-8 is a chemoattractant for neutrophils. Bacteria also block complement activation thus C3b- and MAC-mediated inactivation is avoided. Overall, the TTSS system blocks phagocytosis and suppresses immune system thereby ensuring bacterial survival in the lymphoid tissue. In the intestine, bacteria cause abscess in the Peyer's patches, induce damage to the epithelial lining.

The enterotoxin, Yst, promotes fluid secretion from cells. All three components of Yst; Yst-a, Yst-b, and Yst-c stimulate the membrane bound guanylate cyclase leading to increased accumulation and activation of intracellular cyclic guanosine monophosphate (cGMP), followed by an activation of

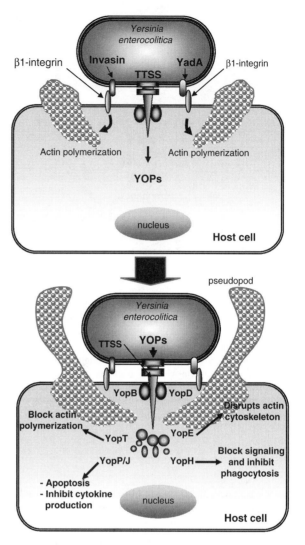

Fig. 13.2 Cellular mechanism of interaction of *Yersinia enterocolitica* with macrophage. After binding to macrophage, bacteria deliver YOPs through type III secretion system (TTSS) into the host cell cytosol

cGMP-dependent protein kinase, culminating in the final biological event, i.e., inhibition of Na^+ absorption and stimulation of Cl^- secretion.

Symptoms

Children are more susceptible than adults to foodborne yersiniosis showing acute enteritis. Symptoms include severe abdominal pain at the lower quadrant of the abdomen mimicking appendicitis, diarrhea, nausea, vomiting, and fever. It also causes enterocolitis, mesenteric lymphadenitis, and terminal ileitis. Symptoms generally appear within 24–30h after consumption of the contaminated food and last for 3–28 days for infants and 1–2 weeks for adults. The disease can be fatal in rare cases. Severity of infection is pronounced in immunocompromised host or individuals with underlying diseases resulting in septicemia, pneumonia, meningitis, and endocarditis, and can be fatal. Yersiniosis can develop skin infection and arthritis. *Yersinia* is also known to cause nosocomial infection in hospital patients exhibiting symptoms of diarrhea. *Yersinia* infection can result in sequelae in some patients leading to arthritis or Reiter's syndrome.

Prevention and Control

Yersinia enterocolitica is a psychrotroph, and therefore refrigeration cannot be used to control their growth. Good sanitation at all stages of handling and processing of food, and proper heat treatment are important to control the occurrence of foodborne yersiniosis. Consumption of raw milk or meat cooked at low temperatures should be avoided. *Yersinia enterocolitica* is susceptible to heat and pasteurization and can be easily destroyed by ionizing radiation, UV-radiation, and other food preservation procedures.

Yersinia produces two types of β-lactamases (enzymes that hydrolyse the β-lactam ring of the β-lactam antibiotics) and are thus resistant to penicillin group of antibiotics; however newer β-lactam antibiotics (ceftriaxone, ceftazidime, moxalactam) are found to be effective. *Yersinia enterocolitica* is also sensitive to imipenem and aztreonam antibiotics. Broad-spectrum cephalosporins also are effective against extraintestinal infections.

Detection

Culture Methods
Enrichment medium such as irgasan–ticarcillin–potassium chlorate (ITC) broth has been used to increase bacterial numbers. *Yersinia* are isolated on several commonly used enteric media: MacConkey agar, Hektoen-Enteric (HE) agar, and xylose–lysine deoxycholate (XLD) agars. Other selective media include cefsulodin–irgasan–novobiocin (CIN) and virulent yersinia enterocolitica (VYE) medium. Of all the selective media, CIN agar has been found to be the most effective. A modified CIN agar incorporating esculin was able to differentiate virulent *Y. enterocolitica* (esculin nonhydrolyzing and produced red colonies) from avirulent environmental *Y. enterocolitica* or other *Yersinia* species that produced dark colonies.

Biological activity of enterotoxin (Yst) is determined by the suckling mice assay or by rabbit ileal loop assay (see Chap. 5).

Serodiagnosis
Agglutination-based serodiagnosis of *Yersinia* using host serum has been used. However, it was found to be an inconsistent diagnostic tool because of the crossreaction of antiserum with several other pathogens. Though ELISA improved the sensitivity, crossreactions, and the false-positive rate were high. Indirect immunofluorescence assay has been used with biopsy specimens. It appears that culture confirmation, in conjunction with serodiagnosis may be used to correctly diagnose a patient suffering from yersiniosis.

Molecular Detection Method
Standard PCR method that target virulence genes namely *yadA* and *virF* located on pVY, and the 16S rRNA genes are used. Real-time quantitative (Q-PCR)

employing *ail* gene as target has been used for detection of *Y. enterocolitica* in pig feces and meat.

Molecular Typing Methods

Serotyping based on O (somatic) and H (flagellar) antigens, multilocus enzyme electrophoresis (MLEE), phage typing, and DNA-based schemes have been used to study typing, taxonomy, and epidemiology of *Yersinia enterocolitica*. Among the DNA-based schemes, pulsed-field gel electrophoresis (PFGE), ribotyping, amplified fragment length polymorphism (AFLP), randomly amplified polymorphic DNA (RAPD), and DNA sequencing have been used that produced reproducible typing information.

Yersinia pestis

Introduction

Yersinia pestis causes plague, which is either bubonic or pneumonic (pulmonary). Plague was described as early as 430 BC in Athens, Greece, and is called an old-world disease. Plague is often referred to as "Black Death." *Y. pestis* has been responsible for three pandemics and over 200 million deaths within the last 1,500 years. Plague can be transmitted through contact with wild rodents and their fleas which act as vector (Fig. 13.3). *Yersinia pestis* has high affinity for lymphoid tissues and causes acute inflammation, abscess, and swelling of lymph nodes. These inflamed and pus emitting lymph nodes are called buboes (hence bubonic). In recent years, *Y. pestis* has gained significant interest as a possible agent of biological warfare. Pulmonary (pneumonic) plague is of major concern because infection can be acquired directly by inhalation of infectious aerosols generated by the infected person.

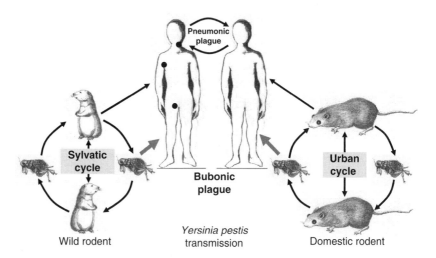

Fig. 13.3 Pathways showing transmission of *Yersinia pestis* to humans. Flea can transmit *Y. pestis* to wild rodents (sylvatic cycle), domestic rats (urban cycle), or humans. Infected flea also help maintain both sylvatic and urban cycles. Flea-bite or direct contact of rodents with human may result in the development of bubonic plague. Human-to-human transmission via aerosol/droplets may cause pneumonic plague

Biology

Yersinia pestis is a nonmotile, nonspore-forming, coccobacillary organism measuring 0.5–0.8 µm by 1–3 µm. The optimum growth temperature is 28 °C with a range of 4–40 °C. *Yersinia pestis* is a facultative intracellular organism and it has three biotypes: antiqua, medievalis, and orientalis. Biotypes can be distinguished by their ability to utilize certain substrates. Antiqua is glycerol, arabinose, and nitrate-positive. Medievalis is glycerol and arabinose-positive but nitrate-negative. Orientalis is glycerol-negative and arabinose and nitrate-positive. Existence of another biotype called microtus has been proposed. Microtus may have lost genes essential for virulence determinants and host adaptation. Microtus is glycerol-positive and arabinose and nitrate-negative.

Pathogenesis

Yersinia pestis virulence genes are located on the chromosome and on three plasmids (pVY, pMT1, pPCP1). The virulence factors are F1 (fraction 1 antigen) capsule, plasminogen activator (Pla) protein, low calcium response (Lcr) stimulon, pesticin, and a murine toxin (Ymt). These proteins are used as markers for biotyping of *Y. pestis*. The gene for Lcr is present on pVY; F1 and the murine toxin (Ymt), a phospholipase D are present on pMT1; and Pla and pesticin are located on pPCP1. Chromosome-encoded invasin and pVY-encoded YadA are inactive in *Y. pestis*. Pla serves as an adhesin, invasin, and proteolytic factor and helps in systemic dissemination of bacteria. F1 blocks phagocytosis.

Bacteria multiply rapidly inside the mid gut (stomach) of the rat flea (*Xenopsylla cheopis*), aided by phospholipase D and form a mass at around 26 °C. During feeding of blood (37 °C) on the host (rat or humans), the fleas regurgitate the bacilli to the wound. Wild animals, particularly rats are the major reservoir and can be infected asymptomatically. In rural areas, humans coming in contact with rodents or fleas (vector) carrying *Y. pestis* become infected. Cat-to-human transmission is possible and human-to-human transmission is mostly by the airborne route or by human adapted flea (*Pulex irritans*) (Fig. 13.3). The bubonic form is most common and starts with the bite of an infected flea. The incubation period is 2–8 days. Clinical manifestation starts with fever, chills, headache, and leads to swelling and enlargement of inguinal (groin) and submaxillary lymph nodes called "buboes." Inflammation or cellulitis may develop around the buboes. Large carbuncles may also develop. Gastroenteritis characterized by abdominal pain, nausea, vomiting, and diarrhea may develop in some cases. Septicemia, disseminated intravascular coagulation and shock can develop in some patients. Secondary pneumonic plague may result in bronchopneumonia with production of bloody purulent sputum. Pneumonic plague is highly contagious and can be transmitted easily by the airborne route. In the primary septicaemic plague, buboes are absent but bacteremia develops and the mortality rate is 30–50%. In the secondary pneumonic phase, bacteria disseminate to the respiratory tract and develop severe bronchopneumonia, cavitation, and formation of bloody purulent (containing pus) sputum. This form of plague is highly contagious and fatal.

Treatment and Prevention

Early intervention with antibiotics within 1–3 days of exposure may be effective. Streptomycin, gentamicin, chloramphenicol, and tetracycline are reported to be effective. Antibiotic-resistant strains have been reported recently from Madagascar where plague is endemic or could be engineered to be used as agents of bioterrorism. Use of ciprofloxacin as a prophylactic antibiotic has been found to be effective but is not good for treatment. Treatment with other fluoroquinolones has been reported to be promising.

A killed "whole cell" vaccine is available but require multiple boosters. Efficacy of the vaccine against pneumonic plague is questionable. Other preventive measures include public education about the transmission of the disease by rodents and fleas. Access of rodents to food for human consumption should be prevented. Flea-bite should be avoided by wearing protective clothing or by using insect repellant. Insecticides can be used to eliminate fleas from home or pet populations.

Detection of *Y. pestis*

Body fluids like blood, sputum, bubo aspirates, or cerebrospinal fluid can be stained for bipolar staining (Giemsa or Wayson's) to directly visualize the organism under a microscope. The samples can be streaked onto blood agar plates, brain heart infusion (BHI) agar plates, or MacConkey agar plates and incubated at 28 °C for 48 h (Note: at 37 °C, the organism develops an envelope and becomes highly virulent). The colonies are characteristically opaque, smooth, and round shaped with irregular edges. The bacteria can be extremely slow growing; thus, culture plates should be examined for 1 week before discarding. Biochemical characterization of *Yersinia* is reliable and can be done with diagnostic kits (Table 13.1). Serological tests such as passive hemagglutination (PHA) test could be used for diagnosis. Other immunoassys such as ELISA, direct immunofluorescence assay, and dipstick assays that target F1 antigen or Pla protein were found to be very useful rapid tools. A real-time 5′ nuclease PCR that targets the *pla* gene of *Y. pestis* has been developed for analysis of human sputum or respiratory swabs samples.

Summary

The genus *Yersinia* consists of 11 species, of which *Y. enterocolitica, Y. pseudotuberculosis*, and *Y. pestis* are pathogenic to humans. The former two are enteropathogenic and responsible for gastroenteritis and the latter one is responsible for plague. *Y. enterocolitica* and *Y. pseudotuberculosis* are transmitted through food, and pig serves as the primary reservoir while *Y. pestis* infection is transmitted by flea-bite and rodents act as the intermediate host. All three pathogens carry chromosomal- and plasmid-encoded virulence factors, which are required for adhesion, invasion and colonization of intestinal epithelial cells and lymph nodes, growth inside macrophages, macrophage apoptosis, and serum resistance. In *Y. enterocolitica*, chromosome-encoded virulence gene products include invasin, attachment invasion locus (Ail), siderophores (yersiniabactin), and an enterotoxin (Yst), which are important during initial colonization and invasion of intestinal M-cells and enterocytes in the intestine. The pYV plasmid-encoded virulence factors include *Yersinia* outer membrane proteins (YOPs) that are responsible for bacterial virulence protein translocation to the host cell cytosol, resist macrophage and PMN-mediated inhibition, and serum resistance. Expression of YOPs is temperature dependent and occurs mostly at $37\,°C$ which is critical for bacterial pathogenesis while inside a host. Upon entry into the digestive tract, bacteria invade M-cells overlying Peyer's patches, multiply in the lymphoid follicle, and are engulfed by macrophages. *Y. enterocolitica* is resistant to phagocytic killing by PMN and macrophages, and these cells help disseminate the organism to mesenteric lymph nodes, liver, and spleen.

Y. pestis causes bubonic and pneumonic form of plague by colonizing the lymphoid tissues of the gastrointestinal and the respiratory tracts. The organisms acquired by either flea-bite or by aerosol are transported to the lymph nodes by macrophages where they survive and resist the killing by macrophages and PMN by producing several virulence factors: F1 (fraction 1 antigen) capsule, plasminogen activator (Pla), and low calcium response (Lcr) stimulon. In the bubonic form, the submaxillary lymph nodes are enlarged and appear as buboes. In this form, fever, chill, headache, and septicemia develop and the mortality rate in untreated cases is 30–50%. In the secondary pneumonic phase, bacteria disseminate to respiratory tract and develop severe bronchopneumonia with bloody purulent sputum. This form of plague is highly contagious and invariably fatal, if not treated.

Further Readings

1. Bhaduri, S., Wesley, I.V., and Bush, E.J. 2005. Prevalence of pathogenic *Yersinia enterocolitica* strains in pigs in the United States. Appl. Environ. Microbiol. 71:7117–7121.

2. Bottone, E.J. 1997. *Yersinia enterocolitica*: the charisma continues. Clin. Microbiol. Rev. 10:257–276.

3. Bottone, E.J. 1999. *Yersinia enterocolitica*: overview and epidemiologic correlates. Microbes Infect. 1:323–333.

4. Cornelis, G.R. 2002. *Yersinia* type III secretion: send in the effectors. J. Cell Biol. 158:401–408.

5. Cornelis, G.R., Boland, A., Boyd, A.P., Geuijen, C., Iriarte, M., Neyt, C. et al. 1998. The virulence plasmid of *Yersinia*, an antihost genome. Microbiol. Mol. Biol. Rev. 62:1315–1352.

6. Fredriksson-Ahomaa, M., Stolle, A., and Korkeala, H. 2006. Molecular epidemiology of *Yersinia enterocolitica* infections. FEMS Immunol. Med. Microbiol. 47:315–329.

7. Grassl, G.A., Bohn, E., Muller, Y., Buhler, O.T., and Autenrieth, I.B. 2003. Interaction of *Yersinia enterocolitica* with epithelial cells: invasion and beyond. Int. J. Med. Microbiol. 293:41–54.

8. Heise, T. and Dersch, P. 2006. Identification of a domain in *Yersinia* virulence factor YadA that is crucial for extracellular matrix-specific cell adhesion and uptake. Proc. Natl Acad. Sci. USA 103:3375–3380.

9. Navarro, L., Alto, N.M., and Dixon, J.E. 2005. Functions of the *Yersinia* effector proteins in inhibiting host immune responses. Curr. Opin. Microbiol. 8:21–27.

10. Pujol, C. and Bliska, J.B. 2005. Turning *Yersinia* pathogenesis outside in: subversion of macrophage function by intracellular yersiniae. Clin. Immunol. 114:216–226.

11. Robins-Browne, R.M. 1997. *Yersinia enterocolitica*. In Food Microbiology: Fundamentals and Frontiers. Edited by Doyle, M.P., Beuchat, L.R., and Montiville, T.J., ASM, Washington, DC, pp 192–215.

12. Schubert, S., Rakin, A., and Heesemann, J. 2004. The *Yersinia* high-pathogenicity island (HPI): evolutionary and functional aspects. Int. J. Med. Microbiol. 294:83–94.

13. Singh, I. and Virdi, J.S. 2004. Production of Yersinia stable toxin (YST) and distribution of yst genes in biotype 1A strains of *Yersinia enterocolitica*. J. Med. Microbiol. 53:1065–1068.

14. Smego, R.A., Frean, J., and Koornhof, H.J. 1999. Yersiniosis I: microbiological and clinicoepidemiological aspects of plague and nonplague *Yersinia* infections. Eur. J. Clin. Microbiol. Infect. Dis. 18:1–15.

15. Virdi, J.S. and Sachdeva, P. 2005. Molecular heterogeneity in *Yersinia enterocolitica* and '*Y. enterocolitica*-like' species – implications for epidemiology, typing and taxonomy. FEMS Immunol. Med. Microbiol. 45:1–10.

16. Zhou, D., Han, Y., Song, Y., Huang, P., and Yang, R. 2004. Comparative and evolutionary genomics of *Yersinia pestis*. Microbes Infect. 6:1226–1234.

14

Vibrio cholerae, V. parahaemolyticus, V. vulnificus

Introduction

Vibrios are inhabitants of estuarine and fresh waters and some species are pathogenic to humans, and marine vertebrates and invertebrates. In humans, some species of vibrios can cause gastroenteritis following ingestion of contaminated food or water and septicemia when pre-existing cuts or abrasions on skin come in contact with contaminated water or seafoods. Vibrios are of significant concern in both developed and developing countries because of their continued burden of disease resulting from contaminated water and fish products. Three major species; *Vibrio cholerae, V. parahaemolyticus*, and *V. vulnificus* are responsible for the majority of human infections; however, several other species are responsible for sporadic infections.

Classification

The genus *Vibrio* contains 63 species and at least 11 of them are pathogenic to humans including *V. cholerae* (O1 and O139), *V. parahaemolyticus, V. vulnificus, V. mimicus, V. hollisae, V. fluvialis, V. alginolyticus, V. damsela, V. furnissii, V. metschnikovii*, and *V. cincinnatiensis*. Among these, the first three species cause most human infections.

Biology

Vibrio species are Gram-negative curved rods with size ranging from 1.4 to 2.6 μm long with 0.5 to 0.8 μm width. They are motile and possess a single polar flagellum. They are facultative anaerobes, most are oxidase-positive, and utilize D-glucose as main carbon source. *Vibrio* species produce many extracellular enzymes: amylase, gelatinase, chitinase, and DNase. Some *Vibrio* species are halophilic (tolerance up to 10% NaCl) and sodium ions stimulate their growth. Vibrios grow well in neutral to alkaline pH 9.0 and are acid-sensitive. The optimum pH range is 8–8.8 and the optimum growth temperature is 20–37 °C. Water temperatures in either side of the range severely affect the bacterial growth. Nutrient deficiency, salinity, and changes in temperature promote stress resulting in viable but nonculturable state (VBNC) especially for *V. cholerae* and *V. vulnificus*. Two chromosomes, usually one large and the other small, are present in Vibrios, which provide diversity in gene structures and gene contents. Chromosomes 1 and 2 in *V. cholerae* are 3.0 and 1.1 Mb, respectively, in *V. parahaemolyticus* 3.3 and 1.9 Mb, and in *V. vulnificus* 3.3 and 1.85 Mb.

Source and Transmission

Vibrio species are isolated from fresh, brackish, and marine waters. Vibrios are found as free-living in water or are associated with inanimate surfaces or aquatic organisms, aquatic animals (seabirds), sewage water, sediments, seafood, fish, and shellfish. Vibrios are also associated with chitinous zoo plankton and shellfish so that they can survive longer than as a free-living organism. Bivalve shellfish such as clams, oysters, and mussels accumulate bacteria because of their filter-feeding strategy. Natural disasters like flood, cyclone, and hurricane lead to the failure of sewage system and results in con-tamination of aquatic environment. Food, washed in contaminated water can transmit bacteria. Water temperature, nutrient availability, salinity, and asso-ciation with marine organisms influence the bacterial loads in water. Bacterial counts are very high during spring–summer months and counts diminish (or absent) at temperature below 10 °C. Water salinity of 5–30% favors growth and survival.

Vibrio cholerae

Introduction

One of the most important members of the genus *Vibrio* is, *Vibrio cholerae*, and the disease caused by toxigenic strains of the two serotypes (O1 and O139) of this organism is known as cholera. Filippo Pacini, an Italian physician first discovered *V. cholerae* in 1854 as the causative agent of cholera. Outbreaks of cholera dates back to 460–377 BC during the times of Hippocrates. In the modern history, epidemic and pandemic cholera occur with global implica-tions and frequent outbreaks are currently reported in Asia, Africa, and South and Central America. According to the World Health Organization (WHO), in 2005, there were 131,943 cases of cholera including 2,272 deaths in 52 different countries. Moreover, cholera outbreak has serious economic impacts worldwide. For example, the 1991 outbreak in Peru resulted in the loss of 500 million dollars in yearly fish exports.

Biology

V. cholerae is a Gram-negative rod or curved shaped bacterium (0.7–1.0 × 1.5–3.0 μm). It is a facultative anaerobe and produces pale-yellow, translucent colonies that are about 2–3 mm in diameter on a special medium known as thiosulfate citrate bile salts sucrose (TCBS) agar. *V. cholerae* is able to grow at a temperature range of 15–45 °C, a pH range of 6–10 and a salt (NaCl) concentration of 6%. There are 206 known serotypes and of which two major serotypes, O1 and O139 are responsible for the disease cholera. A major difference between O1 and the O139 is the presence of a thin capsule in O139 and its absence in O1. This difference can be observed during bacterial growth on solid agar media, where O1 produces translucent while O139 produces opaque colonies. Furthermore, the lippolysaccharide (LPS) of O1 serotype is smooth while it is semirough in O139. The O139 LPS has a highly substituted core oligosaccharide and shorter side chains of O antigen which are responsible for rough attribute. The LPS and capsule of O139 also share a unique sugar, 3,6-dideoxyhexose (colitose). The serotype O1 is subclassified into the Classical or E1 Tor biotype based on a set of phenotypic traits. The two biotypes are further classified as Inaba, Ogawa, and Hikojima subserotypes. All strains that cause cholera carry virulence genes for cholera toxin (CT) and toxin-coregulated pili (TCP). Other factors including outer membrane porins, biotin and purine biosynthetic enzymes, iron-regulated outer membrane proteins (IrgA), and O antigen of LPS are also known virulence factors in *Vibrio*. The non-O1/O139 also cause diarrhea but generally milder than O1 and these serotypes are common in the US. *V. cholerae* has a single polar flagellum, which helps the bacterium to reach to the intestinal mucosa and aids in colonization. Flagellin mutants are nonmotile and are less virulent. *V. cholerae* maintains two lifestyles; one in the aquatic environment as a free-living or attached to zooplanktons and other inside the host gastrointestinal tract. A 53 kDa colonization factor, *N*-acetylglucosamine (GlcNAc)-binding protein A encoded by *gbpA* has been shown to be involved in bacterial colonization in vitro and in vivo.

Gastroenteritis

V. cholerae is the most studied organism that is responsible for acute secretory diarrhea known as cholera. The infectious dose is 10^4–10^{10} cfu g^{-1} and the disease is spread through contaminated water and food and transmission is through the fecal–oral route. Individuals with the blood group O are more susceptible to cholera than the other blood groups. The serotype O1 is responsible for most fatal form of cholera and the non-O1/O139 are usually avirulent or mildly virulent. There are two biotypes; classical and E1 Tor biotypes and they share common LPS O antigen. A strain of *V. cholerae* O139 Bengal, was responsible for several outbreaks in India and Bangladesh and originated around 1992. It is capable of causing serious diarrhea. Following ingestion, bacterial adherence and colonization of small intestinal mucosa are facilitated by TCP, flagella, and neuraminidase. Bacteria then produce several toxins such as CT, ZOT (zona occludin toxin), ace (accessory cholera enterotoxin), and HlyA which act on mucosal cells. Toxins alter the ion balance by affecting the ion transport pumps for Na$^+$, Cl$^-$, HCO$_3^-$, and K$^+$ in cell and results in extensive fluid and ion losses.

The symptoms of diarrhea appear within 6 h to 5 days and last for 2–12 days. Diarrhea appears as "rice-water" with a fishy odor. The patients show high pulse rate, dry skin, lethargy, low urine volume, abdominal cramps, nausea, and vomiting. The watery diarrhea causes severe dehydration, loss of electrolytes, and ions causing hypertension that can be life threatening. Infants and children are highly susceptible. Recovering patients develop immunity against cholera.

Pathogenic Mechanism

Adhesion

Bacterial adherence and colonization is mediated by long filamentous pili called TCP because the genes encoding these pili are regulated similarly to genes encoding for the cholera toxin. TCP mutants are avirulent in humans. TCP pilin, subunit of TCP pili is encoded in *tcpA* and is located in the TCP pathogenicity island. Other colonization factors include mannose–fucose resistant cell associated hemagglutinin (MSHA), mannose-sensitive hemagglutinin (MSHA), and some outer membrane proteins (OMPs). TCP and other colonization factors are regulated by regulatory proteins (ToxR/ToxS and ToxT). Immediately adjacent to *tcp* cluster, the *acf* gene for accessory colonization factor (ACF), a lipoprotein, is located. The exact role of ACF is not known but it is believed to be involved in the bacterial colonization.

Cholera Toxin

Cholera toxin (CT) is the most important virulence factor in *V. cholerae*. Mutants are avirulent or may cause milder diarrhea because of the presence of other toxins. CT is the best studied bacterial toxin. It is an A–B type "ADP-ribosylating toxin." The A subunit is a 27-kDa protein encoded by *ctxA* and the B subunit consists of five identical proteins of 11.7 kDa and it is encoded by *ctxB*. A and B subunits are secreted into the periplasm, where they are assembled. Virulence genes are located in a lysogenic filamentous bacteriophage, CTXΦ (Fig. 14.1).

The B subunit of toxin first binds to host mucosal cell by binding to the Ganglioside GM_1 receptor (Fig. 14.2). It is a sialic acid containing oligosaccharide covalently attached to the ceramide lipid. It is found on the surface of many cells. The toxin is internalized and the A subunit is detached. The A subunit has the enzymatic activity, it ADP-ribosylates the Gs proteins (composed of three subunits: α, β, γ) also known as "GTP hydrolyzine proteins." Gs proteins regulate the activity of host cell adenylate cyclase and serve as "off" and "on" switch. Binding of A subunit to Gs subunit α, locks it in the "on" position and stimulates the production of the cyclic adenylate cyclase (cAMP). cAMP activates the protein kinase A, which in turn causes phosphorylation of protein especially the CFTR (cystic fibrosis transmembrane conductance regulator) protein in the ion pump and thus alters the function of sodium and chloride ion

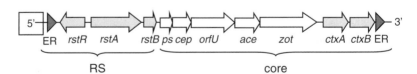

Fig. 14.1 Schematic of CTXΦ phage genome. Genes encoding cholera toxin, *ctxAB* appear at the 3′ (*right*) end of the genome. The *rstR, rstA,* and *rstB* genes constitute the RS region (*shaded arrows*) that is responsible for the site-specific integration and replication of the phage. The Core region of the phage encodes phage coat and morphogenesis proteins. ER represents End Repeat, a 14 bp sequence that flanks either end of the phage genome and represents the site specificity of the phage (adapted from Uma, G. et al. 2003. Curr. Sci. 85:101–108)

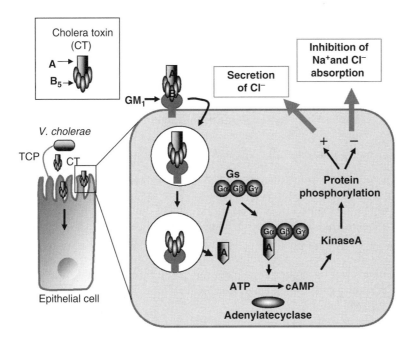

Fig. 14.2 Schematic drawing showing the action of cholera toxin (CT) on enterocyte. The cells of *V. cholerae* first binds the cells using pili (toxin-coregulated pili: TCP) or other colonization factor and produce CT. The CT is an A–B type toxin composed of one A subunit and five B subunits. The B subunits bind to GM_1 receptor and the CT is transported inside the cell. The A subunit ADP-ribosylate Gs protein (GTP hydrolyzing protein) and increases the catalysis of ATP (adenosine triphosphate) to form cyclic adenosine mono-phosphate (cAMP). cAMP mediate phosphorylation of CFTR protein involved in ion transport (pump), thus affecting ion losses (Cl^-, HCO_3^-, Na^+) and fluid flow

transport, resulting in the increased Cl^- and HCO_3^- secretion by crypt cells, and decreased absorption of Na^+ and Cl^- by absorptive cells (Fig. 14.2).

Regulation of Cholera Toxin Production
CT production is regulated by transmembrane proteins ToxR/S and TcpP/H. ToxR, a 32-kDa transmembrane protein binds to 7 bp DNA sequence located in the upstream of *ctxAB* and increases the expression of CT. ToxR and TcpP/H activate ToxT which in turn activates CT and TCP expression. ToxR and ToxT also regulate the expression of ACF, outer membrane proteins OmpU and OmpT, and other lipoproteins. The quorum-sensing regulatory proteins LuxO and HapR also control CT and TCP expression.

Other Toxins

V. cholerae also produces two other toxins; zonula occludin toxin (ZOT) and ace (accessory cholera enterotoxin) toxin. The ZOT (44.8 kDa) disrupts the "tight junction" that binds mucosal cells together and preserve the integrity of the mucosal membrane. Normally, the tight junction – maintains cellular integrity and prevents the loss of ions or water molecules. ZOT induces a reorganization of F-actin and decreases the G-actin and affects the cytoskeletal rearrangements possibly mediated by protein kinase C. Consequently tight junction looses its barrier function and enhances the pericellular permeability. ZOT also disrupts the ion balance and promotes diarrhea.

Ace toxin is responsible for diarrhea in animals but probably has no role in human diarrhea. A hemolysin called HlyA or E1 Tor hemolysin is a 65-kDa toxin and is responsible for enterotoxicity. It binds to cholesterol and oligomerizes in the membrane forming a pore of 1.2–1.6 nm. Other virulence factors like siderophores also help bacterial iron acquisition from host.

Immune Response to CT

V. cholerae induces a strong immune response by production of the secretory immunoglobulin A (sIgA). sIgA acts as an opsonin and opsonization may aid in the transcytosis of *V. cholerae* through M-cells by interacting with the CT receptor ganglioside GM_1 on M-cells. CT aids in intestinal dendritic cell (DC) maturation through production of prostaglandin E_2 (PGE_2) and nitric oxide (NO). The CT also influences the immune response of lymphocytes and monocytes by altering the expression of several genes. CT induces cAMP, which regulates the expression of genes that regulate various functions including immune response.

Vibrio parahaemolyticus

Biology

Vibrio parahaemolyticus is a halophilic enteropathogen and primarily causes gastroenteritis; however, it also causes extraintestinal infections like eye and ear infections, and wound infections affecting extremities. *V. parahaemolyticus* grows at a minimum temperature of 15 °C and a maximum temperature of 44 °C. *V. parahaemolyticus* expresses peritrichous flagella or lateral flagella (encoded by the *laf* gene) when grown in solid media exhibiting swarming phenotype. It expresses single polar flagellum (*fla*) exhibiting swimming motility when grown in liquid medium. All strains of *V. parahaemolyticus* produce H_2S when grown in triple sugar iron (TSI) medium. Urease production is an unusual phenotype for *V. parahaemolyticus* and a majority of clinical and environmental isolates are urease-negative; however urease-positive clinical isolates exists. There is a strong correlation between urease production and the presence of *trh* (see below) gene among clinical isolates.

V. parahaemolyticus strains are classified based on their somatic (O) and capsular (K) antigen patterns and the predominant serotype is O3:K6, which is distributed globally. The O3:K6 serotype is thought to be originated from Japan and was responsible for a major outbreak in Calcutta (India) in 1996. The serovariants of O3:K6 exist including O4:K12, O4:K68, O1:K41, O1:K25, O1:KUT, and several others which are responsible for regional outbreaks such as those occurred in the South and South East Asia. A recent outbreak (2006) occurred in the US was caused by the serotype O4:K12 due to the consumption of raw shellfish. Three costal states (New York, Oregon, and Washington) were involved with 177 cases without any fatalities.

V. parahaemolyticus is distributed in marine costal waters worldwide. The spread and dissemination of *V. parahaemolyticus* depend on the water temperature, zooplankton blooms, and dissolved oxygen. Countries located in the temperate climate experience higher numbers of outbreaks during summer months than the tropical countries. Countries located in the tropical zone maintain a warmer temperature thus conducive for year-round outbreaks. The outbreaks of gastroenteritis are associated with consumption of contaminated seafoods including raw oyster and other shellfish.

Pathogenesis

Vibrio parahaemolyticus is infectious and a dose of about $2 \times 10^5 - 3 \times 10^7$ cfu is required to cause disease. The incubation period is about 15 h (range, 4–96 h) and the disease may last for 2–3 days. Bacteria colonize and produce toxins and the pathogenesis depends on the production of a set of toxins that cause cell damage resulting in pore formation and loss of fluids and electrolytes. Bacteria also induce a strong inflammatory response in the intestine, which is more severe than the infection caused by the *V. cholerae*. Though the *V. haemolyticus* infection induces a strong immune response, the detailed mechanism of pathogenesis is still unclear.

The symptoms include acute abdominal pain, nausea, vomiting, headache, low grade fever, and diarrhea (watery or bloody). The stool is described as "meat washed" due to the presence of blood. The disease is usually self-limiting.

Toxins

V. parahaemolyticus produces four hemolysins: a thermostable direct hemolysin (TDH), a heat-labile TDH-related hemolysin (TRH), a thermolabile hemolysin (TLH), and δ-VPH (22.8 kDa). Toxin production in *V. parahaemolyticus* was originally detected on Wagatsuma agar (special kind of blood agar medium) and this phenomenon was called Kanagwa phenomenon (KP) and later the toxin was termed as TDH. *V. parahaemolyticus* acquired the *tdh* gene through horizontal gene transfer. TDH is a 21-kDa protein that causes a zone of β-hemolysis on Wagatsuma blood agar plate. TDH is heat-stable (100 °C for 10 min) and is produced by KP$^+$ strain. The KP$^-$ strain generally carries the heat-labile TRH toxin. TDH acts as a porin and allows the influx of ionic species; Ca^{2+}, Na^+, Mn^{2+} from enterocytes. The porin channels increase with increased concentrations of TDH and increase ionic influx, cell swelling and death of cells due to ionic imbalance. It also disrupts the epithelial barrier function by affecting the tight junction proteins (such as claudins and occludins as well as cytosolic zonula occludin protein 1 (ZO-1), ZO-2, ZO-3, cingulin, and 7H6). Toxin action can also be detected by a ligated mouse ileum model. During gastroenteritis, a strong humoral immune response to TDH and LPS has been reported and the predominant immunoglobulin is determined to be IgM.

TRH is heat-labile (inactivated at 60 °C for 10 min) and induces fluid accumulation when tested in a rabbit ileal loop model. TRH also induces chloride secretion and its involvement in diarrhea has been strongly suggested.

TLH has a phospholipase A2/lysophospholipase activity and consists of two molecular weight species of 47.5 and 45.3 kDa, and is present in all *V. parahaemolyticus* strains. Role of this hemolysin in pathogenesis is unknown.

The majority of clinical isolates obtained from the Pacific Northwest of US belonged to serotype O4:K12 contained both TDH and TRH, while the serotype

O3:K6 that caused recent outbreaks in the Gulf of Mexico and the eastern coasts of the US had only TDH.

V. parahaemolyticus have two sets of genes for synthesis of type III secretion system (TTSS), which are necessary to inject virulence proteins inside the host cells to induce cell apoptosis. The first cluster (TTSS1) is located in the large chromosome (3.3 Mb) and the second one (TTSS2) is located in the small chromosome (1.9 Mb).

V. parahaemolyticus has the ability to detach from damaged/sloughed mucosal cells and reattach to new mucus surface. The detachment factor is a "zinc and calcium-dependent protease" and is also called "hemagglutinin protease" or HAP because it has "hemagglutination activity."

Vibrio vulnificus

Introduction

Vibrio vulnificus is considered the most infectious and invasive of all the human pathogenic Vibrios and it is the leading cause of seafood-related mortality in the US with 40 deaths per year. *V. vulnificus*-related infections are thought to be very high in Japan because of the warmer coastal water and increased consumption of raw seafoods. Consumption of raw oyster and other seafoods are implicated in outbreaks. *Vibrio vulnificus* is responsible for septicemia, wound infection, and gastroenteritis in humans. Immunocompromised individuals or persons with underlying conditions such as diabetes, chronic renal disease, and cirrhosis of liver due to chronic alcoholism, are highly susceptible to infection. Gastroenteritis symptoms are associated with abdominal pain, vomiting, and diarrhea. Septicemia and wound infection progress very rapidly and a patient may die within 24 h of exposure to bacteria due to endotoxic shock. *V. vulnificus* produces several virulence factors: capsules, siderophores, and toxins such as hemolysins, collagenase, protease, elastase, DNase, mucinase, hyaluronidase, fibrinolysin, lipase, and phospholipase.

Pathogenic Mechanism

Capsular Polysaccharide

Vibrio vulnificus produces a capsular polysaccharide (CPS) which is the primary virulence factor helping bacteria to avoid phagocytosis by macrophages. Presence of CPS correlates with the opaque colony phenotype and is pathogenic. Translucent colony is avirulent. CPS synthesis is encoded by four genes; *wcvA, wcvF, wcvI*, and *orf4*. Mutation in any of these genes results in the loss of capsule biosynthesis, bacterial colony becomes translucent, and the bacterium becomes avirulent.

Acquisition of Iron

Vibrio vulnificus expresses phenolate, hydroxamate, and vulnibactin siderophores to scavenge iron from the host transferrin, hemin, and lactoferrin. Host with high levels of iron such as those are suffering from chronic hemochromatosis or cirrhosis of liver are highly susceptible to *V. vulnificus* infection. *hupA* and *fur* genes regulate iron acquisition. Avirulent strains are unable to acquire iron from host.

Flagella and Motility

It is an important virulence factor encoded in *flgC* gene and is responsible for cytotoxicity in host cells. The mutant showed defective motility and was attenuated for infection in suckling mice. This may be because decreased adherence could have prevented the delivery of toxic factors to the host cells. Also mutation in flagellar protein biosynthesis gene, *fliP* affected the virulence.

Hemolysin

Three types of hemolysin/cytolysins have been reported to be produced by *Vibrio vulnificus*. The most widely studied hemolysin (VVH) is a water-soluble polypeptide (51 kDa) that binds to cholesterol on the membrane and forms small pores in the erythrocyte membrane. It can induce apoptosis by elevating cytosolic free Ca^{2+}, release of cytochrome C from mitochondria, activate caspase-3 and degrade poly-ADP ribose-polymerase (PARP) and induce fragmentation of DNA. Toxin increases vascular permeability and causes skin damage. Two other hemolysins are less characterized and they are encoded by *hlyIII* and *trkA*.

V. vulnificus also produces a RTX (repeat in toxin) family of toxin. They share a repeated nine amino acid sequence motif among several Gram-negative hemolysins. It causes depolymerization of actin, pore formation in red blood cells, and necrotic cell death in cultured mammalian cells. However, its significance in pathogenesis is not conclusive.

Metalloprotease

Metalloprotease is a 45-kDa Zinc containing protease: The N-terminal 35-kDa part mediates proteolysis and the C-terminal 10-kDa part binds to protein substrates on erythrocytes and blood cells. Protease has two functions: membrane permeability enhancement and tissue hemorrhage.

Septicemia

It is a systemic disease generally occurs in host with underlying conditions such as liver or kidney disease, malignancies, and diabetes. Infection is also severe in people with high levels of iron in serum caused by liver cirrhosis or genetic disorder, hemochromatosis. From intestinal tract, bacteria invade epithelial cells; first binds to epithelial cells with the aid of pili, produce hemolysin, which induce apoptosis and facilitate bacterial invasion and translocation to blood circulation. Bacteria acquire iron from host cells using siderophores and grow and cause septicemia. Bacterial polysaccharide capsule activates complement cascade; however, escapes neutrophil or macrophage-mediated phagocytosis. From blood circulation bacteria can invade cutaneous tissue with the help of toxins such as hemolysins, collagenase, protease, lipase, and phospholipase. These toxins also aid in the development of edematous hemorrhagic skin lesions. Induction of proinflammatory cytokines such as TNF-α, IL-1β, and IL-6 can cause septic shock. Septicemia is manifested by fever, prostration, hypotension, chills, nausea, and occasional vomiting, diarrhea and abdominal pain. Mortality rate is 40–60% and is related to underlying conditions.

Wound Infection

Wound infection is associated with recreational or occupational activities in seawater and seafood industry. *V. vulnificus* is thought to be associated with most Vibrio-related wound infections. Other species involved are *V. parahaemolyticus, V. alginolyticus* and *V. cholerae* non O1. They produce collagenase and metalloprotease that allow them to colonize in the wound. Protease evokes two types of reactions; it increases vascular permeability, and hemorrhagic actions. Protease is bifunctional: The N-terminal portion mediates enzymatic activity. It degrades Type IV collagen located in the vascular basement membrane and cause tissue damage leading to hemorrhage. The C-terminal end binds to mast cell receptor, activates and aids in the release of histamine and bradykinin. These mediators increase the membrane permeability and cause wound edema. Pre-existing cuts, skin lesions, or injuries, trauma resulting from activities from handling seafoods, marine animals or from recreational activities are predisposing factors. Infection results in wound edema, vesicle formation, cellulitis, erythema, and tissue necrosis. Severe infection may require hospitalization and amputation of extremities. Antibiotic therapy is needed to clear the infection. Fatality rate for wound infection is 20–30% and is related to underlying conditions.

Control and Prevention of Infection by *Vibrio* species

Vibrios are becoming one of the most dangerous emerging foodborne pathogens because of increased popularity of seafoods. Aquaculture is one of the fastest growing industries worldwide. The risk of contamination of seafoods is much greater in coastal and fresh waters than in open sea. Seafood safety may include harvesting from unpolluted water or when the water temperature is low during the winter months. In addition, improved handling and processing, and packaging and shipment should be adopted to prevent bacterial growth. Workers safety should be addressed especially for those who have inflicted with cut wounds or abrasions (lacerations) in skin. They should take precautions to avoid contact with water or seafoods. Consumer education should be part of the seafood safety program: the danger of eating raw or undercooked seafoods especially for persons with underlying conditions with liver disease, diabetes, and kidney disease should be highlighted.

Hallmark of cholera is profuse diarrhea. Loss of water and electrolytes lead to dehydration. Fluid therapy is the most effective treatment to prevent dehydration. Rehydration with salts and glucose in water is very effective since glucose stimulates water uptake. Fluid could be administered by oral or intravenous route, and the latter route is needed for patients in comatose condition or showing signs of advanced form of dehydration. Antibiotics including tetracycline, cotrimoxazole, erythromycin, doxycycline, chloramphenicol, and furazolidone can be used to treat cholera.

Cholera vaccine is used to prevent infection in the population in the endemic zone. sIgA generated from vaccination prevents bacterial colonization on the mucosal surface. Since there are no crossprotection between serotype O1 and O139, bivalent vaccine are needed to provide protection against both serotypes. Earlier, injectable heat-killed bacterial cells were used but this vaccine exhibited toxic side effects due to the presence of endotoxin (LPS) and thus currently discontinued. Several strategies are undertaken to

develop safe yet protective vaccine against cholera. O139 capsular polysaccharide (CPS) conjugated to tetanus toxoid or nontoxic diptheria toxin
mutant or chicken serum albumin induced protective immunity against *V.
cholerae* in experimental animal challenge study. Oral vaccination with
killed bacteria together with purified B subunit of cholera toxin was used
and the protection was about 60%. A live oral vaccine with inactivated *ctxA,
zot, ace* and *cep* (core encoded protein; related to pilin synthesis) genes has
been used and it is found to be very promising with the protection rate of
about 84% in adult volunteers.

Detection of *Vibrio* species

Culture and Serological Methods
Selective enrichment is performed in alkaline peptone water (APW) containing
1–3% NaCl and colonies are isolated by streaking enriched samples onto the
selective thiosulfate citrate bile salt agar (TCBS). Bacteria have been also isolated on cellobiose polymyxin B colistin (CPC) and mannitol–maltose agar.
In addition, *V. vulnificus* is also isolated using sodium dodecyl sulfate-
polymyxin B-sucrose agar (SPS), *Vibrio vulinficus* agar (VVA), and modified
CPC (mCPC) agar. Biochemical characterizations are performed to determine
the species of *Vibrio*. *V. parahaemolyticus* has been tested for Kanagwa
phenomenon (hemolytic activity) by growing them on Wagatsuma agar
containing high-salt (7%) with blood for thermostable direct hemolysin
(TDH) activity. Identification is further accomplished by serotyping for
somatic O antigen and capsular K antigens. Immunological assays including
ELISA have been used that target intracellular and TDH antigens.

Molecular Techniques
A single gene or multigene-specific PCR assays have been developed that
use specific sequence in 16S rRNA, *tdh, trh, gyrB, toxR, ctxB, ctxAB*, and
tcpA for detection of *Vibrio* spp. A detection limit of 10^1–10^2 cfu has been
reported when used in a multiplex format targeting two to three genes. For
genomic typing and identification; ribotyping, restriction-fragment length
polymorphism (RFLP), amplified restriction fragment length polymorphism
(AFLP), randomly amplified polymorphic DNA, and enterobacterial intergenic
consensus sequence-PCR (ERIC-PCR) have been used. Multilocus enzyme
electrophoresis and multilocus sequence typing of housekeeping genes have
been performed for identification and typing purposes.

Summary

Though the cholera caused by *Vibrio cholerae* is considered an old-world
disease, it is continued to be a serious problem in the developing countries.
The infections caused by other Vibrios are also increasing worldwide especially in developed countries and recognized as emerging. *Vibrio cholerae* is
known for its epidemic and pandemic outbreaks and countries in Asia, Africa
and South and Central America have experienced the most outbreaks. The
water or fecal–oral transmission mode spreads *Vibrio cholerae*. Upon entry
into the intestine, the bacterium produces several adhesion factors including
toxin-coregulated pili (TCP), flagella, neuraminidase and accessory colonization factor (ACF) for colonization. The bacterium produces cholera toxin

(CT) and zona occludin toxin (ZOT), which affect the ion transport pumps for Na^+, Cl^-, HCO_3^-, and K^+ and junctional integrity and results in extensive fluid and ion losses. Diarrhea appears within 6 h to 5 days and last for 2–12 days. *Vibrio parahaemolyticus* and *V. vulnificus* infections are associated with seafoods harvested from estuarine or fresh waters. They produce several heat-stable (TDH) and heat-labile hemolysins (TRH) and phospholipases which are responsible for membrane pore formation, apoptosis, and fluid loss resulting in diarrhea. Besides, these organisms also cause septicemia and wound infections which could be fatal especially when infected by *V. vulnificus*. *V. vulnificus* is the most invasive of all Vibrios and in addition to hemolysins, it produces collagenase, metalloprotease, lipase, and phospholipases which promote rapid tissue destruction resulting in death within 24 h after the start of infection.

Further Readings

1. Albert, M.J. and Nair, G.B. 2005. *Vibrio cholerae* O139 Bengal-10 years on. Rev. Med. Microbiol. 16:135–143.
2. Faruque, S.M. and Nair, G.B. 2002. Molecular ecology of toxigenic *Vibrio cholerae*. Microbiol. Immunol. 46:59–66.
3. Harwood, V.J., Gandhi, J. P., and Wright, A.C. 2004. Methods for isolation and confirmation of *Vibrio vulnificus* from oysters and environmental sources: a review. J. Microbiol. Methods 59:301–316.
4. Gulig, P.A., Bourdage, K.L., and Starks, A.M. 2005. Molecular pathogenesis of *Vibrio vulnificus*. J. Microbiol. 43:118–131.
5. Kirn, T.J., Jude, B.A., and Taylor, R.K. 2005. A colonization factor links *Vibrio cholerae* environmental survival and human infection. Nature 438:863–866.
6. Levin, R.E. 2005. *Vibrio vulnificus*, a notably lethal human pathogen derived from seafood: a review of its pathogenicity, subspecies characterization, and molecular methods of detection. Food Biotechnol. 19:69–94.
7. Levin, R.E. 2006. *Vibrio parahaemolyticus*, a notably lethal human pathogen derived from seafood: a review of its pathogenicity, characteristics, subspecies characterization, and molecular methods of detection. Food Biotechnol. 20:93–128.
8. Lynch, T., Livingstone, S., Buenaventura, E., Lutter, E., Fedwick, J., Buret, A.G., Graham, D., and DeVinney, R. 2005. *Vibrio parahaemolyticus* disruption of epithelial cell tight junctions occurs independently of toxin production. Infect. Immun. 73:1275–1283.
9. Miyoshi, S. 2006. *Vibrio vulnificus* infection and metalloprotease. J. Dermatol. 33:589–595.
10. Nair, G.B., Ramamurthy, T., Bhattacharya, S.K., Dutta, B., Takeda, Y., and Sack, D.A. 2007. Global dissemination of *Vibrio parahaemolyticus* serotype O3:K6 and its serovariants. Clin. Microbiol. Rev. 20:39–48.
11. Tantillo, G.M., Fontanarosa, M., Di Pinto, A., and Musti, M. 2004. Updated perspectives on emerging vibrios associated with human infections. Lett. Appl. Microbiol. 39:117–126.
12. Thompson, F.L., Iida, T., and Swings, J. 2004. Biodiversity of vibrios. Microbiol. Mol. Biol. Rev. 68:403–431.
13. Uma, G., Chandrasekaran, M., Takeda, Y., and Nair, G.B. 2003. Recent advances in cholera genetics. Curr. Sci. 85:101–108.
14. World Health Organization. 2006. Cholera cases and deaths notified to WHO, 2005. Wkly Epidemiol. Rec. 81:297–308.
15. Zhang, X.H. and Austin, B. 2005. Haemolysins in *Vibrio* species. J. Appl. Microbiol. 98:1011–1019.

15

Shigella species

Introduction

Shigella was first discovered in 1896 by a Japanese microbiologist, Kiyoshi Shiga, who was investigating an outbreak of *sekiri* (means "dysentery" in Japanese). He isolated a bacillus from stool sample and called it *Bacillus dysenteriae*, now it is known as *Shigella dysenteriae* type 1. The toxin produced by this organism is called, Shiga toxin. *Shigella* are a member of the *Enterobacteriaceae* family and causes shigellosis characterized by bacillary dysentery (mucoid bloody stool). *Shigella* are commonly found in water contaminated with human feces and fecal–oral route is the primary mode of transmission. It is responsible for a worldwide health problem; however, it is a serious concern in the developing countries. It is estimated that *Shigella* spp. account for 1.1 million deaths and 165 million cases of dysentery annually worldwide. Shigellosis has a high morbidity and mortality rate in children under 5 years of age. Malnourished children are highly susceptible and *Shigella* infection further promotes impaired nutrition, recurring infection and retarded growth. Antibiotic resistant strains are continuously emerging thus treatment regiments become very difficult against shigellosis. CDC estimates annually there are 18,000 reported cases of shigellosis in the US. Since the disease sometimes appears as mild and not reported, thus the number could be as high as 360,000 cases per year.

Biology

Shigella are a Gram-negative, nonsporulating rod. It is a facultative anaerobe, and is nonmotile. *Shigella* are generally catalase-positive, and oxidase- and lactose-negative. They ferment sugars, usually without forming gas. The strains grow between 7 and 46 °C, with an optimum temperature of 37 °C. The cells are not as fragile as once thought. They survive for days under harsh physical and chemical exposures, such as refrigeration, freezing, 5% NaCl, and media with a pH of 4.5. *Shigella* are sensitive to heat treatments and are killed by pasteurization. *Shigella* can multiply in many types of food when stored at their growth temperature ranges. *Shigella* are biochemically and serologically related to enteroinvasive *E. coli* (EIEC), which also induce diarrhea/dysentery (see Chap. 10).

Classification

The genus *Shigella* contains four species (serogroups): *Shigella dysenteriae* (serogroup A), *S. flexneri* (serogroup B), *S. boydii* (serogroup C), and *S. sonnei* (serogroup D) and each species has several serotypes. Of four species of *Shigella*, *S. dysenteriae* type 1 is known to cause deadly epidemics; *S. flexneri* and *S. sonnei* are responsible for endemic outbreaks; and *S. boydii* causes rare disease. *S. boydii*-related outbreak is mostly limited to Indian subcontinent. *Shigella dysenteriae* and *S. flexneri* are responsible for shigellosis in developing countries and *S. sonnei* causes sporadic outbreaks in industrialized countries transmitted by contaminated water or undercooked food. In the US, *S. sonnei* accounts for two-thirds of shigellosis, while *S. flexneri* the rest. However, *S. flexneri* is the most extensively studied species among the shigellae and the molecular study of this pathogen is beginning to shed light on its mechanism of infection and the disease process.

Source and Transmission

Drinking water and food contaminated with human feces serve as the primary source of shigellae. Vegetables can be contaminated in the field by farm workers defecating in the fields, exposure to sewage, or the polluted or contaminated irrigation water. *Shigella* infection is a major problem in the developing countries due to inadequate sanitary facilities, and flies can transmit bacteria from human feces in their legs and wings and can contaminate food. Children and infants are highly susceptible to shigellosis and moreover the chance of contracting the infection is very high because of lack of hygienic practices. Adults are less susceptible, because of improved hygienic practices and immunity. In daycare center, person-to-person contact results in the infection. Homosexuals, migrant workers, and travelers to endemic zone are often infected with *Shigella* and also become the source of infection for others.

Pathogenesis

Shigella are highly infectious and a dose of 10–100 cells can cause infection. The incubation period is 1–4 days or as many as 8 days. *Shigella* are resistant to low pH (pH 2.5 for 2 h) thus facilitate survival during transit through stomach. The primary site of infection is the large intestine (colon and rectum). *Shigella* are nonmotile and cannot persist in small intestine because of increased ciliary movement of the epithelial cells, fast flow of liquids, and increased intestinal peristaltic movement. The invasive traits are expressed at 37 °C but not at 30 °C. Thus shigellae cells growing at 30 °C need a few hours of conditioning at 37 °C before they can invade intestinal epithelial cells. Shigellae colonize and invade mucosal epithelial cells and the stages of pathogenesis (Fig. 15.1) include (1) invasion, (2) intracellular multiplication, (3) inter- and intracellular (cell-to-cell) movement, and (4) the host cell killing. These events ultimately provoke pronounced epithelial inflammation, and ulcerative lesions with mucopurulent bloody stool that represent characteristics clinicopathological symptoms of dysentery. The pathogenic mechanism is complex, involving an array of gene products that facilitate bacterial adhesion, invasion, cell-to-cell movement, cell death, and inflammatory immune response, which are discussed below (Fig. 15.1).

All pathogenic isolates carry the 220-kb virulence plasmid, and the major virulence genes required for bacterial invasion and the protein export system are located in a 30-kb pathogenicity island (PAI) (Fig. 15.2). The invasion-associated virulence genes are referred as invasion plasmid antigens (*ipa*), and the gene products are designated as IpaB, IpaC, IpaD, and IpaA. Expression of *ipa* genes are regulated by *virB* located in the PAI. The other genes present in the 30-kb PAI include *icsA* required for actin-based intracellular bacterial motility, *icsB* for growth inside cytoplasm, *ipg* for bacterial invasion, and *spa* (surface presentation antigens) and *mxi* (membrane excretion proteins), consisting of genes coding for the components of type III secretion system (TTSS) (Table 15.1).

Several virulence genes are located in the chromosome and they are involved in the regulation of plasmid encoded virulence genes. For example, the *virR* gene controls the temperature-dependent expression of Spa/Mxi proteins.

Shigellae produces three types of enterotoxins: *Shigella* enterotoxin 1, found in *S. flexneri*; *Shigella* enterotoxin 2, found in many shigellae but

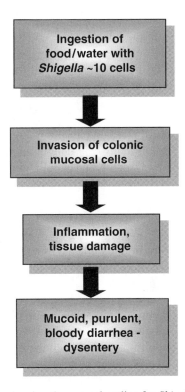

Fig. 15.1 Flow diagram showing the general outline for *Shigella* pathogenesis

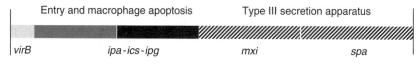

Fig. 15.2 Schematic representation of virulence gene cluster in the 30-kb pathogenicity island of 220-kb *Shigella* plasmid

Table 15.1 Virulence factors acting as effectors secreted via type III secretion system (TTSS) of *Shigella*

Virulence factors	Size (kDa)	Functions
IpaA	70	Membrane ruffling, loss of actin stress fibers
IpaB	62	Hemolysin, release from endosome, lysis of protrusion in cell–cell spreading, CD44 stimulation, translocater for TTSS
IpaC	42	Membrane ruffling, actin polymerization, β1-integrin stimulation, translocater for TTSS
IpaD	37	Regulation of TTSS translocater
IcsB	52	Escape from autophagy
IpgB1	23	Membrane ruffling, actin polymerization
IpgD	66	Phosphatidylinositol (4,5) bisphosphate phosphatase activity
IpaH9.8	62	E3 ubiquitin ligase activity, suppression of inflammatory response
OspF	28	MAPK phosphatase activity, suppression of inflammatory response
OspG	23	Kinase activity for ubiquitin E2 molecule, suppression of inflammatory response
VirA	45	Degradation of microtubule, membrane ruffling, intracellular bacterial motility

not all; and Shiga toxin (Stx), which is produced by *S. dysenteriae* type 1. The gene for *Shigella* enterotoxin 1 is encoded in the chromosome, *Shigella* enterotoxin 2 in the virulence plasmid, and the Stx in the chromosome (a bacteriophage-borne).

Invasion

In the intestine (Fig. 15.3), *Shigella* are unable to adhere the epithelial cells, since the bacteria do not possess any adhesion factor or flagella. Instead, *Shigella* deliver the effectors via TTSS such as IpaA, IpaB, IpaC, IpgB1, IpgD, and VirA, and these proteins are involved in the induction of membrane ruffles to form macropinocytosis to allow bacterial entry into cells. *Shigella* entry is exclusively limited to M-cell-mediated pathway. M-cells are distributed in the

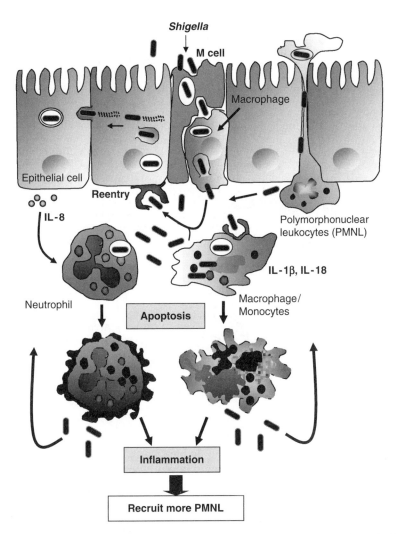

Fig. 15.3 Schematic diagram showing *Shigella* entry through mucosal membrane in the intestine and the induction of inflammation (*See Color Plates*)

Peyer's patches and solitary lymphoid nodules in the small and large intestine. M-cells have endocytic activity for foreign antigens including microbes and transport them to antigen presenting cells such as macrophages and dendritic cells present underneath the M-cells (Figs. 15.3 and 15.4). Polymorphonuclear leukocytes (PMNL) attracted to the lumen of intestine upon bacterial infection can also transport bacteria across the mucosal barrier. Once *Shigella* pass through the epithelial barrier via the M-cells, they infect resident macrophages and subsequently escape from the phagosome into the cytoplasm, where they multiply and induce rapid cell death. *Shigella* are released from dead macrophages and actively enter into the enterocytes through the basolateral surface by inducing macropinocytosis by a process called trigger mechanism (Fig. 15.4). Once inside the epithelial cell, the bacterium is trapped inside the vacuole. *Shigella* disrupt the vacuolar membrane and escapes into the cytoplasm, in which *Shigella* multiply and move within the cytoplasm as well as into the

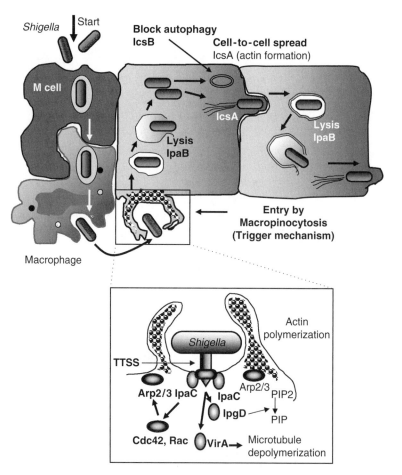

Fig. 15.4 Schematic diagram showing various stages of cellular mechanism of *Shigella* infection. The steps include; entry, lysis of vacuole, replication, blockage of autophagy, actin polymerization, and cell-to-cell movement. *Shigella* entry into epithelial cells is mediated by trigger mechanism (macropinocytosis), and the details of which is presented inside the box (this is also presented in Fig. 4.3 in Chap. 4)

adjacent cells by inducing actin polymerization at one pole of the bacterium. Proinflammatory cytokines produced by macrophages and epithelial cells recruit PMNL, and degranulation of PMNL invokes tissue damage which in turn destabilizes intestinal barrier integrity, thus further promoting additional *Shigella* entry of into the subepithelial region. Cytokines released from infected epithelial cells also recruit more inflammatory cells which exacerbate the infection by inducing tissue destruction, mucus secretion, and blood and pus formation.

The large 220-kb plasmid of *Shigella* encoding the invasion-associated proteins (Table 15.1) is essential for bacterial invasion of epithelial cells. IpaB–C complex interacts with β1-integrin, while IpaB interacts with CD44, those located at the basolateral surface of polarized epithelial cells. The IpaB binding to CD44 plays an important role in *Shigella* uptake. Interaction of IpaB with CD44 initiates a cascade of signaling event that allows recruitment of ezrin, which forms crosslink with actin and possibly aids in the membrane ruffles, lamellipodia formation, and the formation of macropinocytic pocket for engulfment of bacteria. *Shigella* also inject the invasion-associated proteins (IpaA, IpaB, IpaC, IpgB1, IpgD, and VirA) into the host cell cytoplasm through TTSS needle (60 nm long). IpaC and IpaA initiate signaling cascade to induce cytoskeletal rearrangement and to form focal adhesion-like structure that is required for bacterial entry by membrane ruffling. Rho GTPases (Cdc42, Rac, and RhoA) allow recruitment of actin, vinculin, and ezrin through activation of nucleator proteins Arp2/3 and N-WASP (neuronal Wiskott–Aldrich syndrome protein) to form the membrane ruffling structure. IpgB1 also plays important role in inducing large scale membrane ruffling, and IpaA binds to vinculin and also stimulates RhoA to promote local actin cytoskeleton rearrangements necessary for bacterial entry.

Intracellular Multiplication

Shigella encased inside the phagosome is released with the help of IpaB, which forms membrane pore (Fig. 15.4). IpaB together with IpaC act like a hemolysin that oligomerize in the vacuolar membrane to form pore. *Shigella* thus released from the phagosome multiplies rapidly inside the cytoplasm utilizing cytoplasmic nutrients rich in amino acids. The bacterial generation time inside the cytoplasm is about 40 min. Bacteria growing inside, however, can occasionally be the target for degradation by autophagy. Normally, autophagy mediates the bulk degradation of undesirable cytoplasmic proteins and organelle in the cytoplasm. In addition, autophagy also plays a critical role in host innate defense by degrading intracellular bacteria which multiplies in the cytoplasm. *Shigella* upon release from phagosome secrete IcsB via TTSS and IcsB blocks the recognition of intracellular *Shigella* by autophagy, thus allowing the bacteria to escape autophagic degradation and promote bacterial replication inside the cell. Strains that are lacking *icsB* gene are trapped by autophagosome, which is eventually fused with the lysosome for degradation in the autolysosomal compartment.

Bacterial Movement: Inter- and Intracellular Spreading

Shigella produce IcsA (intracellular spread)/VirG protein, a 116-kDa surface-exposed outer membrane protein, which mediates actin polymerization to aid in bacterial movement inside the cell, thus referring to actin-based bacterial motility (Fig. 15.4). The IcsA accumulates at one pole of the bacteria as they multiply inside the host cells and induces actin polymerization that propels the bacterium forward through cytoplasm at a rate of 10–15 μm min^{-1}. Indeed, IcsA directly binds to nucleation promoting factor N-WASP, one of the WASP family, to polymerize actin with the aid of Arp2/3 complex which initiates actin polymerization. As bacteria propel inside the cell, eventually coming in contact with the cytoplasmic membrane and push the cell membrane to form a protrusion. The protrusion containing *Shigella* is engulfed by the neighboring cell. IpaB aids in the lyses of the double membrane to allow bacterial release into the neighboring cell. Bacteria again initiate actin polymerization and continue cell-to-cell movement without ever leaving the intracellular location, by which *Shigella* can continuously expand their own replicative niches.

Cell Death and Inflammation

After arrival of shigellae in the subepithelial region, they are rapidly phagocytosed by resident macrophages. *Shigella* escape phagocytic vacuole with the help of IpaB and escape degradation by lysosomal antimicrobial system. *Shigella* replicate and induce macrophage apoptosis by two distinct pathways, (1) activation of caspase-1 by IpaB to release IL-1β and (2) by translocation of cytosolic lipid A component of LPS. Apoptosis is generally a noninflammatory process; however, during *Shigella* infection, the bacteria appear to promote inflammation. Caspase-1 also activates proinflammatory cytokines IL-1β and IL-18, which recruit inflammatory cells at the site of infection (Fig. 15.3).

Intracellular growth inside epithelial cells also triggers inflammatory response. During intracellular growth, bacterial peptidoglycan (PGN) composed of diaminopimelate (DAP) containing *N*-acetylmuramic acid and *N*-acetylglucosamine is recognized by Nod1 (nuclear oligomerizing domain protein), and induce inflammatory cytokine, IL-8 production through activation of nuclear transcription factor, NF-κB. In addition, LPS released by intracellular bacteria also provokes sustained activation of NF-κB. Increased IL-8 recruits polymorphonuclear neutrophils (PMN) in the site of infection. Increased PMN activity destabilizes the mucosal membrane integrity, thus permitting entry of more shigellae during this early phase of infection. In the later stages of infection, a large number of activated infiltrating neutrophils aid in the resolution of disease by actively phagocytosing the shigellae. However, it is rather mysterious why bacteria would induce such a strong inflammatory response to promote its own clearance by PMN.

Killing of mucosal epithelial cells is mediated by bacterial growth inside and is not due to the production of Shiga toxin (Stx) since a toxin-negative mutant also kills epithelial cells. Intracellular bacterial growth result in blockage of protein synthesis, depletion of essential small molecules, interference of respiration, drop in ATP levels, increase in pyruvate levels, and disruption of energy metabolism and apoptosis.

Shiga Toxin and Hemolytic Uremic Syndrome

Shigella dysenteriae type 1 produces Stx and the gene (*stxAB*) for which is encoded in the chromosome (bacteriophage-borne). It is an A–B type toxin of approximate molecular mass of 70 kDa. The A subunit is a 32-kDa toxin and the B subunit is a pentamer, each consisting of 7.7 kDa. Stx is released upon lysis of the cell. The B subunit of Stx binds to the receptor, globotriaosylceramide (Gb$_3$), a glycolipid and allows the entry of A subunit into cell. The A subunit inhibits protein synthesis in the 28S RNA of the 60S host cell ribosome. Shiga toxin exhibits varieties of functions; (1) it acts as an enterotoxin and induces fluid accumulation; (2) as a neurotoxin, it blocks nerve impulses and elicits paralysis; and (3) as a cytotoxin, it kills cells by inhibiting protein synthesis, and by triggering apoptosis.

The major site of action of Shiga toxin is on the kidney tubule which is rich in Gb$_3$ receptors. The toxin damages the tubule resulting in acute kidney failure and blood is excreted into urine causing hemorrhagic uremic syndrome (HUS).

The Stx also causes damage to the colonic endothelium of blood vessels causing bloody stool. Other sequelae of HUS are the development of Reiter's syndrome, or arthritis due to an autoimmune disease. In addition, release of LPS induces increased production of IL-1 and TNF-α, which provokes toxic shock syndrome. Increased cytokine production also induces vascular damage and the kidney failure.

Regulation of Virulence Genes

The plasmid encoded virulence gene expression in *Shigella* is temperature dependent with maximum expression occurring at 37 °C and virtually no expression is seen at 30 °C. These virulence genes are regulated by *virF*, which is not transcribed at 30 °C, because a repressor gene, *virR* is active at 30 °C and furthermore, it inhibits the expression of *virF*. Another transactivator of virulence genes is *virB*, which is controlled by *virF* and the temperature. The *virB* is located directly downstream from *ipaA* and is responsible for expression of Ipa proteins necessary for bacterial invasion and cell-to-cell movement. There is another important regulatory gene, *mxiE*, located in the 30-kb PAI on the plasmid regulates the expression of genes that are necessary for intracellular survival. When *Shigella* enter into epithelial cells, the MxiE protein acts as a transcriptional activator for expression of a subset of genes such as *ipaH9.8*, *ospF*, and *ospG*. Indeed, under intracellular environment, MxiE induces the expression of several virulence-associated genes, carrying so-called MxiE boxes (GTATCGTTTTTTANAG) located between position −49 and −33 with respective transcriptional start site in the upstream region of the MxiE target genes. This strategy allows the intracellular *Shigella* to induce expression of specific genes that are necessary for survival inside the cell. Furthermore, MxiE regulated genes (*ipaH9.8, ospF*, and *ospG*) also are involved in modulating host innate immune response.

Resistance Against Infection

Host defense against mucosal pathogens like *Shigella* is conducted by the innate immune response, which provides an early defense against infection. Phagocytic cells including, macrophages, monocytes, dendritic, and neutrophils provide the bulk of resistance. Upregulation of proinflammatory cytokines including IL-1β, IL-6, IL-8, TNF-α, and TNF-β, allow recruitment of neutrophils. Invading intracellular bacteria can be recognized by various innate immune systems. For instance, invasion of epithelial cells by *Shigella* is recognized by pattern recognition molecules such as Nod1, which ultimately activate NF-κB and produces IL-8, thus leading to recruitment of neutrophils. In addition, IFN-γ production and complement activation as byproducts are also involved in the innate immunity. Defense against *Shigella* is also mediated by antimicrobial peptides, designated LL-37 belonging to the cathelicidin family, and β-defensin-1 secreted by neutrophils and macrophages. However, *Shigella* deliver the effectors such as IpaH9.8, OspG, and OspF via TTSS to host cells and downregulate the host innate immune response including the expression of antimicrobial peptides in human colonic tissues to overcome the host innate defense. As mentioned above, survival of shigellae inside the intestinal epithelial cell is also promoted by several strategies such as escape from phagocytic vacuoles by IpaB, escape from autophagic compartments by IcsB, and intracellular movement by IcsA. In addition, bacterial LPS may play a pivotal role in bacterial survival. It protects against complement-mediated lysis and induces apoptosis in phagocytes. Glucosylation of LPS also promotes bacterial invasion and evasion of innate immunity.

Shigella are present in both extracellular and intracellular locations during infection, and thus humoral and cellular immune systems are thought to be important for eradicating the pathogen. Adaptive humoral immune response has been observed in experimental animal model to produce immunoglobulins (IgG, IgM, and sIgA) directed against the LPS. In addition, activated T-cells are also induced upon *Shigella* infection of animals.

Symptoms

The symptoms appear within 12h to 7 days, but generally in 1–3 days. In case of mild infection, symptoms last for 5–6 days, but in severe cases, symptoms can linger for 2–3 weeks. Typical symptoms of dysentery include anorexia, fever, colitis, mucopurulent bloody stool, abdominal cramp, and tenesmus (a sense of incomplete evacuation of bowel with rectal pain). Inflammation in lamina propria and mucosal layer results in edema, erythema, and mucosal hemorrhage. In adults, the disease is self-limiting and resolve within 5–7 days. Some individuals may not even develop symptoms. An infected person sheds the pathogens long after the symptoms have stopped. Generally, children are more susceptible to shigellosis than adults and it is fatal in children especially when children suffer from malnourishment. Children with malnourished conditions show dehydration, megacolon, rectal prolapse, and intestinal perforations and infection is life-threatening. Neurological disorder including lethargy, headache, and convulsions are seen. In some patients, HUS may develop resulting in kidney failure.

Prevention and Control

Foodborne shigellosis, at least in the developed countries, is caused by contamination of foods by food handlers shedding the pathogen in the feces and having poor personal hygiene. Therefore, food handlers should be forbidden to prepare or serve ready-to-eat foods. In addition, proper education of the food handlers about the importance of good personal hygiene and the need to not handle food if one suspects having a digestive disorder is important. Use of rigid sanitary standards to prevent cross-contamination of ready-to-eat food, use of properly chlorinated water to wash vegetables to be used for salads, and refrigeration of foods are necessary to reduce foodborne shigellosis.

In the developing countries, water and contaminated food are the major source and the disease is primarily associated with unsanitary living conditions. Many communities do not have any sanitary toilet facilities thus outdoor open-air toilet practices allow direct bacterial transmission to food and water supplies. Children under the age of 5 are most susceptible and this group has the least awareness of sanitary and hygienic practices. Thus they need most attention if the *Shigella*-induced disease need to be controlled in the developing countries. Hand washing should be practiced to prevent spread of infection.

The most important therapy is rehydration and the electrolyte containing fluid is administered orally or intravenously. Antibiotic treatment is controversial since *Shigella* develops resistance against antibiotics. Treatment of shigellosis varies depending on the severity of infection. However, for severe cases antibiotics are normally administered along with the fluid therapy to prevent dehydration. The most common antibiotics used are: ampicillin, trimethoprim/sulfamethoxazole, nalidixic acid, or ciprofloxacin.

Vaccination strategy seems to be a probable solution for developing countries. The vaccines must be effective against 17 epidemiologically important serotypes of *Shigella* including *S. dysenteriae* type 1, 15 *S. flexneri* serotypes, and *S. sonnei*. Several vaccines are currently under development that include the use of attenuated wild type strain without virulence genes, and conjugate vaccine that include *Shigella*-O polysaccharide conjugated to a carrier protein. Many of these vaccines are now gone through phase I and II trials and some are even in the phase III clinical trials and should be available for use in near future.

Diagnosis and Detection

Animal and Cell Culture Models

Shigella does not cause infection in adult laboratory rodents when administered orally. In guinea pigs, there is no infection, when administered orally – but it can cause infection when applied to eye causing severe inflammatory reaction called "sereny kerato conjunctivitis test." Rabbit ileal loop assay (RIL) has been used to test the diarrheagenic action of toxins. Recently, a newborn mouse model has been developed to study shigellosis showing inflammation and tissue damage in the large intestine. Monkeys have been used primarily to study bacterial colonization; however, their use is restricted because of ethical considerations and the cost.

Various cultured cell lines are used to study bacterial invasion and bacterial cell-to-cell spread. Plaque assay has been used to study bacterial ability to move from cell-to-cell by forming a focal lesion on a cultured cell monolayer (see Chap. 5).

Bacterial Culture Methods

Shigellae can be isolated from stool sample; however bacteria remain viable for a short period outside the human body, therefore stool should be tested immediately or stored in appropriate media. Furthermore, stool should be collected in the early phase of infection, before the antibiotic treatment has begun to ensure isolation of bacteria. Shigellae have been isolated by streaking the diluted stool sample or rectal swab on MacConkey, Salmonella-Shigella, Xylose–Lysine Deoxycholate, and Hektoen Enteric agars. Isolated colonies can be tested by slide agglutination test for serotyping and for presumptive identification. Shigellae have been isolated from food samples in a similar way.

Immunological Methods

Immunoassays including enzyme immunoassay (EIA), latex agglutination (LA) test, and dipstick immunoassays are used for detection of *Shigella* species. Commercially available LA test designated, Wellcolex Color *Shigella* test

(WCT-Shigella) has greater than 90% accuracy in detection of *Shigella*. Similarly commercial EIA kits are available for *S. dysenteriae* (Shigel-Dot A), *S. flexneri* (Shigel-Dot B), *S. boydii* (Shigel-Dot C), and *S. sonnei* (Shigel-Dot D). These EIA assays are performed on the membrane and are also known as dot blot assay and have greater than 94% success rate.

Molecular Techniques

Conventional or nested PCR methods have been developed to detect various species of *Shigella* and the gene targets include *ipaH, virA, ial* (invasion-associated locus), LPS, and plasmid DNA. The detection limit for the majority of these assays is from 1 to 1×10^4 cfu g^{-1} of food samples.

Summary

Shigella is highly infective and causes disease worldwide. In particular, it causes epidemic in developing countries and is mostly associated with unhygienic and poor sanitary living conditions. *S. dysenteriae* type 1 causes epidemic outbreak and *S. flexneri* and *S. sonnei* cause endemic outbreaks. *Shigella* expresses multiple virulence factors encoded in a large plasmid. The major virulence genes are encoded in a 30-kb pathogenicity island as well as are scattered on the large plasmid. The virulence proteins are required for bacterial invasion, intracellular growth, cell-to-cell spread, killing of host cells, and evasion of host innate defense systems. The virulence proteins are delivered to the host cell by TTSS. After being ingested with contaminated drink or food, bacteria colonize in the colon and rectal mucus membrane. The infection dose is very low, 10–100 cells, because, in part, for their ability to survive gastric acid (pH 2.5). Bacteria cross the epithelial barrier by passing through naturally phagocytic M-cells and after arrival in the subcellular location; they are engulfed by macrophages and dendritic cells. Bacteria kill these cells by apoptosis and invade epithelial cells in the basolateral side by inducing membrane ruffling and macropinocytosis, a mechanism termed as "trigger mechanism." This process is aided by bacterial ability to deliver many virulence proteins called effectors by TTSS. Bacteria are then released from phagosome with the aid of IpaB, move inside cell by inducing actin polymerization with the help of IcsA and infect neighboring cell. Infection results in the release of high levels of proinflammatory cytokines that recruit PMN. Bacteria-induced epithelial cell damage and the activation of PMN results in massive inflammation characterized by ulceration and hemorrhage and patients show signs of mucopurulent bloody stool, abdominal cramp and tenesmus. Children under the age of 5 are most susceptible and the infection could be fatal in malnourished children.

Further Readings

1. Levine, M.M. 2006. Enteric infections and the vaccines to counter them: future directions. Vaccine 24:3865–3873.
2. Niyogi, S.K. 2005. Shigellosis. J. Microbiol. 43:133–143.
3. Ogawa, M. and Sasakawa, C. 2006. Intracellular survival of *Shigella*. Cell. Microbiol. 8:177–184.

4. Phalipon, A. and Sansonetti, P.J. 2007. *Shigella*'s ways of manipulating the host intestinal innate and adaptive immune system: a tool box for survival? Immunol. Cell Biol. 85:119–129.

5. Philpott, D.J., Edgeworth, J.D., and Sansonetti, P.J. 2000. The pathogenesis of *Shigella flexneri* infection: lessons from *in vitro* and *in vivo* studies. Philos. Trans. R. Soc. Lond. B 355:575–586.

6. Sansonetti, P.J. 2001. Rupture, invasion and inflammatory destruction of the intestinal barrier by *Shigella*, making sense of prokaryote–eukaryote cross-talks. FEMS Microbiol. Rev. 25:3–14.

7. Sur, D., Ramamurthy, T., Deen, J., and Bhattacharya, S.K. 2004. Shigellosis: challenges & management issues. Indian J. Med. Res. 120:454–462.

8. Suzuki, T., Nakanishi, K., Tsutsui, H., Iwai, H., Akira, S., Inohara, N., Chamaillard, M., Nunez, G., and Sasakawa, C. 2005. A novel caspase-1/Toll-like receptor 4-independent pathway of cell death induced by cytosolic *Shigella* in infected macrophages. J. Biol. Chem. 280:14042–14050.

9. von Seidlein, L., Kim, D.R., Ali, M., Lee, H., Wang, X., Thiem, V.D., Canh, D.G., Chaicumpa, W., Agtini, M.D., Hossain, A., Bhutta, Z.A., Mason, C., Sethabutr, O., Talukder, K., Nair, G.B., Deen, J.L., Kotloff, K., and Clemens, J. 2006. A multi-centre study of *Shigella* diarrhoea in six Asian countries: disease burden, clinical manifestations, and microbiology. PLoS Med. 3:1556–1569.

10. Warren, B.R., Parish, M.E., and Schneider, K.R. 2006. *Shigella* as a foodborne pathogen and current methods for detection in food. Crit. Rev. Food Sci. Nutr. 46:551–567.

Glossary

Terminology	Definition
Achlorhydria	Reduced or absence of hydrochloric acid from the gastric juice
Adenitis	Inflammation of a gland
Akinetic mute	A state in which a person is not able to move or make sounds.
Ataxia	Defective muscular coordination caused by BSE, Tetanus, Botulism
Angiogenesis	Generation of new capillary blood vessels or of vascularisation of a tissue
Atrophy	Reduction in the size of a cell, tissue, or organ
Bradycardia	Slow heart rate – below 60 beats per min
Carbuncle	A Staphylococcal skin infection (by *S. aureus* or *Yersinia pestis*) characterized by tender pea sized red nodule and it may ooze pus.
Cathepsin	Intracellular proteolytic enzymes of animal cells or tissues, such as cathepsin B, a lysosomal thiol proteinase, cathepsin C, dipeptidyl peptidase, cathepsin D, that has pepsin like specificity, cathepsin G, similar to chymotrypsin, cathepsin H, that has aminopeptidase activity, cathepsin N, that acts on N terminal peptides of collagen.
Cellulitis	An acute pus forming inflammation of the deep subcutaneous tissues and sometimes muscle, which may be associated with abscess formation
Cholecystitis	Inflammation of gall bladder
Demyelination	Loss of myelin sheath on nerve cell with intact axons or fibre tracts. Demyelination of the peripheral nervous system seen with Guillain-Barre syndrome
Edema	Fluid accumulation and swelling
Edrophonium chloride	Dimethylethyl (3-hydroxyphenyl) ammonium chloride, anti cholinesterase drug, a competitive antagonist of skeleatal muscle relaxants
Furuncle	Boil (painful red skin swelling)
Ganglioside	It is a sialic acid containing oligosaccharide covalently attached to ceramide lipid on the epithelial cell
Hemochromatosis	A rare genetic disorder (disease) that results in the over abundance of iron in the body tissues
Hyperplasia	The abnormal increase in cell numbers in tissue
Lymphadenitis	Inflammation of lymph nodes

Terminology	Definition
Lysostaphin	A mixture of three proteolytic enzymes produced by *Staphylococcus epidermidis* and two of which lyse bacterial cell wall
Morbidity	A diseased condition or state, the incidence of a diseases in a population
Mortality	The death rate. The ratio of the total number of deaths to the total population
Myalgia	Muscle pain
Myonecrosis	Necrosis of muscle- caused by alpha toxin producing *Clostridium perfringens*
Polyps	A mass of tissue bulges or projects outwards resulting from granuloma formation
Prolapse	Falling or dropping down of an internal organ like – uterus or rectum
Septicemia	Presence and persistence of pathogenic microorganisms or their toxins in the blood during systemic disease
Tachycardia	The excessive rapid heartbeats - usually above 100 per minute
Tenesmus	A sense of incomplete evacuation of bowel with rectal pain – associated with bacillary dysentery
Xenobiotic	A completely synthetic chemical compound which inhibits microbial growth and it is believed to be resistant to environmental degradation

Index

Printed in the United States of America